Symbolic Mathematics for Chemists

T0337985

Symbolic Mathematics for Chemists

A Guide for Maxima Users

Fred Senese
Frostburg State University
MD, USA

Registered Offices
John Wiley & Sons, Inc., 111 River Street, Hoboken, NJ 07030, USA
John Wiley & Sons Ltd, The Atrium, Southern Gate, Chichester, West Sussex, PO19 8SQ, UK

Editorial Office
The Atrium, Southern Gate, Chichester, West Sussex, PO19 8SQ, UK

For details of our global editorial offices, customer services, and more information about Wiley products visit us at www.wiley.com.

Wiley also publishes its books in a variety of electronic formats and by print-on-demand. Some content that appears in standard print versions of this book may not be available in other formats.

Library of Congress Cataloging-in-Publication Data

Names: Senese, Fred, author.
Title: Symbolic mathematics for chemists : a guide for Maxima users / Fred Senese.
Description: Hoboken, NJ : John Wiley & Sons, 2019. | Includes bibliographical references and index. |
Identifiers: LCCN 2018024356 (print) | LCCN 2018033308 (ebook) | ISBN 9781119273233 (Adobe PDF) |
    ISBN 9781119273264 (ePub) | ISBN 9781118798690 (pbk.)
Subjects: LCSH: Chemistry–Mathematics. | Logic, Symbolic and mathematical–Data processing.
Classification: LCC QD39.3.M3 (ebook) | LCC QD39.3.M3 S46 2018 (print) | DDC 542/.8553–dc23
LC record available at https://lccn.loc.gov/2018024356

Cover design by Wiley
Cover image: © Billion Photos/Shutterstock; @ leminuit/Getty Images

Set in 10/12pt Warnock by SPi Global, Pondicherry, India
Printed and bound in Singapore by Markono Print Media Pte Ltd

10 9 8 7 6 5 4 3 2 1

To my dearest wife, Hazel, without whose tolerance, patience, support and love this book would not have been possible.

Contents

Preface

Maxima is a free open source symbolic math engine, similar to commercial systems like Mathematica, Matlab, and Maple. It can be used to symbolically solve problems in algebra, trigonometry, calculus, and differential equations. It is available for Windows, Macintosh, Linux, and Android platforms, and is directly downloaded by over 100 000 users each year. Several different graphical user interfaces are available, including wxMaxima, XMaxima, emacs, and TeXmacs. Maxima is also a component of several computer algebra systems, including Sage, SMath Studio, and the Euler Math Toolbox.

This Maxima primer focuses on problem solving, model building, and data analysis in chemistry. It can be used as a companion text for physical chemistry, courses in mathematical chemistry, or any chemistry course that requires computation and data analysis. It is not intended to be a replacement for the Maxima manual, nor is it intended to be a chemistry textbook.

Structure of the Book

Chapter 1 (Fundamentals) introduces wxMaxima, the graphical user interface for Maxima, and gives a quick tour of the General Math pane, includes dialogs for basic plotting, algebra, and calculus.

Chapter 2 (Storing and Transforming Data) explores Maxima's data types, including numbers, strings, lists, and matrices. It also shows how Maxima can import and export data from files.

Chapter 3 (Plotting Data and Functions) shows how to use Maxima to draw scatter plots, histograms, three-dimensional surface plots, and contour plots.

Chapter 4 (Programming Maxima) focuses on writing programs that perform iterative calculations and conditionally execute commands.

Chapter 5 (Algebra) shows you how to rewrite, factor, expand, and extract pieces of expressions and equations in Maxima. It also demonstrates the solution of equations and systems of equations, and interpolation of data using polynomials and cubic splines.

Chapter 6 (Differentiation, Integration, and Minimization) uses Maxima to perform the basic operations of calculus, including computation of limits, differential expansions, derivatives, and integrals. We'll also see how to minimize and maximize functions, find points of inflection, and compute power series and Taylor series expansions for functions.

Chapter 7 (Matrices and Vectors) applies Maxima to algebraic and differential vector and matrix calculations.

Chapter 8 (Error Analysis) uses Maxima to estimate errors in datasets and propagate them through calculations. It also shows how statistics and assumptions about the distribution of errors can be used to objectively test hypotheses about the data.

Chapter 9 (Fitting Data to a Straight Line) applies linear least-squares fits to datasets, and shows you how to assess the quality of the fit.

Chapter 10 (Fitting Data to a Curve) fits nonlinear models to data, and shows how errors in the fit parameters can be estimated using the jackknife and bootstrap methods.

Chapter 11 (Differential Equations) demonstrates symbolic, power series, and numerical solution of differential equations, as well as graphical visualization of the solutions with direction fields.

Chapter 12 (Operators and Integral Transforms) shows how quantum mechanical operators can be defined directly in Maxima. It also introduces Maxima's powerful Fourier transform and fast Fourier transform functions.

Features of the Worksheets

A comprehensive set of worksheets form the core of the book. The worksheets address the full range of computations that students encounter in an undergraduate physical chemistry course.

The worksheets themselves are not printed in their entirety in the book. They are available for download at http://booksupport.wiley.com

Each worksheet begins with clearly defined goals and learning objectives. These will be listed both in the book and at the beginning of the worksheet, along with a detailed abstract that provides motivation and context for the material. Prerequisite and follow-up worksheets are described and linked in the abstract. Users should not have to refer to the book while using the worksheets. The presentation will be practical and conversational; rigor will be retained without burdening students with fussy details.

The worksheets are not computer programs. They do not simply plot a graph or print the answer for a textbook problem. Each worksheet is a cohesive and complete guided inquiry that uses symbolic math to illuminate a topic in chemistry.

Students will have different levels of comfort with symbolic math, and the worksheets are designed with this in mind. Step-by-step instructions and clear, detailed examples are given for beginners. Troubleshooting hints and case studies provide practical experience and foster critical thinking for those who have mastered the basics. Proficient users are offered avenues for further exploration.

The worksheets do not simply present information, like a textbook; they engage students directly by asking them to write symbolic mathematics themselves. Students aren't simply asked to tweak the values of a few variables and observe the effect on a calculated result or graph. The focus is on critical thinking, creative problem solving, and the ability to connect concepts.

Each worksheet includes summary problems that ask students to integrate the ideas and techniques presented. The worksheets end with suggested projects for more proficient users. The projects offer new contexts for application of what has been learned, along with a bibliography for more advanced study.

Conventions Used in This Book

- "Maxima" refers to Maxima 5.37.1 running through the wxMaxima 15.08.1 graphical user interface.
- Code in text is typeset in `Courier` font. Options for commands are set in *`italicized Courier`*.

- Maxima input lines are printed with gray background and numbered lines; and output is printed in white, with Maxima's output labels shown in red. For example:

  (%i1) `2+2;`

  (%o1)   4

- In key sequences, spaces are shown by either Space or by explicit space characters ␣.
- When keys are to be pressed simultaneously, they will be separated by a plus sign; for example Ctrl + c means "press the Ctrl and c keys at the same time".
- The format for especially important commands are marked with a gear.
- Maxima tries to do calculations exactly with integers and rational numbers whenever it can. By default, it prints warnings when a floating point number is converted into a rational number. The printing of these messages has been switched off in the text using *ratprint*: `false`.

Installing Maxima

Maxima for desktops can be downloaded for free at http://sourceforge.net/projects/maxima/files/. Click on the directory for your operating system (`Maxima-Linux`, `Maxima-Windows`, or `Maxima-MacOS`) and download the installer for the latest version.

- The Windows download is a single self-contained .exe installer; just click on the file to start the installation.
- The MacOS X download is a single .dmg disk image[1]. Double-click on the file to mount it as a disk. A window showing the contents of the disk should appear. Drag and drop the Maxima, wxMaxima, and Gnuplot applications into the `Applications` directory.
- For installation instructions under Linux, see http://maxima.sourceforge.net/download.html. You must install both the `maxima` and `maxima-exec-clisp` packages.

A portable distribution of Maxima (which can be installed on a flash drive) is available. Search for "portable Maxima" to find the latest version.

An Android version of Maxima is available from Google Play at https://play.google.com/store/apps/details?id=jp.yhonda&hl=en. See https://sites.google.com/site/maximaonandroid/ for details.

Acknowledgements

I'd like to thank Frostburg State University for supporting this work, and my colleagues in the Department of Chemistry for their support and encouragement. Jerry Simon in particular made suggestions that improved the book. I would also like to thank the reviewers of the initial proposal for the book for their frank comments, and my editors Sarah Higginbotham, Sarah Keegan, Lesley Jebaraj, and Jenny Cossham for their patience and guidance. Ann Seidel did a wonderful job in creating online versions of the worksheets. I dedicate this book to my dear friends Corrie Haldane and Michelle Kaseler, who helped and supported me in innumerable ways during the writing of this book.

---

1 Unfortunately Maxima binaries are only being distributed for Intel-based Macs at the time this book was written. You can build Maxima on other Macs with MacPorts. See http://www.macports.org for instructions.

1

Fundamentals

The most powerful single idea in mathematics is the notion of a variable.

– K. Dewdney [1]

Maxima is a general-purpose computer algebra system (CAS) that can be used to perform numerical and symbolic calculations, simplify complicated expressions, solve algebraic and differential equations, analyze and plot data, differentiate and integrate functions, compute statistics, perform matrix and vector operations, and more. Maxima in various forms has been in use for half a century. Development on Maxima's predecessor Macsyma began in the late 1960s at MIT. In 1998, the U.S. Department of Energy released Macsyma's source code into the public domain, and the program became Maxima. Today, Maxima is actively supported and updated by an international community of users and developers [2].

Maxima differs in several important ways from commercial CASs like Mathcad, MATLAB, Maple, and Mathematica:

- Maxima is open-source software, distributed under the the Free Software Foundation's *GNU General Public License* [3].
- Maxima is available for Windows, Mac OS, Linux, and Android tablets and phones. Commercial software is usually available on a limited number of platforms. For example, *Mathcad* works only under Windows; *Mathematica* and *Maple* are not available for Android tablets or phones, or on non-Intel-based Macs.
- Most commercial CASs provide a single interface, while Maxima is a computational "back end" that can use several different interfaces. Think of Maxima as an automobile engine; it runs unseen under the hood of the car. It is controlled through a graphical "front end" or user interface, analogous to the steering wheel, dashboard, and pedals of the car. Some front-ends and software systems that use Maxima as a back end are wxMaxima [4], XMaxima [5], Euler Math Toolbox [6], Sage [7], TeXmacs [8], SMath Studio [9], and Cantor [10]. Websites can also perform symbolic calculations and plot graphs interactively using Maxima as a back end [11].

1.1   Getting Started With wxMaxima

Maxima comes bundled with two front-ends: Xmaxima and wxMaxima. wxMaxima is the more powerful of the two. It displays typeset mathematical expressions and plots, and it provides menus, toolbars, and dialogs to access most important Maxima functions. It also saves an entire session with Maxima including text, headings, and calculations in a single printable document. In this book, we'll focus exclusively on Maxima running under wxMaxima.

Symbolic Mathematics for Chemists: A Guide for Maxima Users, First Edition. Fred Senese.
© 2019 John Wiley & Sons Ltd. Published 2019 by John Wiley & Sons Ltd.
Companion website: http://booksupport.wiley.com

In wxMaxima, commands are typed into input cells. To create a new cell, click on an empty space anywhere. A horizontal line will appear. This is the *cell cursor*; typing will create a new input cell at that point. You can click between cells or before the first cell to move the cell cursor to at any point in the worksheet.

Start wxMaxima and type 2+2 ; . To execute the command, press ⇧ + Enter . The result will like this:

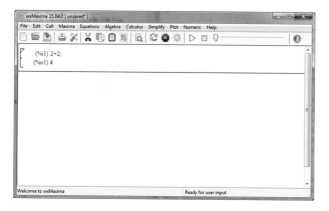

Commands end with either a semicolon (;) or a dollar sign ($). The semicolon shows output from the command; the dollar sign hides it. wxMaxima adds the semicolon automatically to the last command in a cell if you forget to type it.

 Linebreaks, tabs, and spaces are ignored in Maxima commands. You'll get an error message if you try to separate commands only with a linebreak instead of a semicolon or dollar sign.

1.1.1 Input Cells

Let's take a closer look at the way wxMaxima displays a command and its output in input cells:

Maxima labels both input and output within the cell. The first command is labeled with (%i1). Each additional command you type will be labeled with %i followed by the number of the command. The output is labeled with (%o1). Every output is labeled with %o followed by the number of the corresponding input command.

Notice the glyph on the left edge of the cell (Figure 1.1). It shows the division between input and output in the cell, and also provides a way to hide the output (click on the triangle on top of the glyph). Clicking on the glyph itself selects the cell. Dragging the mouse over multiple glyphs selects them all. You can then cut, copy, or delete the selected cell or cells:

- To copy, press Ctrl + c or click on the *Copy Selection* icon on the tool bar (it's just to the right of the scissors icon).

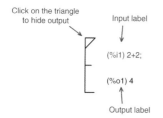

Figure 1.1 An input cell in wxMaxima divides the cell into input and output sections, with a glyph on the left that provides a way to hide output or select the cell for copying, cutting, or deleting.

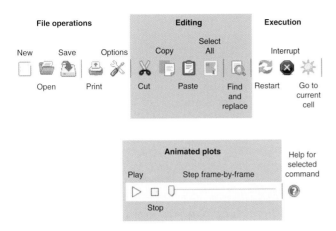

Figure 1.2 The wxMaxima toolbar.

- To delete, press the [Delete] key.
- To cut (which copies the cell and then deletes it), press [Ctrl]+[x] at the same time or click on the scissors icon on the toolbar.

Cells that have been cut or copied can be pasted back into the worksheet. Click on the space where you'd like to insert the cell or cells and press [Ctrl]+[v].

Cells are executed by pressing [⇧]+[Enter]. While a cell is executing, the glyph is outlined in black, and the message *Maxima is calculating* appears in the status bar at the bottom of the window.[1] The black outline will not go away if you've made an error in the cell.

You can place multiple commands into a single input cell, as long as each ends with a semicolon or a dollar sign. This is useful in showing intermediate results or for grouping commands. Press [Enter] after each command to keep the input readable. Maxima executes commands within a cell the order that you type them, and it numbers the output for each line automatically.

1.1.2 The Toolbar

The toolbar provides shortcuts for storing, loading, editing, and managing sessions in wxMaxima (Figure 1.2).

The crossed tool icon is used to set program options and flags, and also to set styles for fonts and colors. We'll look at it in more detail in Section 1.1.3.

The triangle, square, and slider icons are used to play, stop, and step through animated plots. This is an advanced feature that we'll use in one of the worksheets for Chapter 4.

The blue question mark icon on the right will display the Maxima manual in a separate window. To see the help page for a particular command, select the command in your worksheet and click on the Help icon (the question mark on the far right side of the toolbar). Clicking the icon without first selecting a command brings up the wxMaxima manual. You can also press the [F1] key or select [Help] ⟩ [Maxima Help] to bring up help pages for a selected command.

1.1.3 The Menus

Many Maxima functions can be accessed through the menus above the toolbar. The menus also let you configure wxMaxima and control the execution of cells.

---

1 Cells usually execute so quickly that you won't notice this, but complex calculations can take some time. If you want to interrupt a calculation, select [Maxima] ⟩ [Interrupt] or press [Ctrl]+[g].

`File   Edit   Cell   Maxima   Equations   Algebra   Calculus   Simplify   Plot   Numeric   Help`

- The `File` menu has items for creating, opening, saving, and printing *wxMaxima* documents. `File` 〉 `Load Package...` can load special-purpose code *packages* into Maxima. `File` 〉 `Load Batch File...` can load and execute a series of commands that have been previously saved as a file with the `.mac` extension.[2] `File` 〉 `Export` can save your *wxMaxima* file as an HTML file or as a LATEX file.[3]
- The `Edit` menu has items for undoing and redoing changes; cutting, copying, and pasting commands and output; finding and selecting text; zooming in and out; and setting preferences. The `Edit` 〉 `Configure` item is equivalent to the crossed-tool icon on the toolbar. Select `Edit` 〉 `Configure` 〉 `Options` to change wxMaxima's default behavior. Most of the options are self-explanatory.
  - Check **Match parenthesis in text controls** so wxMaxima will automatically type a closing parenthesis after you type an open parenthesis, and so that clicking on a closing parenthesis automatically highlights the matching opening parenthesis.
  - Check **Show long expressions** if you'd like to see results when Maxima displays a message like `<< Expression too long to display! >>`.
  - Check **Enter evaluates cells** if you'd rather not execute cells with `⇧`+`Enter` (the default). This isn't recommended, because it will keep you from breaking long commands up into a series of easy-to-read lines.

  Select `Edit` 〉 `Configure` 〉 `Style` to change the fonts and colors that wxMaxima uses.
- The `Cell` menu is used to evaluate, copy, insert, and hide different types of cells. In addition to input cells containing Maxima commands, you can organize your wxMaxima document by inserting text, titles, section titles, subsection titles, images, and page breaks.
- The `Maxima` menu is used to interrupt or restart Maxima. Several useful floating toolbars can be shown using `View`. You can also use the menu to show and delete functions and variables that have been defined in the current session, and change the way math is displayed in output.
- The `Equations`, `Algebra`, `Calculus`, `Simplify`, and `Plot` menus start dialogs that build Maxima commands.
- The `Numeric` menu has items that are useful for setting the precision of calculations and displayed numeric results.
- The `Help` menu provides context-sensitive help. Selecting a command and then selecting `Help` 〉 `Maxima Help` brings up the manual page for that command. `Help` 〉 `Example...` can provide examples of many commands. `Help` 〉 `Apropos...` shows all commands that are similar to a selected command. `Help` 〉 `Tutorials` loads a web page that links several tutorials on Maxima for beginners. `Help` 〉 `About` displays the current versions of *wxMaxima* and Maxima.

1.1.4   Command History

Unlike Excel or Mathcad, Maxima doesn't automatically execute and update individual cells. A cell won't be executed until you type `⇧`+`Enter` into a selected cell. This lets you execute cells out of order if you like. If you do so, Maxima will update the input and output labels to show the order of execution.

You can execute all of the cells in a worksheet top to bottom by typing `Ctrl`+`R` (or selecting `Cell` 〉 `Evaluate All Cells` from the menu). Maxima will remember the results of previous calculations when you do this; if you want a fresh start, select `Maxima` 〉 `Restart Maxima` first.

---

2 See Section 1.4 for more about packages and batch files.
3 LATEX is a complete mathematical typesetting system. This book and the PDF versions of the worksheets were written with LATEX.

wxMaxima stores your command history; you can repeat or edit commands in the history. To view the command history, select View ⟩ History (or type Alt + ⇧ + I). A list of commands you've typed appears in a pane on the right side of the screen. You can drag this pane outside the wxMaxima window if you like. Double-click on a command in the history pane; it will be inserted into your session.

You can refer to previous inputs and outputs using the (%i) and (%o) labels. For example,

```
(%i1)    2+2;
```

```
(%o1)    4
```

```
(%i2)    4+4;
```

```
(%o2)    8
```

```
(%i3)    %o1 + %o2
```

```
(%o3)    12
```

You can also retrieve the result of the previous cell simply by typing a percent sign (%), which is shorthand for the most recent calculated result:

```
(%i4)    %;
```

```
(%o4)    12
```

This is very useful in passing results from one command to the next in the same cell. For example, if we want to compute the square root of two, we can use the `sqrt` function (which computes the result symbolically) and then pass the result to the `float` function, which computes the numerical result:

```
(%i1)    sqrt(2);
(%i2)    float(%);
```

$$(\%o1)   \sqrt{2}$$
```
(%o2)    1.414213562373095
```

 Use % within a cell to pass results between commands. It is good practice to avoid using % (or the %i or %o labels) to access previous cells because executing cells out of order in the worksheet can lead to confusion.

When you're finished working you can save a worksheet using File ⟩ Save or File ⟩ Save As and open it again later with File ⟩ Open . You can save the file as a wxMaxima document (with a `.wxm` extension), a plain-text XML file (with a `.wxmx` extension), or as a batch file that you can execute from inside other worksheets (with a `.mac` extension).

1.1.5 Basic Arithmetic

Basic arithmetic operations in Maxima are typed and executed the same way they are in Excel and on most scientific calculators. An asterisk (*) is used for multiplication, a slash (/) is used for division, and a caret (^) or double asterisk ** is used for exponentiation.

You must explicitly type * for multiplication. An expression like 2(3 + 4)(5 − 6) is typed as 2*(3 + 4)*(5 − 6).

Expressions in parentheses are evaluated first. Exponents are evaluated next, followed by left-to-right multiplication and division, and finally left-to-right addition and subtraction. For example, to compute (1 + 2) / (3 + 4), you must type (1 + 2)/(3 + 4), not (1 + 2)/(3 + 4).

E-notation can be used to enter numbers in scientific notation. For example, to compute the size of an atomic mass unit, you can type either of the following expressions:

```
(%i1)   1/(6.022*10^23);
(%i2)   1/6.022e23;
```

$$(\%o1) \quad 1.6611295681063124 \, 10^{-24}$$

$$(\%o2) \quad 1.6611295681063124 \, 10^{-24}$$

The e means *times ten to the power that follows*. The e can be upper or lower case.

By default, Maxima provides exact symbolic results rather than approximate numerical output. It uses rational numbers (fractions involving integers) rather than decimal numbers because arithmetic with integers and rational numbers is exact. Expressions like 1/10 or $\sqrt{2}$ (typed as sqrt(2)) and mathematical constants like π (typed as %pi) or e (typed as %e) will be left as is:

```
(%i1)   sqrt(2)*%pi/3;
```

$$(\%o1) \quad \frac{\sqrt{2}\,\pi}{3}$$

The float function can be used to force a numerical result:

```
(%i2)   float(%);
```

$$(\%o2) \quad 1.480960979386122$$

You can set *all* subsequent results to display as decimal numbers by setting the *numer* system variable to true. Setting *numer* to false will turn this off again.

```
(%i1)   numer : true$
(%i2)   (-16+sqrt(256 - 4*2*8))/2;
(%i3)   numer : false$
(%i4)   (-16+sqrt(256 - 4*2*8))/2;
```

$$(\%o2) \quad -1.071796769724491$$

$$(\%o4) \quad \frac{8\sqrt{3}-16}{2}$$

The menu item Numeric ⟩ Toggle Numeric Output can also be used to switch between numeric and symbolic output.

Table 1.1 Common mathematical functions.

`abs(x)`	Absolute value of x
`acos(x)`	The arccosine (inverse cosine) of x
`asin(x)`	The arcsine (inverse sine) of x
`cos(x)`	The cosine of x
`exp(x)`	e^x, the exponential function
`log(x)`	$\ln(x)$, the natural logarithm of x
`max(x,y,...)`	Maximum value of x, y, ...
`min(x,y,...)`	Minimum value of x, y, ...
`round(x)`	The closest integer to x
`sin(x)`	The sine of x
`tan(x)`	The tangent of x
`x!`	The factorial of x

If you want to display numerical results for one command only, and leave symbolic output as the default, you can use *numer* or `float` as a "switch":

```
(%i1)    (2+sqrt(2))/(3+sqrt(3)), numer;
(%i2)    (2+sqrt(2))/(3+sqrt(3))
```

$$(\%o1) \quad 0.72150822153305$$

$$(\%o2) \quad \frac{\sqrt{2}+2}{\sqrt{3}+3}$$

1.1.6 Mathematical Functions

We've already seen that square roots can be calculated with the `sqrt` function. Many other mathematical functions are available in Maxima. You can see a complete list by pressing F1 (or selecting Help ≫ Maxima Help from the menu) to bring up the Maxima manual, and then clicking on the Mathematical Functions link. A few common mathematical functions are listed in Table 1.1; there are many more.

Trigonometric functions like `sin`, `cos`, and `tan` take arguments in radians, not degrees. You must convert degrees to radians using

$$\text{radians} = \text{degrees} \times \frac{\pi}{180} \qquad \text{Degree to radian conversion} \qquad (1.1)$$

Maxima refers to the natural logarithm $\ln(x)$ as `log(x)`, and there is no built-in base-10 logarithm function. You can calculate base-10 logs by dividing the natural log of a number by the log of 10. For example, the base-10 log of 2 is

```
(%i1)    log(2)/log(10);
(%i2)    float(%);
```

$$(\%o1) \quad \frac{\log(2)}{\log(10)}$$

$$(\%o2) \quad 0.30102999566398$$

 Worksheet 1.0: Maxima as a Scientific Calculator

In this worksheet, we'll use Maxima to do basic arithmetic with scientific notation and to calculate basic trigonometric and logarithmic functions.

1.1.7 Assigning Variables

A variable is a symbol that represents a changeable quantity. In Maxima, you can assign numbers, expressions, or even equations to variables. An assignment statement has the form `variable : value`. For example,

```
(%i1)   line:  y=m*x + b;
(%i2)   x :    2.3;
(%i3)   m :  2;
(%i4)   b :  3;
(%i5)   y :  m*x + b;
(%o1)   y = mx + b
(%o2)   2.3
(%o3)   2
(%o4)   3
(%o5)   7.6
```

In the first line, the equals sign relates the two sides of an equation `y = m*x + b`, while the assignment operator `:` assigns a name `line` to the equation. The variable names x, m, and b are assigned numerical values, while y is assigned an expression.

The order of variable definitions is important. The values of variables are substituted into expressions at the time the variable is defined, but not afterwards. The variable `line` contains the equation for a line, but since x, m, b, and y were not yet defined, the equation is stored in terms of the *names* of the variables. But y was defined after values were assigned to x, m, and b, so Maxima stores its computed value.

At this point the variable `line` still has `y = m*x+b` as its value. But if we redefine it, the values of its variables are substituted:

```
(%i6)   line;
(%i7)   line:  y=m*x + b;
(%o6)   y = mx + b
(%o7)   7.6 = 7.6
```

Once a variable is assigned, we can find its value simply by typing its name. If Maxima simply echoes back the name, the variable hasn't been given a value.

There are a few restrictions on variable names:

- The variable name must begin with either a letter or a % sign.
- Variable names are case-sensitive, so `Energy`, `energy`, and `ENERGY` will all be considered different variables.

- If you'd like to use a Greek letter for a variable name, spell it out: a variable named `alpha` will be displayed in expressions as α. If your wxMaxima installation supports Unicode, typing Esc a Esc will type α, Esc p Esc will type π, and so on. Greek letters can also be inserted into variable names like ΔG using the Greek letters pane; select View ⟩ Greek letters and then just select the symbol you want to insert.
- Variable names can be subscripted. End the variable name with either an underscore before a single character or with the subscript enclosed in square brackets. For example, variables named `xi[e]` or `xi_e` display as ξ_e in expressions.
- Special characters can be included in variable names by preceding them with a backslash. For example, type `V\'_m` to name a variable V'_m. This also makes it possible to put spaces in variable names.

For a list of all the variables that have been defined so far, use the `values` command (or select Maxima ⟩ Show Variables from the menu).

```
(%i8)    values;
```

$$(\%o8) \quad [line, x, m, b, y]$$

The variable names are returned as a Maxima list, a sequence of comma-separated items enclosed in square brackets. Many Maxima functions return lists; we'll take a closer look at them in Section 2.3.

Variables persist until you restart Maxima. When variables are no longer needed, we can clear them by name using the `kill` function, or by selecting Maxima ⟩ Delete Variables from the menu.

```
(%i9)    kill(line);
(%i10)   values;
```
$(\%o9)$ *done*
$(\%o10)$ $[x, m, b, y]$

To clear *all* variables, type `kill(all);` or select Maxima ⟩ Clear Memory from the menu.

```
(%i11)   kill(all);
(%i12)   values;
```
$(\%o0)$ *done*
$(\%o1)$ $[]$

Notice that `kill(all)` resets the input and output counts.

Maxima uses special system variables to control the way built-in functions work and expressions are displayed. We've already seen how the *numer* system variable controls whether or not results are displayed as decimal numbers.[4] Another useful system variable is *ratprint*, which controls whether or not Maxima displays warnings about conversions of decimal numbers to rational numbers. By default, it is true; if we find we're being inundated by such warnings, we can set it to false:

```
(%i1)    ratprint;
(%i2)    ratprint: false;
```
$(\%o1)$ *true*
$(\%o2)$ *false*

---

4 See Table 2.1 for a list of system variables that control the display and calculation of decimal numbers.

1.1.8 Defining Functions

Variables can be related to each other with functions; for example, the equation $y = f(x)$ says that y is a function f of x.

You can define your own functions using the : = function declaration operator.

```
F(x, y, ...) := expression
```
Define a function named F with arbitrary dummy arguments x, y, ... that are substituted into *expression*.

For example, we've seen that `log(x)` in Maxima is actually $\ln(x)$. Let's define a function `log10(x)` as the base-10 logarithm of any number x, and then use it to compute the base-10 logarithm of 2:

```
(%i1)    log10(x) := log(x)/log(10);
(%i2)    log10(2), numer;
```

$$(\%o1) \quad \log 10(x) := \frac{\log(x)}{\log(10)}$$

```
(%o2)    0.30102999566398
```

The first line defines the function with a single dummy argument x. The expression `log(x)/log(10)` on the right-hand side is not evaluated during the definition; if we had defined x as a variable previously, the function definition would not have been affected. Once the function has been defined, we can call it with any expression, value, or variable substituted for the dummy argument. In the second line, we compute the base-10 log of 2, and use the *numer* switch to ask for a numerical result.

The function can now be used like any of the built-in functions. For example, once we've defined `log10`, we can use it to define a function that estimates the pH of a solution from its hydrogen ion concentration H:

```
(%i3)    pH(H) := -log10(H);
(%i4)    pH(1e-5), numer;
```

```
(%o3)    pH(H) := −log10(H)
(%o4)    5.0
```

A function must have at least one argument; defining y := x + 1 would give an error. Also take care to use : = to define functions, *not* : or =!

To see a list of functions that you have defined, select Maxima ⟩ Show Functions. The list is also stored in a system variable called `functions`. You can delete function definitions by selecting Maxima ⟩ Delete Function... or by typing `kill(F)`, where F is the function name.

Functions with more than one argument can be defined. For example, we could compute the pH of a buffer solution made from a monoprotic acid HA with a dissociation constant of K_a using the Henderson–Hasselbalch equation,

$$pH = pK_a + \log\frac{[A^-]}{[HA]} \qquad \text{Henderson–Hasselbalch equation} \qquad (1.2)$$

If the buffer is prepared by adding V_b liters of NaOH solution to V_a liters of HA solution, and the formalities of the NaOH and HA solutions are C_b and C_a, respectively, then

$$\frac{[\text{A}^-]}{[\text{HA}]} \approx \frac{\dfrac{V_b C_b}{(V_a + V_b)}}{\dfrac{(V_a C_a - V_b C_b)}{(V_a + V_b)}} = \frac{V_b C_b}{V_a C_a - V_b C_b} \tag{1.3}$$

and we can approximate the pH of the buffer with a function pH_buffer:

```
(%i1)   log10(x) := log(x)/log(10);
(%i2)   pH_buffer(Ka, Ca, Cb, Va, Vb) := -log10(Ka) + log10(Vb*
        ➥ Cb/(Va*Ca-Vb*Cb));
```

$$(\%o1) \quad \log10(x) := \frac{\log(x)}{\log(10)}$$

$$(\%o2) \quad pHbuffer(Ka, Ca, Cb, Va, Vb) := -\log10(Ka) + \log10\left(\frac{Vb\,Cb}{Va\,Ca - Vb\,Cb}\right)$$

For example, with $K_a = 1.75 \times 10^{-5}$, $C_a = 1$ M, $C_b = 1$ M, $V_a = 0.025$ L, and $V_b = 0.0125$ L, we have

```
(%i3)   pH_buffer(1.75e-5, 1, 1, 0.025,0.0125), numer;
(%o3)   4.756961951313706
```

When a function is defined with the : = operator, the expression on the right-hand side is not evaluated. This can cause unexpected results if you're using previously defined expressions in the function definition. For example, the following code works as expected:

```
(%i1)   f(x) := 2*x$
(%i2)   f(a);

(%o2)   2 a
```

but this code doesn't:

```
(%i1)   y : 2*x$
(%i2)   f(x) := y$
(%i3)   f(a);

(%o3)   2 x
```

The x in the expression for y is not recognized as a dummy argument for the function, because y was defined outside the function body.

Maxima does provide a way to define a function with the function body evaluated. Use the define function:

```
define( f(x1, x2, ... ), expression)
```
Defines a function f with arguments x1, x2, ... and a function body *expression*, which is evaluated.

```
 Proof that dH = dq at constant pressure, with only expansion work being done:

 (%i1)  H=U+P*V
        diff(%)                        /* definition of enthalpy */;
        %, diff(U = q + w)             /* infinitesmal changes */;
        %, del(w) = -P*del(V)          /* First Law */;
                                       /* if the system is in mechanical equilibrium with the surroundings,
                                          and does only expansion work */;
        %, del(P) = 0                  /* constant pressure */;
 (%o1)  H = P V + U
 (%o2)  del(H)= P del(V)+del(U)+V del(P)
 (%o3)  del(H)= P del(V)+del(w)+del(q)+V del(P)
 (%o4)  del(H)=del(q)+V del(P)
 (%o5)  del(H)=del(q)
```

Figure 1.3 Annotating a worksheet with text blocks and comments.

Replacing $f(x) := y$ with $define(f(x), y)$ solves the problem:

```
(%i1)    y: 2*x$
(%i2)    define(f(x), y)$
(%i3)    f(a);
```

(%o1) $2\,a$

(M) Worksheet 1.1: Variables and Functions: Calculating pH

In this worksheet, we'll program functions in Maxima to compute the pH of buffer solutions and mixtures of acids and bases.

1.1.9 Comments, Images, and Sectioning

You can annotate your wxMaxima documents with text cells. Select Cell ⟩ Insert Text Cell or press Ctrl + 1 . Currently wxMaxima does not allow text formatting within the cell.[5]

To add a comment to a line of code in an input cell, place text between /* and */ delimiters. For example, in Figure 1.3, a text block is used to introduce a derivation, and each line is commented. Notice that the end-of-line semicolon or dollar sign must come after the comment.

You can also structure your documents with title cells (Cell ⟩ Insert Title Cell), section cells (Cell ⟩ Insert Section Cell), subsection cells (Cell ⟩ Insert Subsection Cell), and page breaks (Cell ⟩ Insert Page Break). The title, section, and subsection cells are collapsible; clicking on the open square in the left margin of all cells that are nested under the heading. Sections and subsections are automatically numbered.

Images stored as portable network graphics (PNG) files can be inserted into your document with Cell ⟩ Insert Image... .

1.2 A Tour of the *General Math* Pane

wxMaxima provides floating "panes" that are helpful for beginners. The panes let you insert commands without having to know their syntax. As you become more comfortable with Maxima, though, you'll find that learning and typing the commands directly gives you many more options.

---

5 Starting a text cell off with Tex: does let you insert LaTeX mathematical typesetting code into a text cell. The session can then be exported as a .tex file and then typeset with LaTeX. You'll see examples of this in the text cells of the worksheets that come with this book.

Selecting View ⟩ General Math (or typing Alt + ⇧ + m) brings up the *General Math* pane:

You can drag the pane out of the wxMaxima window if you like.

Clicking on buttons on the pane inserts code into your worksheet. It's a good idea to move the cell cursor to the spot you want the new cell to be inserted before clicking on any of the pane's buttons.

Most of the buttons operate on the last calculated result. You can also select an expression or result for them to operate on before clicking the button.

1.2.1 Basic Plotting

A dialog for building two dimensional plots can be started using either the Plot 2D... button on the bottom of the General Math pane or the Plot ⟩ Plot 2D... menu item. The dialog lets you specify the expression to plot and the names and ranges of the *x* and *y* variables.

For example, to plot $y = \sin(x)$ with x ranging from -2π to 2π, fill in the Plot 2D dialog as follows:

- The *Expression(s)* field can be used to enter a function of the *x* variable to plot. By default, it is %, which is the last result evaluated in Maxima.
- The variable names and ranges of the *x* and *y* axis variables can be changed, if you like; leaving the range of the *y* variable at its default (with zeros for the *From* and *To* fields) tells Maxima to plot between the minimum and maximum values of *y*. Checking *logscale* makes the scale on the axis logarithmic.
- The *Ticks* field gives the number of points computed to generate the curve. You'll only need to adjust it if the plotted curve isn't smooth.
- The *Format* field's `default` option displays the plot in a pop-up window; the only other useful option is `inline`, which instead inserts the plot in place in your wxMaxima session.
- Maxima uses a program called Gnuplot to generate plots. You can send commands directly to Gnuplot using the *Options* field. For example, the command `set grid;` shows a grid

on the plot, and the command `set size ratio 1;` makes the scales of the *x* and *y* axes the same. We'll look at the many options that can be set directly in Maxima without using Gnuplot commands in Chapter 4.

- The *File* option lets you save the plot to an encapsulated Postscript file (the ending `.eps` will be appended to the file name typed into the field). We'll see how to save plots in several other formats in Chapter 4.

The dialog inserts the `plot2d` function:

```
(%i1)   plot2d([sin(x)], [x,-2*%pi,2*%pi])$
```

which displays the plot in a pop-up window:

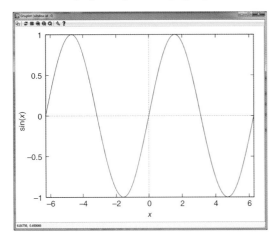

 Maxima will not let you continue unless you close the pop-up plot window.

1.2.1.1 Plotting Multiple Curves

Notice that each of the arguments for `plot2d` are lists, enclosed in square brackets. If you type a series of comma-separated expressions into the *Expression(s)* field in the dialog, each of the expressions will be plotted as a separate curve. For example, typing `sin(x), cos(x)` into *Expression(s)* field (or `[sin(x),cos(x)]` as the first argument of `plot2d`) gives us

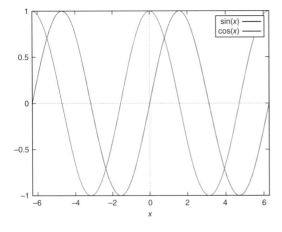

1.2.1.2 Parametric Plots

The Special button on the *Plot 2D* dialog brings up dialogs for parametric or discrete plots. A parametric plot shows x and y as functions of a third parameter t. For example, let's plot the curve defined by the parametric equations

$$x(t) = \frac{3}{2}\cos(t) - \cos(5t), \quad y(t) = \frac{3}{2}\sin(t) - \sin(4t), \quad 0 \le t \le 2\pi$$

Parametric equations to draw a starship

(1.4)

Choose Special ⟩ Parametric plot on the *Plot 2D* dialog to bring up this window:

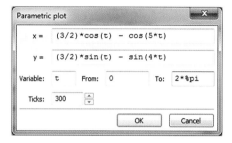

Press OK, and then OK to generate the following code and plot:

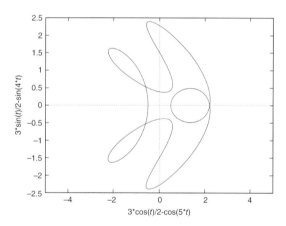

1.2.1.3 Discrete Plots

The Special ⟩ Discrete plot option lets you plot lists of discrete x and y data. For example, fill in the dialog with

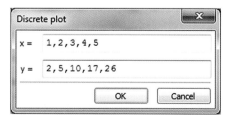

In the *Plot 2D* dialog, set the range for x from 0 to 7. Press OK, and then OK again on the *Plot 2D* dialog to generate the following plot:

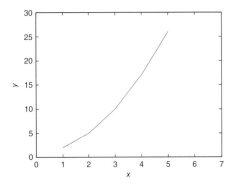

The discrete option looks like [discrete, xdata, ydata] where xdata and ydata are lists of x and y coordinates.[6]

Notice that plot2d automatically draws lines between the points. There is a *style* option for plotting points without connecting them, but you'll have to manually add it to the plot2d code the dialog generates:

```
(%i1)   plot2d([['discrete, [1,2,3,4,5], [2,5,10,17,26]]], [x
        ➥ ,0,7],[style, points])$
```

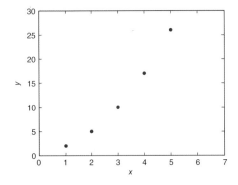

You can connect the points with lines and also draw the points with the *linespoints* style:

```
(%i1)   plot2d([[discrete, [1,2,3,4,5], [2,5,10,17,26]]], [x
        ➥ ,0,7],[style, linespoints])$
```

---

6 We'll see in Section 1.3 that a single quote ' suppresses the execution of the command that comes after it. The keywords discrete and parametric doesn't really need to have a single quote in front of them.

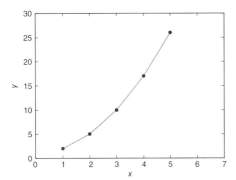

We'll see many more options for formatting and labeling graphs in Chapter 3.

1.2.1.4 Three-Dimensional Plots

The Plot 3D... button plots functions of x and y. It produces a dialog that is very similar to the *Plot 2D* dialog. For example, to plot the function $z = x\sin(y)$, fill in the *Plot 3D* dialog with

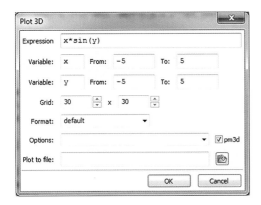

- The Grid field sets the number of points computed in the x and y directions.
- The Format field determines whether the plot appears in a gnuplot pop-up window (with "default" or "gnuplot") or embedded in the worksheet (with "inline").
- The Options field can be used to send commands directly to gnuplot. We'll see in Chapter 4 that this is rarely necessary; the `plot3d` function provides options of its own. The pm3d option controls surface coloring in gnuplot; you'll usually want to leave it on.
- The Plot to file field can save the plot to a PostScript file.

The dialog inserts a call to the `plot3d` function:

```
(%i1)    plot3d(x*sin(y), [x,-5,5], [y,-5,5])$
```

The graph appears in a gnuplot pop-up window when you execute the command:

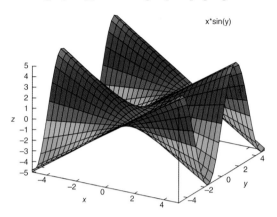

You can view the plot in the pop-up window from different angles by dragging the mouse over it. The mouse wheel pans across the surface. This won't work with embedded plots.

We'll see many options for formatting two- and three-dimensional plots in Chapter 4.

 Worksheet 1.2: Basic Plotting

In this worksheet, we'll graph functions with the Plot 2D... and Plot 3D... buttons on the General Math pane. We'll also see how to plot parametric equations and discrete data.

1.2.2 Basic Algebra

1.2.2.1 Equations
Maxima equations are two expressions related by an equals sign (=). For example, the ideal gas equation of state

$$PV = nRT \qquad \text{Ideal gas equation of state} \tag{1.5}$$

would be typed as

```
(%i1)   P*V=n*R*T;
```

$$(\%o1) \quad PV = nRT$$

1.2.2.2 Substitutions
Let's transform the ideal gas equation into the hard sphere gas equation of state, which considers the excluded volume b of the molecules:

$$P(V - nb) = nRT \qquad \text{Hard sphere equation of state} \tag{1.6}$$

We want to substitute $V - nb$ for V. Click the $\boxed{\text{Subst...}}$ button. The *Substitute* dialog appears.

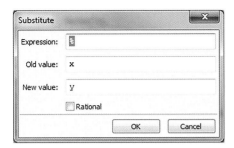

You can leave *Expression* as %, and the last result (P*V=n*R*T) will be substituted. If you had selected $P V = n R T$ from the previous output before clicking Subst..., it would have been automatically filled in as the expression.

Type V for the *Old value* and V-n*b for the *New value*. The code

(%i2)   **subst(V-n*b, V, %);**

(%o2) $P (V - b n) = n R T$

is inserted after the last cell executed. It means "Substitute V-n*b for V in the last result." As you become more comfortable with Maxima, you may find that it's easier to type the code directly.

We can rewrite this equation as the van der Waals equation of state

$$\left(P + \frac{n^2 a}{V^2}\right) (V - nb) = nRT \qquad \text{van der Waals equation of state} \qquad (1.7)$$

by substituting P+a*(n/V)^2 for P:

(%i3)   **subst(P+a*(n/V)^2, P, %);**

(%o3) $\left(\frac{a n^2}{V^2} + P\right) (V - b n) = n R T$

The old value must appear exactly in the equation as you've typed it in the dialog, or the substitution won't work. For example, using the dialog to substitute molar volume Vm for V/n in the equation fails:

(%i4)   **subst(Vm, V/n, %);**

(%o4) $\left(\frac{a n^2}{V^2} + P\right) (V - b n) = n R T$

Checking the Rational checkbox in the dialog performs a smarter substitution that recognizes mathematically equivalent forms of an expression:

(%i5)   **ratsubst(Vm, V/n, %);**

(%o5) $-\dfrac{n \left(b Vm^2 - Vm^3\right) P + n (a b - a Vm)}{Vm^2} = n R T$

The ratsubst function recognizes mathematically equivalent forms of the expression being substituted; subst does not.

1.2.2.3 Simplification

Maxima will not divide or multiply common factors out of an equation or expression automatically. We can simplify the equation by dividing both sides by n,

(%i6)   `%/n;`

(%o6) $-\dfrac{n\left(b\,Vm^2 - Vm^3\right)P + n\left(ab - a\,Vm\right)}{n\,Vm^2} = R\,T$

and then pressing the Simplify button:

(%i7)   **ratsimp(%)**

(%o7) $\dfrac{\left(Vm^3 - b\,Vm^2\right)P + a\,Vm - ab}{Vm^2} = R\,T$

Notice that operations on an equation are applied to both sides of the equation. The `ratsimp` function eliminates common factors above and below a fraction bar. We'll see how it can be fine-tuned in Section 5.3.1.

Let's multiply out the factors on the left-hand side of the equation. Click on the Expand button. The `expand` function is inserted into the next cell:

(%i8)   **expand(%);**

(%o4) $Vm\,P - b\,P + \dfrac{a}{Vm} - \dfrac{ab}{Vm^2} = R\,T$

Such expansions are very useful in making approximations. For example, the constants a and b in the van der Waals equation are related to the strength of intermolecular attractions and repulsions, respectively. These forces are weak, so the values of a and b are relatively small. The term in a times b must be very small, and might even be negligible. We can also tell that larger values of a tend to make PV_m smaller, and larger values of b make PV_m larger.

Now let's factor out the left-hand side. Click on the Factor button to insert the `factor` function:

(%i9)   **factor(%);**

(%o5) $\dfrac{\left(V - b\,n\right)\left(P\,V^2 + a\,n^2\right)}{V^2} = n\,R\,T$

Factoring is often useful in quickly finding solutions for an equation. For example, suppose we want to solve the following equation for x:

(%i1)   `x^4-(7*x^3)/2+3*x^2+x/2-1=0;`

(%o1) $x^4 - \dfrac{7\,x^3}{2} + 3\,x^2 + \dfrac{x}{2} - 1 = 0$

Press the Factor button to obtain

(%i2)   **factor(%)**

$$(\%o2) \qquad \frac{(x-2)\,(x-1)^2\,(2\,x+1)}{2} = 0$$

The first, second, and third factors are zero when $x = 2$, $x = 1$, and $x = -\frac{1}{2}$, respectively.
The factor function works on numbers, too:

(%i3)   **factor(2304);**

(%o3) $2^8\, 3^2$

1.2.2.4 Solving Equations

Let's solve Equation 1.7 for P. Click on the $\boxed{\text{Solve}}$ button, and fill in the Solve dialog that appears:

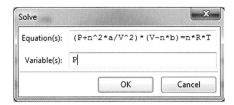

(%i1)   **solve([(P+n^2*a/V^2)*(V-n*b)=n*R*T], [P]);**

$$(\%o1) \qquad \left[P = \frac{n\,R\,T\,V^2 - a\,n^2\,V + a\,b\,n^3}{V^3 - b\,n\,V^2} \right]$$

Notice that each argument of the solve function is enclosed in square brackets. The first argument is a list of equations to solve; the second is a list of variables to solve the equations for. The solve function also returns a list of solutions enclosed in square brackets. In this case we have only one equation, one variable, and one solution. In Chapter 4, we'll see how to solve whole systems of equations for many variables, and we'll see how we can extract and use solutions in the list in further calculations.

1.2.2.5 Simplifying Trigonometric and Exponential Functions

Expressions or equations that include trigonometric functions, logarithms, exponentials, or roots of variables can be further simplified and manipulated using the buttons marked with (r) or (tr). For example, consider the expression

(%i1)   **sqrt(sin(x)^2 + cos(x)^2)+ exp(a*log(x)-b*log(y));**

(%o9) $e^{a\,\log(x) - b\,\log(y)} + \sqrt{\sin(x)^2 + \cos(x)^2}$

Pressing the $\boxed{\text{Simplify (r)}}$ button simplifies the exponential part of the expression by inserting the radcan function:

(%i2)   **radcan(%);**

$$(\%o10) \qquad \frac{\sqrt{\sin(x)^2 + \cos(x)^2}\,y^b + x^a}{y^b}$$

Pressing the [Simplify (tr)] button simplifies the trigonometric part: with the `trigsimp` function:

(%i3)   **trigsimp(%);**

(%o11) $\dfrac{y^b + x^a}{y^b}$

  Worksheet 1.3: Solving Equilibrium Problems

In this worksheet, we'll use the General Math pane to write and solve simple equilibrium problems.

1.2.3 Basic Calculus

In Sections 1.1.7 and 1.1.8, we saw how to use variables and functions in Maxima. In this section, we'll look at how the tiny (infinitesimal) variations in variables and functions can be related. The study of infinitesimal variations is called differential calculus; the summation of those infinitesimal variations is integral calculus.

We'll explore calculus in Maxima in more depth in Chapter 6. For now, let's take a quick look at what basic calculus tools are available on the General Math pane.

1.2.3.1 Limits

Sometimes we can't directly compute a function's value at a point. For example, we can't compute $\sin(x)/x$ at $x = 0$ directly, because division by zero is undefined. But we can find the value by studying what happens as x approaches zero. For very small x, $\sin(x)$ becomes nearly equal to x, so as x approaches zero, $\sin(x)/x$ approaches 1. We say that the limit of $\sin(x)/x$ as x approaches zero is one. This is written mathematically as

$$\lim_{x \to 0} \frac{\sin(x)}{x} = 1$$

The [Limit...] button or [Calculus] ⟩ [Find Limit...] menu item brings up a dialog for computing the limits of an expression. For example, to find the `limit` of the function $\sin(x)/x$ as x approaches 0 from either direction, we fill in the dialog with

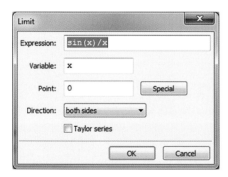

The result is

(%i1)   **limit(sin(x)/x, x, 0);**

(%o1)   1

The *Direction* dropdown lets you compute limits approaching the limit point from either or both directions. For example, computing the limit of $|x|/x$ as x approaches 0 from the right gives

```
(%i1)   limit(abs(x)/x, x, 0, plus);
```

```
(%o1)   1
```

while from the left, we have

```
(%i2)   limit(abs(x)/x, x, 0, minus);
```

```
(%o2)   −1
```

Limits can also be computed using Taylor series (see Section 6.7) by checking the *Taylor series* box in the dialog. For example, taking the limit of the function $(\coth(x) − 1/x)/x$ around $x = 0$ from either direction using a Taylor series inserts the `tlimit` function:

```
(%i1)   tlimit((coth(x)-1/x)/x, x, 0);
```

```
(%o1)   1
        ─
        3
```

1.2.3.2 Differentiation

A derivative dy/dx tells us the slope of a graph of y vs. x at a given point; it means "the tiny change in y per tiny change in x". For it to have any meaning, y must be a function of x.

One chemical application of derivatives is the relationship between potentials and forces. For example, consider the potential energy ϕ for two noble gas atoms separated by a distance r. One approximation for the potential energy is

$$\phi(r) = Ar^{-n} − Br^{-m} \qquad \text{A simple pair potential} \qquad (1.8)$$

where A and B are constants, and $n > m$. The force F acting between the two particles is

$$F = -\frac{d\phi}{dr} \qquad \begin{array}{l}\text{Force is negative} \\ \text{gradient of a potential}\end{array} \qquad (1.9)$$

To compute the derivative in Maxima, press the $\boxed{\text{Diff...}}$ button on the General Math pane, or select $\boxed{\text{Calculus}} \gg \boxed{\text{Differentiate...}}$ from the menu. This starts the *Differentiate* dialog, which can be filled in with

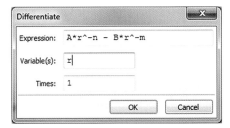

The *Times* field is 1 for a first derivative, 2 for a second derivative, and so on. Pressing OK inserts the `diff` function:

(%i1)   **diff(A*r^-n - B*r^-m,r,1);**

(%o1) $m r^{-m-1} B - n r^{-n-1} A$

1.2.3.3 Series

A power series is an infinite sum of the form:

$$\sum_{i=0}^{\infty} a_i x^i$$

Power series are often written as an alternative form for a function. For example, $\exp(x)$ when $|x| < 1$ can be written as

$$\exp(x) = \sum_{i=0}^{\infty} \frac{x^i}{i!}$$

A Taylor series is a power series expansion of a function $f(x)$ around some particular point $x = c$. It has the following form:

$$f(x) = \sum_{i=0}^{\infty} \frac{(x-c)^i}{i!} \left(\frac{d^i f(x)}{dx^i} \right)_{x=c}$$

Both power series and Taylor series can be computed using the [Series...] button or the [Calculus] [Get Series...] menu item. For example, to see the Taylor series for $\sin(x)$ expanded around $x = 0$,

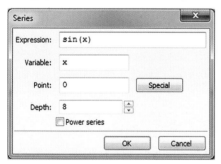

The `taylor` function is inserted:

(%i1)   **taylor(sin(x), x, 0, 8);**

(%o1)   /T/ $x - \dfrac{x^3}{6} + \dfrac{x^5}{120} - \dfrac{x^7}{5040} + \ldots$

To expand $\sin(x)$ as an infinite power series instead, check the *Power Series* box on the dialog. This inserts the `powerseries` function; the `niceindices` function automatically chooses indices like i, j, and k for the expansion:

(%i2)   **niceindices(powerseries(sin(x), x, 0));**

(%o2) $\displaystyle\sum_{i=0}^{\infty} \dfrac{(-1)^i x^{2i+1}}{(2i+1)!}$

1.2.3.4 Integration

The integral of a function $f(x)$ corresponds to the area under the curve when the function is plotted against the integration variable x. When the interval of x isn't specified, the integration is called *indefinite*, and it is essentially the inverse of differentiation. Defining the interval gives us a *definite* integral which can be evaluated either symbolically or numerically.

Suppose we want to integrate the rate law for the reaction $A + B \xrightarrow{k} P$. The rate law is

$$\frac{dx}{dt} = k(a - x)(b - x) \qquad \text{Second-order rate law} \tag{1.10}$$

where a and b are the initial concentrations of A and B, and x is concentration of P formed. We must collect terms in x and t on either side of the equation before integrating:

$$\int \frac{dx}{(a - x)(b - x)} = \int k dt \qquad \text{Integrated second-order rate law} \tag{1.11}$$

Let's evaluate the integral on the left-side of the equation. Press the $\boxed{\text{Integrate...}}$ button on the General Math pane or select $\boxed{\text{Calculus}} \gg \boxed{\text{Integrate...}}$ from the menu to bring up this dialog:

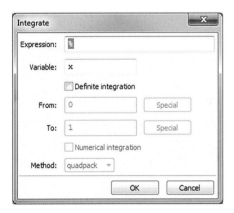

The *Expression* field holds the integrand (the expression to be integrated). Type `1/((a-x)*(b-x)*(c-x))`. Leave the integration variable (under *Variable*) as x. Press the $\boxed{\text{OK}}$ button. The result is a call to the `integrate` function:

```
(%i1)   integrate(1/((a-x)*(b-x)), x);
```

$$(\%o1) \qquad \frac{\log(x - b)}{b - a} - \frac{\log(x - a)}{b - a}$$

This is an indefinite integral. Indefinite integrals always include an integration constant, but Maxima doesn't show the constant.

Let's do the calculation again as a definite integral. Our interval for x can be from 0 to P (the concentration of the product), assuming that A is the limiting reactant (so $P = a$, and $P < b$). Click on the $\boxed{\text{Integrate...}}$ button again. Fill in the integrand `1/((a-x)*(b-x))` again, but this time check *Definite integration* and fill in the integration bounds:

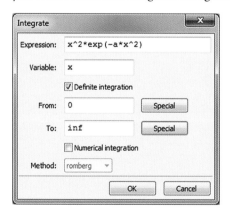

(%i2)   **integrate(1/((a-x)*(b-x)), x, 0, P);**

Is P positive, negative, or zero? positive;
Is P-a positive, negative, or zero? zero;
Is P-b positive, negative, or zero? negative;
Principal Value

(%o2) $\dfrac{\log(b-P)}{b-a} - \dfrac{\log(a-P)}{b-a} - \dfrac{\log(b)}{b-a} + \dfrac{\log(a)}{b-a} + \dfrac{\log(-1)}{b-a}$

The integrate function asks a series of questions about relationships between the integration bounds and variables that appear in the integrand. Answer each of the questions and type ⇧ + Enter at the end of each answer, just as you would in entering commands.

Suppose we wanted to evaluate a definite integral with a constant like π or ∞ as a bound. We've already seen that the codes for these constants are %pi and inf. The codes can be typed directly into the *From* and *To* fields – or you can use the Special button next to these fields to select the constants from a menu. For example, to compute

$$\int_0^\infty x^2 e^{-ax^2}\, dx$$

you would fill out the *Integrate* dialog like this:

The result is

(%i3)   **integrate(x^2*exp(-a*x^2), x, 0, inf);**

Is a positive, negative, or zero? `positive;`

$$(\%o3) \quad \frac{\sqrt{\pi}}{4\,a^{\frac{3}{2}}}$$

Some integrals do not have simple closed-form expressions. For others, Maxima will not provide simple expressions because it looks for solutions over the entire complex domain. For example, Maxima will not compute a simple expression for the integral[7]:

$$\int_0^\infty \frac{x^3}{e^x - 1}\,dx$$

`(%i1)    integrate(x^3/(exp(x)-1),x,0,inf);`

$$(\%o1) \quad \lim_{x \to \infty^-} 6\,li_4\,(e^x) - 6\,x\,li_3\,(e^x) + 3\,x^2\,li_2\,(e^x) + x^3 \log\,(1 - e^x) - \frac{x^4}{4} - \frac{\pi^4}{15}$$

which has a true value of $\pi^4/15$.

We'll see several different strategies for handling integrals like this in Section 6.5.3. One approach is to integrate numerically rather than symbolically. Fill in the Integrate dialog and check *Numerical integration*:

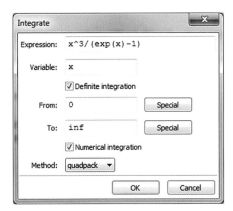

This inserts the rather cryptic code

`(%i2)    quad_qagi(x^3/(exp(x)-1), x, 0, inf);`

$(\%o2) \quad [6.493939402266831, 2.628471501138993 \; 10^{-9}, 165, 0]$

Maxima uses a software package called `quadpack` to integrate functions numerically. Quadpack routines like `quad_qagi` return a list of four numbers: the approximate value of the integral, its estimated absolute error, the number of integrand evaluations, and an error code (which should be zero if all went well). The true value of this integral is $\pi^4/15 = 6.493939402266829$.

---

7 This integral is required to compute the energy density of radiation emitted by a black body. We'll see in Chapter 6 that a simple symbolic result for the integral can be obtained either by a variable change or by using a power series expansion.

Worksheet 1.4: Basic Calculus

In this worksheet, we'll use the General Math pane to integrate and differentiate rate laws, and obtain Taylor and power series expansions for several common functions.

1.2.4 Differential Equations

A differential equation relates an independent variable x to derivatives of a dependent variable y. For example, the one-dimensional Schrödinger equation for a particle with mass m and kinetic energy E is

$$-\frac{h^2}{8\pi^2 m}\frac{d^2\Psi(x)}{dx^2} = E\Psi(x)$$

The Schrödinger equation
(one dimension, zero potential energy)

(1.12)

where Ψ is a function of the variable x and h is Planck's constant. In this case, the function Ψ depends on only one variable, so this is an ordinary differential equation, or ODE.

To solve the equation for Ψ, click on the $\boxed{\text{Solve ODE...}}$ button or select $\boxed{\text{Equation}} \gg \boxed{\text{Solve ODE...}}$ from the menu to bring up the *Solve ODE* dialog.

In the *Equation* field, type `-h^2/(8*%pi^2*m)* diff(Psi(x),x,2) = E*Psi(x)`. Fill the *Function* field with `Psi(x)`, and the *Variable* field with `x`. The result is

```
(%i1)  ode2(-h^2/(8*%pi^2*m)*diff(Psi(x),x,2) =E*Psi(x), Psi(x),
  ➥  x);
```

Is m E positive, negative, or zero? `positive;`

$$(\%o1)\quad \Psi(x) = \%k1\sin\left(\frac{2^{\frac{3}{2}}\pi\sqrt{m}x\sqrt{E}}{h}\right) + \%k2\cos\left(\frac{2^{\frac{3}{2}}\pi\sqrt{m}x\sqrt{E}}{h}\right)$$

Notice that we must type `Psi(x)` rather than `Psi` in the equation to show that it depends on `x`. The `ode2` solver will sometimes need to ask questions about variables that appear in the equation before it can compute a solution.

We'll look at solving differential equations in more detail in Chapter 11.

1.3 Controlling Execution

Maxima has several punctuation marks that give you exceptional control over how individual commands are executed and displayed in a session.

A single quote ′ before a command means "do not evaluate." Unevaluated expressions are called noun in Maxima. They are useful in typing differential equations that include unevaluated derivatives:

```
(%i1)  'diff(x,t,2) = -omega^2*x;
```

$$(\%o1)\quad \frac{d^2}{dt^2}x = -\omega^2 x$$

A pair of single quotes ″ before an expression replaces the expression with its value. Placing the quote–quote before a noun operator forces it to be evaluated, that is, it changes a noun into a verb. For example,

```
(%i1)   a:b$
(%i2)   b:c$
(%i3)   c:d$
(%i4)   a;
(%i5)   ''a;
(%i6)   '' ''a;

(%o4)   b
(%o5)   c
(%o6)   d
```

You can type a comma after an expression and follow it with a switch like *numer* or ratsimp to force the expression to be evaluated in numerical form or rationally simplified form, respectively. You can also add comma-separated variable definitions, substitutions, and switches after an expression, and they will apply only in the context of that expression, without creating variables or affecting other commands. This construct is called an evaluation environment in Maxima.

For example, suppose we want to calculate the volume $V = nRT/P$ of an ideal gas, given the moles of gas n, the temperature T, and the pressure P. We could define n, T, and P as variables, but this would wipe out any previous definitions for n, T, and P, and the new variables would persist through the entire session with Maxima. If we don't want to create those variables, the solution is to use an evaluation environment:

```
(%i1)   V = n*R*T/P, n=1, R=0.08205746, T=273.15, P=1, float;

(%o1)   22.41399519900001
```

Notice that the last thing we added was the float function, which forces the final result to be expressed in terms of decimal numbers. You can add other functions like ratsimp, expand, and factor to evaluation environments, too.

Use evaluation environments when you want to temporarily set variables for the evaluation of a single expression, without setting them globally for the rest of your session. For example, suppose you'd like to find the pH of several different weak acid solutions (each with a different dissociation constant K_a and initial concentration C_a), neglecting the dissociation of water. You could define a function with arguments K_a and C_a (as we did in Worksheet 1.2) or you could use evaluation environments:

```
(%i1)   pH : -log((sqrt(Ka^2+4*Ca*Ka)-Ka)/2)/log(10)$
(%i2)   pH_acetic_acid: pH, Ka=1.75e-5, Ca=0.1, numer;
(%i3)   pH_formic_acid: pH, Ka=6.138e-5, Ca=0.01, numer;
(%i4)   pH_hydrocyanic_acid: pH, Ka=6.08e-10, Ca=0.05, numer;

(%o2)   2.881353542698519
(%o3)   3.122994695104456
(%o4)   5.258587153540856
```

Evaluation environments can be nested inside other commands (including blocks, functions, loops, and other evaluation environments, as we'll see in Chapter 4). When an evaluation environment is part of another expression, though, you need to place it inside a call to the ev function:

ev(*evaluation environment*)
Wrap an evaluation environment to use it inside a function body or a larger expression.

A series of Maxima statements can be made into a single statement by putting them in parentheses and separating them by commas. This is called a block statement, and it too can be used inside a function body.

block(*statement 1, statement 2, ...*)
(*statement 1, statement 2, ...*)
Wrap a series of statements as a single statement for use inside a function body or a larger expression.

The statements inside the parentheses will be executed left to right, and the value of the entire block statement is the value of the final statement in the block. For example,

```
(%i1)    (a: 1, a: a+a, a: a^2);
```

```
(%o4)
```

Understanding nouns, verbs, evaluation environments, and blocks is crucial to mastering Maxima! We'll introduce these constructs in Worksheet 1.5, and explore them in more depth in Chapter 4.

 Worksheet 1.5: Controlling Execution

In this worksheet, we'll see how to suppress and force execution of commands with the quote and quote–quote operators, and practice using evaluation environments to substitute values and expressions into equations and plots.

1.4 Using Packages

Maxima has add-on packages for specialized tasks. Most packages load automatically when you use their functions, you don't have to worry about which package to load to access a particular function.

A few packages don't autoload. You will have to load them yourself, using the load command.

load(*package*)
Execute all commands in *package*.

For example, the distrib package defines useful functions associated with probability distributions. One such function is random_normal(m,s), which returns a random number from a normal distribution with mean `m` and standard deviation `s`. The function won't work unless you load the `distrib` package[8]:

```
(%i1)    random_normal(0,1);
```

```
(%o1)    random_normal(0, 1)
```

```
(%i2)    load(distrib)$
(%i3)    random_normal(0,1);
```

```
(%o3)    −0.96652166545112
```

It's easy to write your own packages. Save a worksheet that defines your functions and variables as a Maxima batch file by selecting File ⟩ Save As and setting *Save as type"* to `Maxima batch file (*.mac)`. For example, in a new worksheet, type

```
(%i1)    log10(x)  := log(x)/log(10);
```

and save the file as a Maxima batch file `log10.mac`. Now open a new worksheet and select File ⟩ Load Package... and browse to the file you just saved. Select the file, and click Open . The load command will be inserted into the new worksheet:

```
(%i1)    load("K:/work/SMC-Maxima/Fundamentals/log10.mac")$
```

Now the `log10` function is available, as you can see by printing the `functions` system variable:

```
(%i2)    functions;
```

```
(%o2)    [log10(x)]
```

---

8 Loading the `stats` package automatically loads the `distrib` package, among others, so if you are doing statistics in a session you can `load(stats)` to get this function.

2

Storing and Transforming Data

You can have data without information, but you cannot have information without data.
— Daniel Keys Moran

Much of chemistry is concerned with the collection of data, but the data themselves aren't of direct interest. We are more interested in *information*—the patterns of behavior data may reveal when it is properly organized, visualized, analyzed, averaged, or reduced. The process of interpreting the data begins when they are stored in a way that allows them to be easily retrieved and used to obtain meaningful results.

This chapter introduces Maxima's main data formats and structures. In Section 2.1, we'll look at how Maxima stores, rounds, and displays numbers, and how we can perform numerical calculations at any desired level of precision. In Section 2.3, we'll see how to build, edit, and do calculations with lists, Maxima's fundamental data type. Section 2.4 deals with indexing, entering, assigning, and editing datasets in the form of matrices. Finally, in Section 2.5 we'll see how to read and write formatted data from external files.

2.1 Numbers

Most computer languages truncate integer calculations, so that 3/2 is computed as 1, rather than 1.5. Maxima is programmed in a language called LISP, which can store integers and ratios of integers to nearly any desired number of digits [12]. Integer arithmetic is exact in LISP, so Maxima computes and displays results in terms of integers or ratios of integers whenever possible.

For computations with decimal numbers, Maxima offers two types of floating point numbers: "floats," which store decimal numbers to 16 significant digits, and "big floats" which can be stored to almost any desired number of digits. Maxima also supports complex numbers (which have both real and imaginary parts).

2.1.1 Floating Point Numbers

Floating point numbers contain a decimal point. They can be typed as ordinary decimal numbers (like 2.1330) or they can be entered in scientific notation.

Like most calculators and spreadsheets, Maxima uses e-notation to represent numbers in scientific notation. For example, you can enter Avogadro's number

```
(%i1)   6.022*10^23;
```

$$(\%o1) \quad 6.021999999999998 \, 10^{23}$$

Symbolic Mathematics for Chemists: A Guide for Maxima Users, First Edition. Fred Senese.
© 2019 John Wiley & Sons Ltd. Published 2019 by John Wiley & Sons Ltd.
Companion website: http://booksupport.wiley.com

in e-notation as

```
(%i2)   6.022e23;
```

$$(\%o2) \quad 6.022\ 10^{23}$$

The "e" stands for "times 10 to the power." The "e" can be upper or lower case.

 The "e" does NOT stand for exponentiation or the mathematical constant e; don't misinterpret a number like $2e3$ as 2^3 or $2e^3$! A common beginner's mistake is to type 6.022×10^{23} as `6.022*10e23`. This actually gives you $6.022 \times 10 \times 10^{23}$, or 6.022×10^{24}.

Notice that the first method for typing Avogadro's number gives an odd result: $6.021999999999998\ 10^{23}$. This is because decimal numbers are stored as binary numbers with a finite number of bits. This limits the number of significant figures in a floating point number to about 16. It introduces a small error into any number that cannot be represented exactly as a binary fraction.

Maxima will automatically try to avoid this issue by displaying results as symbolic constants or ratios of integers (rational numbers) whenever possible. It will only convert results into floating point numbers when forced.

```
(%i1)   %pi;
(%i2)   1/3;
(%i3)   sqrt(2);
(%i4)   (1/10)^20;
```

$$(\%o1) \quad \pi$$
$$(\%o2) \quad \sqrt{2}$$
$$(\%o3) \quad \frac{1}{3}$$
$$(\%o4) \quad \frac{1}{100000000000000000000}$$

These expressions can be converted to floating point numbers using either the `float` function or the *numer* switch, as we saw in Section 1.1.5:

```
(%i5)   float(%pi);
(%i6)   sqrt(2), numer;
(%i7)   1/3, float;
```

```
(%o5)   3.141592653589793
(%o6)   1.414213562373095
(%o7)   0.33333333333333
```

The difference between a number's exact value and its finite stored form is called round-off error. To see the size of the round-off error, we can find the unit round-off error, which is the maximum error that can occur when rounding a number to one. For example, consider the following code:

```
(%i1)   1 + 1e-16;
(%i2)   1 + 1e-15;
```

Table 2.1 System variables for controlling floating point calculations and results.

Variable	Default	
keepfloat	false	Prevents floating point numbers from being converted into rational numbers.
fpprec	16	Number of significant digits for big float numbers.
fpprintprec	0	Number of digits to print in numbers (2–16 in ordinary float numbers; 2 to fpprec for big float numbers).
numer	false	When set to true, *numer* causes functions to be evaluated as floating point numbers, and replaces variables by their values.
ratprint	true	When set to false, *ratprint* suppresses warning messages about the conversion of floating point numbers to rational numbers.

```
(%o1)    1.0
(%o2)    1.000000000000001
```

Using default floating point numbers, the unit round-off error is just under 10^{-15}. Such a small error in an individual result isn't usually a cause for concern. But in longer calculations that destroy significant figures, round-off errors can accumulate enough to make final results inaccurate. Calculations of this sort are called ill-conditioned. Examples of ill-conditioned calculations involve taking the difference between two very large numbers that are nearly the same size, or solving systems of equations when the coefficient for one of the variables is nearly the same size in all of the equations.[1]

Round-off errors can be decreased by extending the floating point precision. Extended precision floating point numbers are called big floats. There are four ways to make a number a big float:

- Type the number in e-notation but use a "b" in place of the "e." For example, a proposed standard value of Avogadro's number [14] could be typed as 6.0221414107040908409072b23.
- Select the number, and then select Numeric ≫ To Bigfloat .
- Type the command bfloat(*number*).
- Use the number in a calculation that involves big floats. Arithmetic between big floats and floats always yields a big float result.

Maxima uses special system variables to control the display and behavior of floating point numbers (see Table 2.1). You can choose the precision of big floats globally by selecting Numeric ≫ Set Precision... , which sets the system variable fpprec to the number of significant digits you want. Only big floats are affected:

```
(%i1)    fpprec: 32$
(%i2)    float(%pi);
(%i3)    bfloat(%pi);

(%o2)    3.141592653589793
(%o3)    3.1415926535897932384626433832795b0
```

---

1 One notorious example of an ill-conditioned calculation caused a failure of the Patriot missile defense system in the first Gulf War in 1992, resulting in the deaths of 28 soldiers [13]. Times and velocities stored by weapons control computers were represented inexactly in memory. The round-off errors in the times and velocities accumulated as the targeting programs ran. After 100 h of continuous operation, the computed targeting window for the missile battery had shifted so much that the computer failed to track an incoming missile.

Big floats are rounded to the specified number of digits:

```
(%i4)    fpprec: 4$
(%i5)    bfloat(%pi);
```

```
(%o5)    3.142b0
```

Let's look at π taken out to 1000 significant figures:

```
(%i1)    fpprec: 1000$
(%i2)    bfloat(%pi);
```

```
(%o7)    3.14159265358979323846264338327[943 digits]30019278766111959
09216420199b0
```

wxMaxima tries to hide most of the digits to keep the output readable. If you really need to see them all, you have to momentarily switch off wxMaxima's default display by placing your `bfloat` calculation between `set_display(ascii)` and `set_display(xml)` commands:

```
(%i3)    set_display(ascii)$
(%i4)    bfloat(%pi);
(%i5)    set_display(xml)$
```

```
(%o6)    3.14159265358979323846264338327950288419716939937510582097
49445923078164\
06286208998628034825342117067982148086513282306647093844460
95505822317253594081\
⋮
78049951059731732816096318595024459455346908302642522308253
3446850352619311881\
71010003137838752886587533208381420617177669147303598253490
4287554687311595628\
63882353787593751957781857780532171226806613001927876611195
909216420199b0
```

The backslashes on the end of each line mean that the line is continued.

Now let's see how unit round-off is affected when using 128-digit big floats. Ordinary floating point numbers are unaffected by setting *fpprec*.

```
(%i1)    fpprec : 128$
(%i2)    1 + 5.5e-16;
```

```
(%o2)    1.0
```

Adding a big float to a number makes the result a big float, too:

```
(%i3)    1 + 5.5b-16;
```

```
(%o3)    1.0000000000000000055b0
```

We can do much better than that. With 128-digit big floats, the smallest possible number that can be added to one and give a number that isn't one is about 4.5×10^{-128}:

```
(%i4)   1 + 4.5b-128;
```

```
(%o4)   1.0000000000000000000000000000000[71digits]0000000
00000000000000000000001b0
```

There are two ways to control the number of digits displayed:

- The *fpprintprec* system variable controls the printing of both big and ordinary floats. You can set *fpprintprec* to anything between 2 and the value of *fpprec* to display numbers rounded to the specified number of digits. For example,

```
(%i1)   fpprintprec: 2$
(%i2)   float(%pi);
(%i3)   fpprintprec : 10$
(%i4)   float(%pi);
(%i5)   fpprintprec : 0$
```

```
(%o2)
(%o4)
```

Setting *fpprintprec* to zero automatically displays numbers to their full precision.
- Use the crossed-tools (Configure wxMaxima) button on the toolbar. The option to set the maximum number of digits displayed appears under the Worksheet tab.

Calculations with big floats require more memory and may be significantly slower than calculations with ordinary floats. This is seldom noticeable in non-repetitive calculations.

2.1.2   Integers and Rational Numbers

Integers are *fixed point numbers*: the decimal point is not shown, but it always just after the last digit. Integers are used to represent counts and to index data in lists and matrices.

We will often need to convert a decimal number like 372.51 into an integer by discarding the digits that come after the decimal point. There are three possibilities:

- Truncate the number at the decimal point with the `floor` function. For example, `floor(372.51)` is 372.
- Round the number with the `round` function; `round(x)` gives us 373.
- Find the smallest integer greater than or equal to the original number. The function we need is `ceiling`; `ceiling(372.51)` displays 373.

Rational numbers are ratios of integers. Numbers like π and $\sqrt{2}$ are *not* rational numbers, because they cannot be written as a simple ratio of integers. Such numbers are said to be *irrational numbers*.

We've already seen how to convert both rational and irrational numbers into floating point numbers using `float` and *numer*.[2] To make floating point results the default, you

---

2 There is a subtle difference between *numer* and `float`. `float` converts rational numbers to decimal numbers, while *numer* additionally causes some mathematical functions to be evaluated in floating point, and it replaces variables with their numeric values if they have been assigned.

can set the *numer* system variable to `true`, or you can toggle *numer* by selecting Numeric〉 〉Toggle Numeric Output from the menu.

To convert numbers from floating point to rational, use the `rat` function:

```
(%i1)    sqrt(2);
(%i2)    float(%);
(%i3)    rat(%);
```

$$(\%o1) \quad \sqrt{2}$$
$$(\%o2) \quad 1.414213562373095$$
$$(\%o3) \quad /R/ \quad \frac{22619537}{15994428}$$

The `rat` function displays a number or expression by expanding it, putting all terms over a common denominator, and then canceling out the greatest common divisor. This is called a *canonical rational expression*, and it is labeled in output with /R/.

The conversion of a decimal number to a rational number is only accurate to within a tolerance given by the system variable *ratepsilon*, which has a default value of 2.0×10^{-15}. Maxima automatically converts decimal numbers into rational numbers to minimize round-off error, and it will warn you whenever such conversions occur:

```
(%i1)    x + 1.1 = 3.5$
(%i2)    solve(%,x);
```

rat: replaced - 2.4 by -12/5 = - 2.4

$$(\%o2) \quad [x = \frac{12}{5}]$$

Warnings about exact conversions can be ignored.[3] If you'd rather not see the warnings at all, set the system variable *ratprint* to false. You can also prevent the conversions in the first place by setting the system variable *keepfloat* to true.

 Worksheet 2.0: Numbers

In this worksheet, we'll display numbers in different formats and precisions, and examine the effects of round-off error.

2.1.3 Complex Numbers

A complex number has two components: a real number x and an imaginary number iy, where i is the square root of -1. The *rectangular form* of the complex number is written as $x + iy$. If $y = 0$, the number is real, so real numbers are just a special case of complex numbers.

In physics and chemistry, complex numbers are often used to represent paired quantities like voltage and current in an electric circuit, or electric and magnetic field strengths in an electromagnetic field. Expressing these quantities as a single complex number $x + iy$ rather than a pair of real numbers (x, y) can simplify mathematical models of the system. Complex numbers and

---

3 Warnings about conversions that aren't exact can be ignored as well, unless your calculation requires extreme precision or is ill-conditioned. In these cases you'll want to use extended precision, and set *ratepsilon* to a smaller value.

functions are absolutely essential in quantum chemistry, where they are used to build wave-functions that correctly represent and distinguish specific states. We also encounter complex numbers in the solutions of many simple algebraic equations, even when the equations themselves involve only real numbers.

In Maxima, *i* is typed as `%i`, so a complex number like $5 + 2i$ would be typed as `5+2*%i`. We can extract the real and imaginary parts with the `realpart` and `imagpart` functions:

```
(%i1)   z1 : 5+2*%i;
(%i2)   realpart(z1);
(%i3)   imagpart(z1);
```

```
(%o1)   2 %i + 5
(%o2)   5
(%o3)   2
```

Complex numbers obey the ordinary laws of algebra. When adding or subtracting complex numbers, the real and imaginary parts are added or subtracted separately:

$$(a + bi) \pm (c + di) = (a \pm c) + (b \pm d)i \qquad \text{Addition and subtraction of complex numbers} \qquad (2.1)$$

```
(%i4)   z2 : 3+4*%i$
(%i5)   z1+z2;
(%i6)   z1-z2;
```

```
(%o4)   6 %i + 8
(%o5)   2 - 2 %i
```

Similarly, multiplying or dividing a pair of complex numbers yields

$$(a + bi)(c + di) = (ac - bd) + (bc + ad)i \qquad \text{Product of two complex numbers} \qquad (2.2)$$

$$\frac{a + bi}{c + di} = \frac{ac + bd}{c^2 + d^2} + \frac{cb - ad}{c^2 + d^2}i \qquad \text{Quotient of two complex numbers} \qquad (2.3)$$

since $i^2 = -1$. Maxima can perform these operations automatically when you multiply or divide two complex numbers, but (as usual) it won't simplify results unless you ask it to.

```
(%i7)   z1*z2;
(%i8)   rectform(%);
```

```
(%o7)   (2 %i + 5) (4 %i + 3)
(%o8)   26 %i + 7
```

```
(%i9)   z1/z2;
(%i10)  rectform(%);
```

$$(\%o9) \quad \frac{2 \%i + 5}{4 \%i + 3}$$

$$(\%o10) \quad \frac{23}{25} - \frac{14 \%i}{25}$$

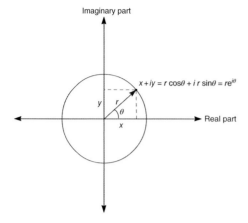

Imaginary part

$x + iy = r\cos\theta + i\,r\sin\theta = re^{i\theta}$

Real part

Figure 2.1 Points in the complex plane can be represented in either rectangular form $(x + iy)$ or polar form $(re^{i\theta})$.

To cast the final result of complex arithmetic into the form $a + bi$, use the `rectform` function (available from the menu as Simplify ⟩⟩ Complex Simplification ⟩⟩ Convert to Rectform , or from the Rectform button on the *General Math* pane). You can also use `rectform` as a switch:

```
(%i11)   (z1+z2)/(z1-z2), rectform;
```

$$(\%o11) \quad \frac{7i}{2} + \frac{1}{2}$$

Complex numbers can be visualized as points in the *complex plane* (Figure 2.1). The real part of the number is its x coordinate in the complex plane, and the imaginary part is y, so $x + iy$ corresponds to the point (x, y) in the complex plane.

We can represent a complex number in either rectangular or polar form. Since $x = r\cos(\theta)$ and $y = r\sin(\theta)$, the polar form is

```
(%i1)   polar_form: x + %i*y, x=r*cos(theta), y=r*sin(theta);
```

$$(\%o1) \quad i\,r\sin(\theta) + r\cos(\theta)$$

which can be simplified to

```
(%i2)   exponentialize(polar_form), ratsimp;
```

$$(\%o2) \quad re^{i\theta}$$

The `exponentialize` function converts trigonometric functions to exponentials by substitution using the Euler's formula [15]:

$$\exp(i\theta) = \cos\theta + i\sin\theta \qquad \text{Euler's formula} \qquad (2.4)$$

...so the simplified polar form can also be obtained by

```
(%i3)   ratsubst(exp(%i*theta), cos(theta)+%i*sin(theta),
      ➥ polar_form);
```

$$(\%o3) \quad re^{i\theta}$$

The angle θ is called the *complex argument* or the *phase angle*. It can be computed using the `carg` function[4]:

```
(%i1)    carg(a+b*%i);
```

```
(%o1)    atan2(b, a)
```

The distance from the origin r is called the *modulus* or *absolute value*. It can be computed with the `cabs` function:

```
(%i2)    cabs(a+b*%i);
```

$$(\%o2) \quad \sqrt{b^2 + a^2}$$

The *absolute square* or *square modulus* $|z|^2$ is just z^2 if z is a real number. But if z is a complex number $x + iy$,

$$|z|^2 = zz^* = (x + iy)(x - iy) = x^2 + y^2 \qquad \text{The square modulus} \qquad (2.5)$$

The absolute square isn't the number times itself; it's the number times its *complex conjugate*, which is the number with the sign on i flipped. On paper, the complex conjugate is usually indicated by a superscript asterisk: $(x + iy)^* = x - iy$.

In Maxima, use the `conjugate` function to write a complex conjugate:

```
(%i1)    z: x+%i*y;
(%i2)    conjugate(z);
```

```
(%o1)    %i y + x
(%o2)    x - %i y
```

and either square the `cabs` function or multiply a complex number by its complex conjugate to compute the square modulus:

```
(%i3)    cabs(z)^2;
(%i4)    conjugate(z)*z;
```

```
(%o3)    y² + x²
(%o4)    y² + x²
```

Maxima can convert between the polar and rectangular forms of complex numbers with `polarform` and `rectform`:

```
(%i5)    polarform(z);
(%i6)    rectform(%);
```

$$(\%o5) \quad \sqrt{y^2 + x^2}\, e^{\%i\, \text{atan2}(y, x)}$$

```
(%o6)    %i y + x
```

---

4 The `carg` function doesn't simplify its results. See Section 5.3.2 for more about simplifying trigonometric functions.

Maxima computes all results in the complex domain, so solving equations will find any complex solutions along with the real solutions. Maxima computes results in the complex domain, so it will find both imaginary and real solutions of an equation. For example,

```
(%i1)   eqn: x^3 - x^2 +4*x -4 =0;
```

$$(\%o1) \quad x^3 - x^2 + 4x - 4 = 0$$

has solutions

```
(%i2)   solve(eqn, x);
```

$$(\%o2) \quad [x = -2\,\%i, x = 2\,\%i, x = 1]$$

Notice that even though all coefficients in the equation are real, it has some imaginary solutions. This can occur even in "physical" equations where only the real solutions have physical meaning.

For example, let's find the molar volume of argon by solving the van der Waals equation at STP:

```
(%i1)   solve((P+a/V^2)*(V-b)=R*T, V), P=1.0, a=1.355, b=0.0320,
    ➡    T=273.15, R=0.082059, numer;
```

$(\%o1) \quad [V{=}7.433445749265302 * (0.86602540378444 * \%i - 0.5) +$
$7.470389034296296 * (-0.86602540378444 * \%i - 0.5) + 7.482138616666667,$
$V = 7.470389034296296 * (0.86602540378444 * \%i - 0.5) + 7.433445749265302 *$
$(-0.86602540378444 * \%i - 0.5) + 7.482138616666667, V = 22.38597340022827]$

The `rectform` function is useful for simplifying the complex solutions:

```
(%i2)   rectform(%);
```

$(\%o2) \quad [V = 0.030221224885867 - 0.03199382333609 * \%i,$
$V = 0.03199382333609 * \%i + 0.030221224885867, V = 22.38597340022827]$

The equation is cubic in V, so we obtain three solutions. The first two solutions are complex; the last is the physically meaningful solution.

 Worksheet 2.1: Complex Numbers and Arithmetic

In this worksheet, we'll see how to define, manipulate, and simplify expressions that involve complex numbers.

2.1.4   Constants

Constants are variables that don't vary. They have fixed values, and are treated differently from variables in operations like differentiation, integration, and equation solving.

Maxima includes built-in mathematical constants like π and e (see Table 2.2). You'll sometimes see them appear in symbolic solutions of equations, limits, integrals, and sums.

Table 2.2 Built-in dimensionless mathematical constants in Maxima.

`%e`	Base for natural logarithms (2.718...)
`%i`	The square root of -1
`%gamma`	Euler's constant (0.5772...)
`ind`	A bounded, indefinite result
`inf`	Real positive infinity
`infinity`	Complex infinity
`minf`	Real negative infinity
`%phi`	The golden mean, $\frac{1+\sqrt{5}}{2}$
`%pi`	3.141592...
`true`	Logical true
`false`	Logical false
`und`	Undefined
`zeroa`	An infinitesimal above zero
`zerob`	An infinitesimal below zero

The indeterminate constants `ind` and `und`, and the infinite constants `inf`, `minf`, and `infinity` are mainly used in reporting and computing limits, integrals, and Taylor series. They'll be treated as simple variables if you try to use them in expressions; they were not intended to be used that way, and sometimes give unexpected results. For example, Maxima correctly refuses to simplify $1/\infty$ (because infinity is not really a number) but it will incorrectly report that infinity plus infinity is two times infinity because it treats infinity as a simple variable:

```
(%i1)    1/inf;
(%i2)    inf+inf;
```

$$(\%o1) \quad \frac{1}{\infty}$$
$$(\%o2) \quad 2\infty$$

A variable can be declared a constant using the `declare` function:

```
(%i1)    Avogadro : 6.02214129e23;
(%i2)    declare(Avogadro,constant);
```

$(\%o1)$ $6.02214129 \; 10^{23}$
$(\%o2)$ *done*

Maxima contains a library of physical constants with attached units. We'll see how to access them in the next section.

2.1.5   Units and Physical Constants

Units and unit conversions are essential in chemical calculations. Though unit conversions are usually simple enough to set up manually, it can be tedious to track them through lengthy computations.

Maxima can track units and convert them on demand.[5] To enter a quantity with units, type a back tick ⌐ ' ⌐ (usually found to the left of the ⌐ 1 ⌐ key on your keyboard), followed by the unit expression. For example,

```
(%i1)   load(physical_constants)$
(%i2)   8.314`J/(mol*K);
```

$$(\%o1) \quad 8.314`\frac{J}{mol\,K}$$

You can see a list of all known units by typing known_units(). The list is rather limited, but we'll see below that it is easy to define your own units and unit conversions. You must be careful not to use any of the known unit names as variable names, or the results can be confusing when doing unit conversions or dimensional analysis.

Maxima doesn't have a preferred system of units, so you must specify the units you'd like. This is done with the unit conversion operator, which is two back ticks ⌐ ' ⌐ ' ⌐. For example, to convert 4 centimeters to meters,

```
(%i1)   4`cm `` m;
```

$$(\%o1) \quad 0.04`m$$

Let's compute the pressure P (in pascals) exerted by a fluid in a drinking straw filled to 15 cm with a water-like liquid (with density 1.0 g cm^{-3}). The pressure is density times acceleration due to gravity (9.8 m s^{-2}) times the height of the fluid column:

```
(%i1)   load(physical_constants)$
(%i2)   density : 1.0 `g/cm 3 ;
(%i3)   %g : 9.81 `m/s 2 ;
(%i4)   height: 15 `cm;
(%i5)   P: density*%g*height `` Pa;
```

$$(\%o2) \quad 1.0` \frac{g}{cm^3}$$

$$(\%o3) \quad 9.81` \frac{m}{s^2}$$

$$(\%o4) \quad 15` cm$$

$$(\%o5) \quad 1471.5` Pa$$

 We used %g rather than g as the variable name for the acceleration due to gravity to avoid redefining the abbreviation for grams. Don't use unit names or abbreviations as variable names!

To see a list of all of the unit conversions Maxima knows, type the system variable known_unit_conversions. The list shows that Maxima recognizes prefixes like m (milli), k (kilo), M (mega), and G (giga), but it won't apply them consistently to all SI base units.

---

5 As of Maxima 5.37.1, you must load the physical_constants package before you can use units and physical constants in Maxima.

You'll often need to add your own unit conversions to the list using `declare_unit_conversion` function. For example, Maxima doesn't know about atmospheres (atm), or torr,[6] so to convert the pressure in atm or torr, we have to declare unit conversions first:

```
(%i6)   declare_unit_conversion(atm = 101325*Pa, torr = atm/760)
          ➡ $
(%i7)   P `` atm;
(%i8)   P `` torr;

(%o7)   0.014522575869726 ` atm
(%o8)   11.03715766099185 ` torr
```

The units have to be declared so that the left-hand side of each conversion is a simple unit, not an expression. For example, `760*torr = atm` doesn't work; `torr = atm/760` does. You may use backticks instead of multiplications and divisions as long as there is a one in front of the left-hand unit; for example, `1` torr = (1/760)` atm` or `1` atm = 760` torr` will work, but `760` torr = 1`atm` does not.

Maxima has a library of physical constants. Physical constant variable names begin with a percent sign, by default.[7] Some useful constants in chemistry are listed in Table 2.3. Maxima displays constants by their names in expressions. To see the value and units, use the `constvalue` function. Constants are stored as rational numbers, so you have to use `float` to print them

Table 2.3 Some useful physical constants available in Maxima's `physical_constants` package.

`%a_0`	The Bohr radius
`%c`	The speed of light
`%%e`	The charge on an electron
`%e_0`	Electric permittivity of free space
`%E_h`	The Hartree energy
`%F`	The Faraday constant
`%h`	The Planck constant
`%h_bar`	The reduced Planck constant $\hbar$
`%%k`	The Boltzmann constant
`%m_e`	The mass of an electron
`%m_u`	Atomic mass unit
`%mu_0`	Magnetic permeability of free space
`%N_A`	The Avogadro constant
`%R`	The ideal gas law constant
`%R_inf`	The Rydberg constant
`%sigma`	The Stefan–Boltzmann constant
`%V_m`	Molar volume of an ideal gas at STP

All constants are in SI units.

----

6 The `maxima-init.mac` file that comes with this book adds these and many more chemically useful units, conversions, functions, and physical constants to Maxima's repertoire on startup.

7 If you like, you can suppress the % sign on the beginning of constant names. Select Edit ⟩⟩ Configure ⟩⟩ Options , and uncheck the box that says "Keep percent sign with special symbols."

as decimal numbers. In the following example, we get the value and units of the ideal gas law constant %R, and convert them to L atm mol^{-1} K^{-1}.

```
(%i1)   load(physical_constants)$
(%i2)   constvalue(%R);
(%i3)   float(%);
```

$$(\%o1) \quad \frac{1039309}{125000} \text{`} \frac{J}{mol\,K}$$

$$(\%o2) \quad 8.314472 \text{`} \frac{J}{mol\,K}$$

Now let's convert R to L atm mol^{-1} K^{-1}. Declare conversions for atm and torr, and use an uppercase L for liters.

```
(%i4)   declare_unit_conversion(atm = 101325*Pa, torr = atm/760,
    ➥     L = liter)$
(%i5)   constvalue(%R) `` L*atm*mol^-1*K^-1$
(%i6)   float(%);
```

$$(\%o5) \quad 0.082057458672588 \text{`} \frac{atm\,L}{mol\,K}$$

Some functions (like sqrt) can handle quantities with units:

```
(%i1)   load(physical_constants)$
(%i2)   x : 3.2`m^2/s^2;
(%i3)   sqrt(x);
```

$$(\%o2) \quad 3.2 \text{`} \frac{m^2}{s^2}$$

$$(\%o3) \quad 1.788854381999831 \text{`} \frac{m}{s}$$

Many other functions (like sin and log) require unitless arguments.

```
(%i4)   sin(x);
(%i5)   log(x);
```

$$(\%o4) \quad \sin\left(3.2 \text{`} \frac{m^2}{s^2}\right)$$

$$(\%o5) \quad \log\left(3.2 \text{`} \frac{m^2}{s^2}\right)$$

You must strip units from measurements before passing them as arguments to these functions. There are two ways to do this. You can use the qty (quantity) function, which returns the number part of a measurement without units, or you can divide the measurement by its units. For example,

```
(%i6)   sin(qty(x));
```

$$(\%o6) \quad -0.05837414342758$$

```
(%i7)   log(x/(1`m^2/s^2));
```

```
(%o7)   1.16315080980568
```

 Unfortunately units in plotting, integrating, differentiating, and equation-solving aren't consistently supported in the current version of Maxima (5.37.1). If you are going to use units in Maxima, attach them to final results for conversion, rather than propagating them through a lengthy calculation.

 Worksheet 2.2: Units and unit conversion

In this worksheet, we'll define units and perform unit conversions with them. We'll also introduce in Maxima.

2.2 Boolean Expressions and Predicates

Boolean expressions have a value of `true` or `false`. Every digital computation uses them in some form, as does any device that makes decisions based on sensor inputs. In Maxima, we'll use Boolean expressions to write programs that perform conditional calculations, or repeat calculations until some desired stopping criterion has been met. We'll also use Boolean expressions in filtering datasets.

2.2.1 Relational Operators

Boolean expressions usually involve comparison of two quantities using *relational operators* like >, <, and =. For example, the Boolean expression 1 < 2 is `true` and 2 < 1 is `false`. Table 2.4 lists several other relational operators.

Maxima does not simplify Boolean expressions to `true` or `false` when they are typed alone. You must evaluate them using either the `is` function, or the *pred* (predicate) switch. For example,

```
(%i1)   1<2;
(%i2)   is(1<2);
(%i3)   1<2, pred;
```

Table 2.4 Relational operators that compare the values of two expressions or variables.

Operator	Meaning	Example
>	Is greater than	i > 0
<	Is less than	i < n
>=	Is greater than or equal to	i >= j + 1
<=	Is less than or equal to	1 <= m
=	Is numerically equal to	i = 2 * k
#	Is not equal to	k # j/2

```
(%o1)   1 < 2
(%o2)   true
(%o3)   true
```

The output can be `true`, `false`, or `unknown`:

```
(%i4)   is(x>z);
```

```
(%o2)   unknown
```

Arithmetic operations are performed before relational operations. For example, typing `x + 1 > y - z` is the same as typing `(x + 1) > (y - z)`.

2.2.2 Logical Operators

Logical operators operate on true and false values, or on Boolean expressions. For example, the `not` operator gives the opposite value:

```
(%i1)   not true;
(%i2)   not false;
```

```
(%o1)   false
(%o2)   true
```

The `and` operator returns `true` only if both of the expressions it connects are true:

```
(%i3)   true and true;
(%i4)   true and false;
(%i5)   false and true;
(%i6)   false and false;
```

```
(%o3)   true
(%o4)   false
(%o5)   false
(%o6)   false
```

The `or` operator returns true if at least one of the expressions it connects are true:

```
(%i7)    true or true;
(%i8)    true or false;
(%i9)    false or true;
(%i10)   false or false;
```

```
(%o7)    true
(%o8)    true
(%o9)    true
(%o10)   false
```

The `logic` package contains additional logical operators including `nand`, `nor`, `implies`, `eq`, and `xor`. We'll look at them briefly in Worksheet 2.4.

Logical and, or, and not operations are performed *after* relational operations; not operations are performed first. For example, not i>j and not j>k is equivalent to (not (i>j)) and (not (j>k)).

2.2.3 Predicates

A Boolean function is called a predicate. Some useful built-in predicates are listed in Table 2.5. Notice that built-in predicates that determine the data type of a single argument have names that end in "p." Predicates that compare two arguments don't follow this convention.

Maxima has three ways to test for equality and equivalence:

1) An equals sign = tests for equality after basic simplification. For example,

```
(%i1)   is(4=4);
(%i2)   is(2+2=4);
(%i3)   is(4=4.0);

(%o1)   true
```

Table 2.5 Predicate functions that can be used in logical tests.

Predicates for determining type	Returns true only if *expression* is a(n):
alphanumericp(*expression*)	Alphabetic character or digit
bfloatp(*expression*)	Big float
constantp(*expression*)	Constant
evenp(*expression*)	Even integer
floatnump(*expression*)	Floating point number
integerp(*expression*)	Integer
listp(*expression*)	List
matrixp(*expression*)	Matrix
numberp(*expression*)	Integer, rational number, floating point number, or big float
nonnegintegerp(*expression*)	An integer greater than or equal to zero
nonscalarp(*expression*)	A list or matrix
oddp(*expression*)	Odd integer
primep(*expression*)	Prime number
ratnump(*expression*)	Rational number
scalarp(*expression*)	Number, constant, or variable, but not a list or matrix
stringp(*expression*)	String

Predicates for comparisons	Returns true only if a and b are:
equal(a,b)	Equal after rational simplification
logic_equiv(a,b)	Logically equivalent[a]
notequal(a,b)	Not equal after rational simplification
sequal(a,b)	Strings of the same length, containing the same characters

[a] You must load the logic package to use this function.

(%o2)   *true*
(%o3)   *false*

The last test is false because integer 4 and real 4.0 aren't literally equal.

2) The equal(a,b) predicate tests for equality after algebraic simplification (using rat-simp). This is more useful for comparing algebraic expressions than the = operator. For example, consider the following two expressions. They aren't literally equal, but they are algebraically equal:

(%i1)   expr1: (P-n^2*a/V^2)*(V-n*b);
(%i2)   expr2: (P*V^3-b*n*P*V^2-a*n^2*V+a*b*n^3)/V^2;

$$(\%o1) \quad \left(P - \frac{a\,n^2}{V^2}\right)(V - b\,n)$$

$$(\%o2) \quad \frac{P\,V^3 - b\,n\,P\,V^2 - a\,n^2\,V + a\,b\,n^3}{V^2}$$

(%i3)   **is**(expr1=expr2);

(%o3)   *false*

(%i4)   **is**(**equal**(expr1,expr2));

(%o4)   *true*

3) The logic_equiv function (from the logic package) tests for equality after logical simplification (using logic_simp). For example, the following two expressions aren't literally or algebraically equal, but they *are* logically equivalent:

(%i1)   statement1: A **or** (B **and** C);
(%i2)   statement2: (A **or** B) **and** (A **or** C);
(%o1)   *A or B and C*

(%o2)   *(A or B) and (A or C)*

(%i3)   **is**(statement1 = statement2);

(%o3)   *false*

(%i4)   **is**(**equal**(statement1,statement2));

(%o4)   *unknown*

(%i5)   **load**(logic)$
(%i6)   **logic_equiv**(statement1,statement2);

(%o6)   *true*

In general, two expressions can be tested for equality after both are reduced to some common "canonical" form. We'll see how to test equality for expressions involving trigonometric and exponential functions in Worksheet 2.4.

In Section 4.3, we'll use Boolean expressions and predicates to make decisions or comparisons in Maxima programs, and in Section 2.3.6, we'll use predicates to filter and manipulate datasets.

 Worksheet 2.3: Boolean Expressions and Predicates

In this worksheet, we'll use relational operators to compare variables and expressions, and determine when expressions are computationally or logically equivalent.

2.3 Lists

Lists are a fundamental data type in Maxima. Maxima's equation-solving and curve-fitting functions return their results as lists, and plotting and data analysis functions take lists as arguments.

A list is an ordered series of items. The items can be anything: variables, strings, numbers, functions, equations, expressions, or even other lists. The items are separated by commas, and the entire list is enclosed in square brackets.

2.3.1 List Assignments

Lists can be stored as variables. For example, here is a list called `conc` that contains four items:

```
(%i1)   conc : [0.001, 0.01, 0.1, 1];
```

```
(%o1)   [0.001, 0.01, 0.1, 1]
```

Notice that when you type the first square bracket `[`, wxMaxima automatically types a closing square bracket for you. An empty list is `[]`.

You can use lists to do multiple assignments:

```
(%i2)   [c1, c2, c3, c4] : conc;
(%i3)   c3;
```

```
(%o5)   [0.001, 0.01, 0.1, 1]
(%o6)   0.1
```

The assignments are carried out in parallel. This makes it easy to swap the values of variables:

```
(%i1)   x : 1$
(%i2)   y : 2$
(%i3)   [x, y] : [y, x];
(%i4)   x;
(%i5)   y;
```

```
(%o3)   [2, 1]
(%o4)   2
(%o5)   1
```

2.3.2 Indexing List Items

Items in the list are indexed starting from one. In the `conc` list, 0.001 is item 1, 0.01 is item 2, and so on. To access an individual item, type the item number in square brackets after the list name. For example, `conc[3]` is 0.1.

List items can be assigned by index. For example, to set the fourth concentration to 0.5,

```
(%i6)   conc[4] : 0.5$
(%i7)   conc;
```

```
(%o3)   [0.001, 0.01, 0.1, 0.5]
```

You can use the `first` and `last` functions to retrieve the first and last items in the list.[8] For example, `first(conc)` is 0.001, and `last(conc)` is 0.5.

You can't use indices that are larger than the length of the list. Use the `length` function to find the length of a list:

```
(%i8)   length(conc);
```

```
(%o4)   4
```

2.3.3 Arithmetic with Lists

Operations on a list are applied to every list item individually:

```
(%i1)   list1: [1,2,3];
(%i2)   list2: list1^2;
```

```
(%o1)   [1, 2, 3]
(%o2)   [1, 4, 9]
```

Mathematical functions often distribute across lists. For example,

```
(%i3)   sqrt(list1);
```

$$(\%o3)   [1, \sqrt{2}, \sqrt{3}]$$

```
(%i4)   sin(list1);
(%i5)   float(%);
```

```
(%o4)   [sin(1), sin(2), sin(3)]
(%o5)   [0.8414709848079, 0.90929742682568, 0.14112000805987]
```

User-defined functions can distribute across lists as well:

```
(%i6)   f(x) := x^2 + x + 1;
(%i7)   f([x,y,z]);
```

---

8 There are also `second`, `third`, `fourth`, `fifth`, `sixth`, `seventh`, `eighth`, `ninth`, and `tenth` functions.

(%o6) $f(x) := x^2 + x + 1$

(%o7) $[x^2 + x + 1, y^2 + y + 1, z^2 + z + 1]$

More generally, we can use the `map` function (⟨Algebra⟩⟩Map to List⟩) to apply any function to each element in a list:

(%i8)   **map**(f, [x,y,z]);

(%o8) $[x^2 + x + 1, y^2 + y + 1, z^2 + z + 1]$

Addition, subtraction, multiplication, and division on two lists performs the operation between corresponding items. The result is also a list:

(%i9)   list1 + list2;

(%o9) $[2, 6, 12]$

(%i10)   list1/list2;

(%o10) $[1, \dfrac{1}{2}, \dfrac{1}{3}]$

The dot product of two lists is the sum of the products of corresponding items. Dot products can be computed simply with the dot operator ⟨ . ⟩; for example, [a, b, c] . [x, y, z] gives a*x + b*y + c*z.

(%i11)   list1 . list2;

(%o11)   36

The `map` function can also be used to apply operations between corresponding elements in two lists. It must be used to apply relational and logical operations. For example, we can set up a list of equations if the left-hand side is in one list, and the right-hand side in another:

(%i12)   **map**("=", [x,y,z], [r***sin**(theta)***cos**(phi), r***sin**(theta) *
    ➡ **sin**(phi), r***cos**(theta)]);

(%o11) $[x = \cos(\phi)\, r\sin(\theta), y = \sin(\phi)\, r\sin(\theta), z = r\cos(\theta)]$

Note that the operator must be placed in double quotes.

You can apply any operation between elements in the same list using the `apply` function (⟨Algebra⟩⟩Apply to List...⟩). For example, to sum the elements in a list,

(%i1)   list: [1,2,3,4,5,6]$
(%i2)   **apply**("+",list);

(%o1)   21

2.3.4 Building and Editing Lists

The `makelist` function can build lists automatically from expressions. An empty list can be built from makelist():

(%i1) **makelist();**

(%o1) []

To build a list that contains n copies of an element, use `makelist(element, n)`:

(%i2) **makelist(x,6);**

(%o2) $[x, x, x, x, x, x]$

To compute a list from an expression, use `makelist(expression, i, i_1, i_N)`, where i is an index that runs from i_1 to i_N. This form can also be inserted using Algebra ⟩ Make list... from the menu.

(%i3) **makelist(x^i, i, 1, 6);**

(%o3) $[x, x^2, x^3, x^4, x^5, x^6]$

If you'd like the difference between successive i values to be something other than one, add the spacing as an additional argument:

(%i4) **makelist(x^i, i, 1, 6, 2);**

(%o4) $[x, x^3, x^5]$

Finally, to compute a list from an expression with values of x that appear in a list, use `makelist(expression, x, list of x values)`:

(%i5) **makelist(x^2 + x + 1, x, [1, 3, 7, 12]);**

(%o5) $[3, 13, 57, 157]$

Several functions are available for adding, deleting, and inserting list items. All of the functions return a new list, without changing the original list. If you do want to change the original list, you must replace the original list with the function results.

2.3.4.1 Adding Items

We'll often want to accumulate the results of calculations in a list. We can join two lists with the `append` function.

(%i1) **append([a,b,c], [d, e, f]);**

(%o1) $[a, b, c, d, e, f]$

The `cons` and `endcons` functions add an item to the beginning or end of a list, respectively.

```
(%i2)   cons(a, [b,c,d]);
```

```
(%o2)   [a, b, c, d]
```

```
(%i3)   endcons(a, [b,c,d]);
```

```
(%o3)   [b, c, d, a]
```

2.3.4.2 Deleting Items

The delete function removes all occurrences of an item from a list.

```
(%i4)   delete(c, [a, b, c, d, c, b, a]);
```

```
(%o4)   [a, b, d, b, a]
```

The unique function lists all unique elements in a list, in the order that they occur:

```
(%i5)   unique([a,b,c,d,c,b,a]);
```

```
(%o5)   [a, b, c, d]
```

The rest function deletes the first *n* elements in a list:

```
(%i6)   [a, b, c, d, e, f];
(%i7)   rest(%, 2);
(%i8)   rest(%, -2);
```

```
(%o6)   [a, b, c, d, e, f]
(%o7)   [c, d, e, f]
(%o8)   [c, d]
```

Notice that if the second argument is negative, the last two elements are deleted instead.

2.3.5 Nested Lists

A nested list is a list of lists. Nested lists are convenient for storing x, y data; for example, a list of N points could be stored as [[x1, y1], [x2, y2], ..., [xN, yN]].

Nested lists can be built with makelist calls. The first argument of makelist must itself be a list. For example, suppose we wanted to make a nested list called data to store x, y values with y being a function $f(x)$. We want $xmin \leq x \leq xmax$, with adjacent x a distance of deltax apart:

```
(%i1)   xmin: 10$
(%i2)   xmax: 12.5$
(%i3)   deltax: 0.5$
(%i4)   f(x) := x^2$
(%i5)   data: makelist([x, f(x)], x, xmin, xmax, deltax);
```

```
(%o5)   [[10, 100], [10.5, 110.25], [11.0, 121.0], [11.5, 132.25], [12.0, 144.0], [12.5, 156.25]]
```

Nested lists are easily converted into matrices using the `apply` function:

(%i6)   M: **apply**(**matrix**, data);

(%o6) $\begin{pmatrix} 10 & 100 \\ 10.5 & 110.25 \\ 11.0 & 121.0 \\ 11.5 & 132.25 \\ 12.0 & 144.0 \\ 12.5 & 156.25 \end{pmatrix}$

The reverse operation (converting a matrix into a nested list) is accomplished with the `args` function:

(%i7)   **args**(M);

(%o7)   [[10, 100], [10.5, 110.25], [11.0, 121.0], [11.5, 132.25], [12.0, 144.0], [12.5, 156.25]]

Nested lists are easy to plot, as we'll see in Chapter 3. In Chapter 4, we'll build more complex nested lists with more than one index using loops.

2.3.6   Sublists

The `sublist` function chooses elements from a list using any condition you like. The condition has to be written as a predicate.

sublist(*list, predicate*)
Choose all items from *list* for which the *predicate* is true. The predicate must be the name of a function, not a function call.

You can use any of the built-in predicates listed in Table 2.5 with `sublist`, or you can define your own. For example, suppose we need to pick out all elements of a list that are greater than zero and less than 10:

(%i1)   pick(x) := **is**(x>0 **and** x<10)$
(%i2)   list: [-1.2, 0, 1.3, 6.7, 11.5]$
(%i3)   **sublist**(list, pick);

(%o3)   [1.3, 6.7]

 Worksheet 2.4: Lists

In this worksheet, we'll build and edit data lists, and use them to compute molecular bond angles and bond lengths. We'll also look at advanced functions for examining, filtering, and comparing data lists.

2.4   Matrices

A *matrix* arranges data by row and column. The row and column arrangement shows relationships between the data. Suppose you had a list of (x, y) data points. To write the data in matrix form, place it in a table with the x values in column 1 and the y values in column 2. Each row in the table corresponds to a different data point.

The items in a matrix don't have to be numbers; they may also be expressions. For example, several methods for solving a system of equations like

$$3x + ay = b$$
$$(b - 1)x + cy = d$$

begin by writing the coefficients and constant terms in the equations as a matrix like

$$\begin{pmatrix} 3 & a & -b \\ b-1 & c & -d \end{pmatrix}$$

We say this is a 2×3 matrix (it has two rows, and three columns).

Many chemical problems are most naturally expressed in terms of matrices, and matrix arithmetic makes their solution straightforward. We'll explore matrix computation in more detail in Chapter 7. In this section, we'll concentrate on storing data in matrix form, with only a brief introduction to matrix arithmetic.

2.4.1   Row and Column Vectors

A matrix with a single column is called a column vector. Lists can be converted into column vectors using the command `transpose(list)`. The transpose function makes the rows of a matrix into columns, and the columns into rows.

```
(%i1)    list: [a,b,c];
(%i2)    column_vector: transpose(list);
```

$$(\%o1) \quad [a, b, c]$$

$$(\%o2) \quad \begin{pmatrix} a \\ b \\ c \end{pmatrix}$$

A row vector is a matrix with a single row. Lists can be converted into row vectors using the command `matrix(list)`.

```
(%i3)    row_vector: matrix(list);
```

$$(\%o3) \quad \begin{pmatrix} a & b & c \end{pmatrix}$$

In the previous section, we saw that the dot operator gives a dot product when used between two lists. Between two matrices, it acts as matrix multiplication.

```
(%i4)    row_vector . column_vector;
```

$$(\%o2) \quad c^2 + b^2 + a^2$$

That's the same result we would have obtained with `list . list`. But note the result when we swap the order of the vectors:

(%i5) `column_vector . row_vector;`

(%o2) $\begin{pmatrix} a^2 & ab & ac \\ ab & b^2 & bc \\ ac & bc & c^2 \end{pmatrix}$

We'll explore this matrix product further in Section 7.2. In this section, we'll concentrate on entering, indexing, and storing data in matrices.

2.4.2 Indexing Matrices

Using several lists with the `matrix` function generates a matrix with each list as a row:

(%i6) `M : matrix([1,2,3], [4,5,6], [7,8,9], [10,11,12]);`

(%o4) $\begin{pmatrix} 1 & 2 & 3 \\ 4 & 5 & 6 \\ 7 & 8 & 9 \\ 10 & 11 & 12 \end{pmatrix}$

Matrix elements need two indices: one for the row, and one for the column. In either program, the indices are separated by a comma, and enclosed in square brackets. For example, `M[3,4]` gives you the element of matrix M that is in row 3 and column 4. You only need to type `M[3,4]`. wxMaxima automatically types the closing bracket for you.

Assigning the value of a matrix element changes the matrix. For example,

(%i7) `M[3,3];`

(%o5) 9

(%i8) `M[3,3] : 10;`

(%o6) 10

(%i9) `M;`

(%o7) $\begin{pmatrix} 1 & 2 & 3 \\ 4 & 5 & 6 \\ 7 & 8 & 10 \\ 10 & 11 & 12 \end{pmatrix}$

Indices that are greater than the number of rows or columns in the matrix will give an error. To find the size of a matrix, use the `matrix_size` function.

`matrix_size(matrix)`
Return a two-item list containing the number of rows and the number of columns in `matrix`.

Remember that when a function returns a list of values, you can assign the results to a list of variables to access them separately:

```
(%i10)   [nrows, ncols] : matrix_size(M);
```

$$(\%o8) \quad [4, 3]$$

2.4.3 Entering Matrices

Matrices can be built directly with the `matrix` function, as we saw above. They can also be built interactively by selecting Algebra ⟩ Enter matrix... . Maxima will then ask for the number of rows and columns in the matrix, the matrix type, and the name of the matrix.

There are four choices for the matrix type.

- A diagonal matrix has nonzero elements only along the diagonal.
- A symmetric matrix has elements $M_{ij} = M_{ji}$.
- An antisymmetric matrix has elements $M_{ij} = -M_{ji}$.
- If you choose a "general" matrix (or if the matrix isn't a square matrix) you will be prompted for all elements.

You can also build a matrix from a nested list. Each row in the matrix is a list; the entire matrix is a list of the row lists. The nested list is converted into a matrix using `apply(matrix, nested list)`.

```
(%i1)   M : apply(matrix, [ [1,2,3], [4,5,6], [7,8,9] ] );
```

$$(\%o1) \quad \begin{pmatrix} 1 & 2 & 3 \\ 4 & 5 & 6 \\ 7 & 8 & 9 \end{pmatrix}$$

When the matrix elements depend in some way on their row and column indices, you can generate a matrix using an *array function*, which is a function of matrix indices that uses brackets instead of parentheses. For example, we can generate a Hilbert matrix H (which has elements $H_{ij} = 1/(i + j - 1)$):

```
(%i1)   h [i, j] := 1 / (i + j - 1);
(%i2)   H : genmatrix(h, 4, 4);
```

$$(\%o1) \quad h_{i,j} := \frac{1}{i + j - 1}$$

$$(\%o2) \quad \begin{pmatrix} 1 & \frac{1}{2} & \frac{1}{3} & \frac{1}{4} \\ \frac{1}{2} & \frac{1}{3} & \frac{1}{4} & \frac{1}{5} \\ \frac{1}{3} & \frac{1}{4} & \frac{1}{5} & \frac{1}{6} \\ \frac{1}{4} & \frac{1}{5} & \frac{1}{6} & \frac{1}{7} \end{pmatrix}$$

You can generate the same matrix more simply using the Algebra ⟩ Generate Matrix from Expression... menu dialog:

Generate Matrix ✕

matrix[i,j]:	1/(i+j-1)
Width:	4
Height:	4
Name:	H

OK   Cancel

2.4.4 Assigning Matrices

You can assign matrices the same way you would assign variables, but you must be careful because assigning one matrix to another simply gives the same matrix a second name:

(%i1) C : **matrix**([1, 2, 2], [2, 4, 3], [2, 3, 5]);

(%o1) $\begin{pmatrix} 1 & 2 & 2 \\ 2 & 4 & 3 \\ 2 & 3 & 5 \end{pmatrix}$

(%i2) D : C;

(%o2) $\begin{pmatrix} 1 & 2 & 2 \\ 2 & 4 & 3 \\ 2 & 3 & 5 \end{pmatrix}$

(%i3) D[1,1] : x$

(%o3) x

(%i4) C;

(%o4) $\begin{pmatrix} x & 2 & 2 \\ 2 & 4 & 3 \\ 2 & 3 & 5 \end{pmatrix}$

Assigning x to D_{11} also assigns x to C_{11}! If you want to make **D** an independent copy of **C**, you will have to use the copy function:

(%i1) C : **matrix**([1, 2, 2], [2, 4, 3], [2, 3, 5])$
(%i2) D : copy(C);

(%o2) $\begin{pmatrix} 1 & 2 & 2 \\ 2 & 4 & 3 \\ 2 & 3 & 5 \end{pmatrix}$

```
(%i3)    D[1,1] : x$
(%i4)    C;
(%i5)    D;
```

$$(\%o4) \quad \begin{pmatrix} 1 & 2 & 2 \\ 2 & 4 & 3 \\ 2 & 3 & 5 \end{pmatrix}$$

$$(\%o5) \quad \begin{pmatrix} x & 2 & 2 \\ 2 & 4 & 3 \\ 2 & 3 & 5 \end{pmatrix}$$

This time **C** is unaffected by changes in the elements of **D**.

2.4.5   Editing Matrices

We'll often want to add, extract, or rearrange data from matrices in a calculation.
To add columns or rows to a matrix, use the addcol and addrow functions:

addcol(*matrix*, *col*$_1$, *col*$_2$, ...)
Return a matrix with one or more lists or matrices appended as columns to the right side of *matrix* as columns.

addrow(*matrix*, *list*$_1$, *list*$_2$, ...)
Return a matrix with one or more lists or matrices appended to the bottom of *matrix* as rows.

The added rows and columns must of course have lengths that match the respective number of columns or rows in the matrix. For example:

```
(%i1)    M: matrix([1, a, a^2], [1, b, b^2], [1, c, c^2]);
(%i2)    M: addcol(M, [a^3, b^3, c^3]);
(%i3)    M: addrow(M, [1, d, d^2, d^3]);
```

$$(\%o1) \quad \begin{pmatrix} 1 & a & a^2 \\ 1 & b & b^2 \\ 1 & c & c^2 \end{pmatrix}$$

$$(\%o2) \quad \begin{pmatrix} 1 & a & a^2 & a^3 \\ 1 & b & b^2 & b^3 \\ 1 & c & c^2 & c^3 \end{pmatrix}$$

$$(\%o3) \quad \begin{pmatrix} 1 & a & a^2 & a^3 \\ 1 & b & b^2 & b^3 \\ 1 & c & c^2 & c^3 \\ 1 & d & d^2 & d^3 \end{pmatrix}$$

```
(%i4)    addcol(M, [5, 6, 7]);
```

```
(%o4)    addrow or addcol: incompatible structure. – an error.
         To debug this try: debugmode(true);
```

To extract columns or rows from a matrix, use the `col` and `row` functions. The first argument is the matrix name, and the second is the index of the row or column you want:

(%i5)   **col(M,4);**

(%o5) $\begin{pmatrix} a^3 \\ b^3 \\ c^3 \\ d^3 \end{pmatrix}$

(%i6)   **row(M,4);**

(%o6) $\begin{pmatrix} 1 & d & d^2 & d^3 \end{pmatrix}$

To delete rows or columns from a matrix, use the `submatrix` function. The indices of rows to delete come *before* the matrix name in the arguments, and the indices of columns to delete come *after* the matrix name. For example,

(%i7)   **submatrix(3,4, M);**

(%o7) $\begin{pmatrix} 1 & a & a^2 & a^3 \\ 1 & b & b^2 & b^3 \end{pmatrix}$

(%i8)   **submatrix(M, 3, 4);**

(%o8) $\begin{pmatrix} 1 & a \\ 1 & b \\ 1 & c \\ 1 & d \end{pmatrix}$

(%i9)   **submatrix(3, 4, M, 3, 4);**

(%o9) $\begin{pmatrix} 1 & a \\ 1 & b \end{pmatrix}$

Maxima has a powerful `subsample` function for selecting rows from matrices that have row elements that satisfy a given condition.

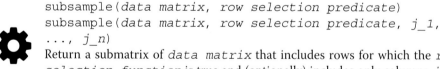

```
subsample(data matrix, row selection predicate)
subsample(data matrix, row selection predicate, j_1,
..., j_n)
```
Return a submatrix of *data matrix* that includes rows for which the *row selection function* is true and (optionally) includes only columns *j_1*, ..., *j_n*.

The row selection predicate takes a list as its single argument. It is applied to each row in the matrix, returning `true` when elements in the list satisfy a certain condition. For example, suppose we only wanted to choose rows from a matrix where the second element is greater than six and the third element is less than 15:

```
(%i1)   M: matrix([1,2,3,4],[5,6,7,8],[9,10,11,12],
          ➥ [13,14,15,16]);
(%i2)   choose(item) := item[2] >6 and item[3] < 15$
(%i3)   subsample(M, choose);
```

$$(\%o1) \quad \begin{pmatrix} 1 & 2 & 3 & 4 \\ 5 & 6 & 7 & 8 \\ 9 & 10 & 11 & 12 \\ 13 & 14 & 15 & 16 \end{pmatrix}$$

```
(%o3)
```

If we wanted to choose columns 2 and 3 from rows that sum to greater than 22, add the column indices you want as extra arguments to `subsample`:

```
(%i4)   choose(item) := apply("+", item) > 22$
(%i5)   subsample(M, choose, 2, 3);
```

$$(\%o5) \quad \begin{pmatrix} 6 & 7 \\ 10 & 11 \\ 14 & 15 \end{pmatrix}$$

The `subsample` function is very useful for excluding outliers from data to be fitted to a curve, and for processing data read from Excel spreadsheets, as we'll see in the next section.

2.4.6   Reading and Writing Matrices From Files

Maxima has several functions for reading and writing numbers, strings, lists, and matrices from files.

We can examine the contents of a text file using the `printfile` function:

```
(%i1)   printfile("k:/work/smc-maxima/data/isotope-data.csv");
```

```
Element, mass number, relative atomic mass,isotopic abundance
hydrogen,1,1.007825032,0.999885
hydrogen,2,2.014101778,0.000115
hydrogen,3,3.016049278,0
helium,3,3.016029319,0.00000134
helium,4,4.002603254,0.99999866
lithium,6,6.015122795,0.0759
lithium,7,7.01600455,0.9241
```

$(\%o1) \quad k : /work/smc - maxima/data/isotope - data.csv$

You must give `printfile` the full path to the file you want to read, enclosed in double quotes.

This file could have been saved by spreadsheet programs like Microsoft Excel or OpenOffice Calc as a CSV file (comma-separated value file).

Use the `read_matrix` function to read data from a CSV file directly into a matrix:

```
(%i2)  M: read_matrix("k:/work/smc-maxima/data/isotope-data.csv
    ➥ ");
```

$$(\%o2)\quad\begin{pmatrix} Element & [mass, number] & [relative, atomic, mass] & [isotopic, abundance] \\ hydrogen & 1 & 1.007825032 & 0.999885 \\ hydrogen & 2 & 2.014101778 & 1.15\,10^{-4} \\ hydrogen & 3 & 3.016049278 & 0 \\ helium & 3 & 3.016029319 & 1.34\,10^{-6} \\ helium & 4 & 4.002603254 & 0.99999866 \\ lithium & 6 & 6.015122795 & 0.0759 \\ lithium & 7 & 7.01600455 & 0.9241 \end{pmatrix}$$

The `read_matrix` function reads strings as lists of separate words (rela-tive atomic mass is read as a list [relative, atomic, mass]). If you don't want this, enclose text in double quotes before you save the spreadsheet as a `.csv` file.

Suppose we want to compute the average atomic mass of hydrogen from this data. First, we extract the atomic masses in column 3 and the isotopic abundances in column 4 for rows that begin with *hydrogen*:

```
(%i3)  choose(item) := item[1] = hydrogen$
(%i4)  H: subsample(M, choose, 3, 4)
```

$$(\%o4)\quad\begin{pmatrix} 1.007825032 & 0.999885 \\ 2.014101778 & 1.15\,10^{-4} \\ 3.016049278 & 0 \end{pmatrix}$$

The average atomic mass is the dot product of the columns in `H_data`:

```
(%i5)  col(H,1) . col(H,2);
```

```
(%o5)  1.00794075382579
```

To write a matrix, list, or nested list to a file, use the `write_data` function:

`write_data(X, file, separator)`
Write a matrix, list, or nested list *X* to a *file*, with values separated by *sep-arator*. The *file* can be a file name (including a complete path) or it can be file handle (see Section 2.5.1). The *separator* can be `comma`, `pipe`, `semi-colon`, `space`, or `tab`. If you want to be able to open the file in Excel, use `comma`.

For example, let's write a matrix A to a comma-separated values file named `tmp.csv`, and print the file back with `printfile`:

```
(%i1)  A : matrix([1,2,3],[4,5,6],[7,8,9]);
(%i2)  write_data(A, "tmp.csv", comma);
(%i3)  printfile("tmp.csv");
```

$$(\%o1) \quad \begin{pmatrix} 1 & 2 & 3 \\ 4 & 5 & 6 \\ 7 & 8 & 9 \end{pmatrix}$$

$(\%o2)$ *done*

    1,2,3
    4,5,6
    7,8,9
$(\%o3)$ *tmp.csv*

2.4.7   Transforming Data in a Matrix

Suppose we want to change individual columns in our data matrix. We might want to convert x to different units, or we might want to replace each of the y values with a calculated expression.

Maxima's `descriptive` package provides a function called `transform_sample` that we can use to do this.[9]

```
transform_sample(data matrix, list of column
variables, list of column expressions)
```
Apply expressions to individual columns in a data matrix. The expressions are written in terms of variables that correspond to each column.

For example, suppose we have the following Excel data for the vapor pressure of pure water at various different temperatures:

```
(%i1)    data: read_matrix("k:/work/smc-maxima/data/vapor-
    ➥ pressure-of-pure-water.csv");
```

$$(\%o1) \quad \begin{pmatrix} [Vapor, Pressure, of, pure, water] & false \\ [T, /,^\circ, C] & [P, /, torr] \\ 0 & 4.579 \\ 5 & 6.543 \\ 10 & 9.209 \\ 15 & 12.788 \\ 20 & 17.535 \\ 25 & 23.756 \\ 30 & 31.824 \\ 35 & 42.175 \\ 40 & 55.324 \\ 45 & 71.88 \\ 50 & 92.51 \end{pmatrix}$$

First, let's delete any rows that don't begin with a number (like the title and column headings), using the `numberp` predicate:

```
(%i2)    chooserow(x) := numberp(x[1])$
(%i3)    data: subsample(data,chooserow);
```

---

9 The `descriptive` package is automatically loaded when you load the `stats` package.

$$(\%o2) \quad \begin{pmatrix} 0 & 4.579 \\ 5 & 6.543 \\ 10 & 9.209 \\ 15 & 12.788 \\ 20 & 17.535 \\ 25 & 23.756 \\ 30 & 31.824 \\ 35 & 42.175 \\ 40 & 55.324 \\ 45 & 71.88 \\ 50 & 92.51 \end{pmatrix}$$

Now lets replace the temperatures t in °C in column 1 with $1/(t + 273.15)$ (the reciprocal kelvin temperature), and the pressures p in torr with $\ln(p/p^\circ)$ (the natural log of the pressure relative to one atmosphere):

```
(%i4)   load(descriptive)$
(%i5)   new_data: transform_sample(data, [t,p],
            [1/(t+273.15), log(p)-log(760)]),numer;
```

$$(\%o4) \quad \begin{pmatrix} 0.0036609921288669 & -5.111837799605419 \\ 0.0035951824555096 & -4.754922657392775 \\ 0.003531696980399 & -4.413137166541147 \\ 0.0034704147145583 & -4.084811202082782 \\ 0.003411222923418 & -3.769119549688235 \\ 0.0033540164346805 & -3.465483303126763 \\ 0.0032986970146792 & -3.17309771116876 \\ 0.0032451728054518 & -2.891490804848345 \\ 0.0031933578157432 & -2.620111622517915 \\ 0.0031431714600031 & -2.358320371365018 \\ 0.0030945381401825 & -2.10600168649715 \end{pmatrix}$$

Ⓜ **Worksheet 2.5: Matrices**

In this worksheet, we'll build, edit, store, and transform data in matrix form. We'll also see how matrices can be used to balance chemical equations.

2.5 Strings

Strings are text enclosed in double quotes . They are most often used to hold file names, to build labels for table rows and columns, to label axes in graphs, to compose error and status messages in programs, and to read and write formatted data in text files.

Strings can be assigned to variables just as numbers can. For example, typing x:"r (nm)" assigns "r (nm)" to the variable x. Note that typing a double quote in wxMaxima automatically inserts a matched pair of double quotes.

A string can contain *almost* any character you can type. If you want to embed a double quote inside a string, type a backslash in front of it: $\boxed{\text{\textbackslash"}}$. A backslash itself would be typed $\boxed{\text{\textbackslash\textbackslash}}$. For example,

```
(%i1)    string1 : "This string contains a backslash (\\), a
            ➥ linebreak (
         ), and a double quote (\")";
(%i2)    string2 : "This string contains special characters like
            ➥ ±, ½, μ, and °.";
```

(%o1)   This string contains a backslash (\), a linebreak ( ), and a double quote (")

(%o2)   This string contains special characters like $\pm$, $\dfrac{1}{2}$, μ, and $^{\circ}$.

Notice that wxMaxima removes the line break in the first string, and types $\frac{1}{2}$ as "(1/2)" in the second string.

If your Maxima installation fully supports Unicode characters, you can enter Greek letters into a string by pressing $\boxed{\text{Esc}}$ before and after a letter; for example, $\boxed{\text{Esc}}$ $\boxed{\text{a}}$ $\boxed{\text{Esc}}$ types α; $\boxed{\text{Esc}}$ $\boxed{\text{D}}$ $\boxed{\text{Esc}}$ types Δ. If wxMaxima alone supports Unicode, you'll have to cut and paste Greek letters into your worksheets. Otherwise you'll have to avoid using such characters in strings.

Strings can't be used directly in computations, even when they contain numbers. For example, `"2"` + `"3"` or `"2+3"` isn't 5. But you can evaluate strings containing numbers and expressions in Maxima using the `eval_string` function.[10] For example,

```
(%i1)    "2^2 + 3*2";
(%i2)    eval_string(%);
```

(%o1) $2\hat{\ }2 + 3 * 2$
(%o2)   10

You can't add strings, but you can concatenate (join) them. The `sconcat` function joins numbers, expressions, and strings into a single string. For example,

```
(%i1)    mean_x : 5.32$
(%i2)    CI_95 : 0.15$
(%i3)    N : 12$
(%i4)    sconcat(mean_x, " ± ", CI_95, " kJ/mol (N=", N, ", 95%
            ➥ CI)");
```

(%o4) $5.32 \pm 0.15\,\mathrm{kJ/mol}\,(N = 12,\ 95\%\ \mathrm{CI})$

2.5.1 Using String Functions to Work with Files

Maxima has several string functions for reading and writing formatted data from external files.[11] The simplest is `with_stdout`, which writes the output of a Maxima command (converted to a string) into a file:

---

10 In older versions of Maxima, you will need to `load(stringproc)` before you can use this function.
11 In older versions of Maxima, you'll have to `load(stringproc)` before using these functions.

Table 2.6 Useful format codes for `printf`.

~h	Big float	~e	Scientific notation
~f	Floating point number	~g	Scientific notation or float, depending on magnitude
~d	Integer	~a	String
~%	New line	~s	String with double quotes added
~r	Convert an integer to words	~t	Tab

The first argument is the file name, and the second is the string.

```
(%i1)    with_stdout("four.txt", string(2+2))$
(%i2)    printfile("pi.txt")$
```

If you want more control over what's being printed into the file, use the open, close, and printf functions. To see how they work, let's create a text file that lists π to 1000 digits. First, we open the file for writing with the `openw` function:

openw(*filename*)
Opens a named *filename* for writing, and returns a "handle" to the file. *file* contains the filename of the file, including its path; the file will be overwritten.

For a file named `pi.txt`,

```
(%i1)    file: openw("pi.txt")$
```

The variable `file` now contains a file handle. We can use the `printf` function to create a formatted string, and print it to the file.

printf(*file handle, string*)
printf(*file handle, format, expression 1, expression 2, ...*)
Prints a *string* or formatted expressions to *file handle*. If the *file handle* is set to `true`, the output is printed in wxMaxima instead of into a file. The *format* is a string containing print codes for the given expressions (see Table 2.6).

To convert the 1000-digit value of π into a string and write it to the file, we set floating point precision to 1000 and use the ~h format code:

```
(%i2)    fpprec: 1000$
(%i3)    printf(file, "~h", bfloat(%pi))$
(%i4)    close(file)$
```

In the last line, we close the file with `close(file)`.
Now let's reopen the file for reading, using the `openr` function for reading files:

```
(%i5)    file: openr("pi.txt")$
```

Now use the `readline` function to read a line from the file, and close the file:

```
(%i6)   readline(file);
(%i7)   close(file)$
```

```
(%o7)   3.14159265358979323846264338327950288419716939937510582...
```

where we've omitted the end of the output for brevity. The `readline` function will return false if there is not another line to read.

In Worksheet 2.6, we'll see how these and other string functions can be used to read and write formatted output from Excel files that have been saved in CSV format.

 Worksheet 2.6: Working with Text in Strings and Files

In this worksheet, we'll use string functions to read and write formatted data in external text files.

3

Plotting Data and Functions

> *There is a magic in graphs. The profile of a curve reveals in a flash a whole situation – the life history of an epidemic, a panic, or an era of prosperity. The curve informs the mind, awakens the imagination, convinces.*
>
> – Henry D. Hubbard [16]

Graphs can reveal patterns or relationships that wouldn't be apparent otherwise. Adding theoretical curves to data plots can help us compare theory with experiment and see the effects of approximations and adjustable parameters.

In Chapter 1, we saw how to make simple graphs in two and three dimensions using the General Math pane. In Chapter 2, we saw how lists and matrices could be used to store and manipulate datasets. In this chapter, we'll use both techniques to visualize patterns in that data with scatter plots, histograms, surface plots, and contour plots.

Section 3.1 focuses on plotting functions and data in two dimensions. We'll see how to format, label, and scale graphs, and also how to plot parametric curves, implicit equations, and histograms. In Section 3.2, we'll build interactive three-dimensional plots of multivariate functions and data using Cartesian, cylindrical, and spherical coordinates, and also plot the contours of three-dimensional surfaces in two dimensions.

3.1 Plotting in Two Dimensions

Suppose we want to plot a damped sine wave `-sin(x)*exp(-x)` with $0 \leq x \leq 10$. To create the plot,

1) Select ⌗Plot⟩⟩Plot 2d⌗ from the menu, or select the ⌗Plot 2D...⌗ button from the *General Math* pane.
2) The Plot 2D dialog appears. Fill it in as shown in Figure 3.1. The command to draw the plot will automatically be inserted in the next cell on the worksheet.

If you want to save the graph as an image, right-click on it. You can save the graph as a PNG image (.png), a JPEG image (.jpg), a Windows bitmap (.bmp), or an X pixmap (.xmp).

3.1.1 Changing Plot Size and Resolution

To change the default size of the embedded plot, click on the crossed tool icon on the wxMaxima toolbar. To change the size for the current session, use *wxplot_size* system variable, which is a two-element list containing the horizontal and vertical sizes of the plot.

Symbolic Mathematics for Chemists: A Guide for Maxima Users, First Edition. Fred Senese.
© 2019 John Wiley & Sons Ltd. Published 2019 by John Wiley & Sons Ltd.
Companion website: http://booksupport.wiley.com

Figure 3.1 Filling in the Plot 2d dialog to plot a function in Maxima.

(%i1) wxplot2d([-sin(x)*exp(−x)], [x,0,10])$

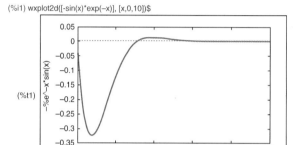

A better way to increase the resolution of a plot is to change the *Format* field to `default` or `gnuplot` (or type `plot2d` in place of `wxplot2d` in the inserted code[1]).

The open-source plotting program Gnuplot draws the plot in a pop-up window.[2]

```
(%i1)   plot2d(-sin(x)*exp(-x), [x, 0, 10])$
```

 wxMaxima won't execute new commands until you close the Gnuplot window.

Try the following while the graph is open in the pop-up Gnuplot window:

- Resizing the pop-up window resizes the graph.
- Move the mouse over the graph. The current x, y coordinates of the cursor will be displayed at the bottom of the window.
- To zoom in, right-click on the upper left corner of the area you're interested in. Drag out a rectangle by moving the mouse to the bottom right corner of the area, and then left-click to stop selecting.
- Mouse over the icons on the toolbar above the graph to see their functions. You can use the icons to copy the graph to the clipboard, replot the graph, toggle the grid on and off, page back and forth between zoom settings you've tried, rescale the plot, and configure the window.

---

1 Putting the letters "wx" in front of the names of Maxima's plotting functions (`contour_plot`, `draw2d`, `draw3d`, `histogram`, `implicit_plot`, `plot2d`, and `plot3d`) places the plot in an embedded PNG file in the worksheet rather than in a pop-up window.
2 Gnuplot is installed as part of the standard Maxima distribution. A call to any plotting function in Maxima creates a list of Gnuplot commands in the file `maxout.gnuplot`, which is placed in your user directory. For more on Gnuplot commands and using Gnuplot as a stand-alone program, see http://www.gnuplot.info.

You can save a plot as a file rather than displaying it:

`[pdf_file, `*`file name`*`]`	Saves the plot into a PDF (Portable Document Format) file.
`[png_file, `*`file name`*`]`	Saves the plot into a PNG (Portable Network Graphics) file.
`[ps_file, `*`file name`*`]`	Saves the plot into a Postscript file.
`[svg_file, `*`file name`*`]`	Saves the plot into a SVG (Scalable Vector Graphics) file.

3.1.2 Plotting Multiple Curves

Study the inserted code in Figure 3.1. The `plot2d` and `wxplot2d` functions take the form:

```
plot2d(curve expression, x range, options)
plot2d([curve_1, curve_2, ...], x range, options)
```
Plot a function or list of functions of one variable *x* over a range of *x* values. Substitute `wxplot2d` for `plot2d` for plots that are placed inline in the worksheet.

All arguments of `plot2d` are lists, enclosed in square brackets and separated by commas. The first argument can list multiple curves. You can enter several functions separated by commas in the Plot 2D dialog, or type them directly into the `plot2d` command yourself (see Figure 3.2).

A curve legend appears in the upper right corner of the plot when there is more than one curve. In Worksheet 3.0, we'll see how the legend can be customized or hidden.

Figure 3.2 Multiple curves on a high-resolution `plot2D` plot in Maxima. The inserted code is `plot2d([-sin(x), exp(-x), -sin(x)*exp(-x)], [x,0,3])`.

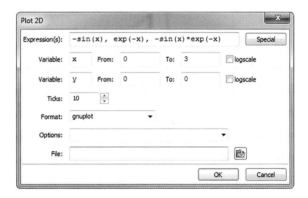

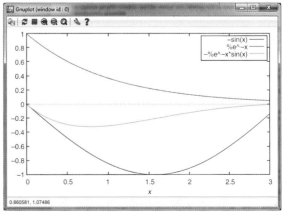

3.1.3 Changing Axis Ranges

The x range argument is required when plotting a function or algebraic expression. It takes the form [x, *minimum value of x*, *maximum value of x*]. A similar option for setting the range of the y axis that takes the form [y, *minimum value of y*, *maximum value of y*]. If you don't supply the *y range*, the plot will automatically use the minimum and maximum y values computed from curve expression over the range of x values.

 Avoid defining variables with the same names as plot2d options. For example, if you defined x before calling plot2d, the x range option won't work!

3.1.4 Plotting Complex Functions

By default, plot2d will plot only real functions, so if your function has both real and imaginary parts, use realpart and imagpart to plot it. For example, let's plot the function $f(x) = \ln \frac{x-1}{x+1}$, which has a nonzero imaginary component if $x < 1$:

```
(%i1)   f(x) :=log((x-1)/(x+1));
(%i2)   plot2d([realpart(f(x)),imagpart(f(x))], [x,-3,3])$
```

$$(\%o1) \quad f(x) := \log\left(\frac{x-1}{x+1}\right)$$

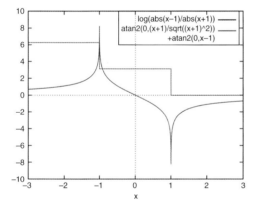

(M) **Worksheet 3.0: Plotting functions**

In this worksheet, we'll use plot2d to plot single and multiple curves with labeled axes and legends, over specific ranges of x and y. We'll also see how to use graphs to locate solutions of equations.

3.1.5 Plotting Data

Plots of experimental (x, y) data points can reveal patterns or relationships that wouldn't be apparent otherwise. They're also useful in comparing experimental data with theoretical curves.

There are several ways to plot x, y data in Maxima. You can use the simple menu dialog we used to plot functions by selecting [Plot ⟩ Plot 2d...] and then [Special ⟩ Discrete plot].

3.1.5.1 Plotting Data in Separate X, Y Lists

If you have more than a few data points, the menu dialog is unwieldy. It is much easier to type the `plot2d` command directly. The *discrete* option specifies a set of discrete points in `plot2d`'s curve list:

> [discrete, *xdata*, *ydata*]
> Specifies a set of discrete points in the list of curves in `plot2d` or `wxplot2d`. *xdata* and *ydata* are lists of the *x* and *y* values. Maxima will choose the *x* and *y* ranges by examining the data in *xdata* and *ydata*. You don't have to add the [x, *minimum value of x*, *maximum value of x*] or [y, *minimum value of y*, *maximum value of y*] options unless you want to plot ranges that are different from the ranges in *xdata* and *ydata*.

By default, Maxima will connect the points with line segments. **To see only isolated points**, add the [style, points] option.[3]

```
(%i1)   xdata : [0.0, 0.9, 1.8, 2.6, 3.3, 4.4, 5.2, 6.1, 6.5,
          ➡ 7.4]$
(%i2)   ydata : [5.9, 5.4, 4.4, 4.6, 3.5, 3.7, 2.8, 2.8, 2.4,
          ➡ 1.5]$
(%i3)   plot2d([discrete, xdata, ydata], [style, points])$
```

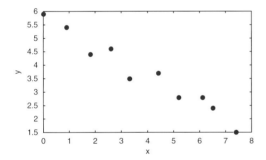

3.1.5.2 Plotting Data as Lists of X, Y Points

If you don't want to use separate lists for the *x* and *y* data, you can enter the points as a nested list of *x*, *y* pairs:

```
(%i1)   xydata : [
            [0.0, 5.9], [0.9, 5.4], [1.8, 4.4], [2.6, 4.6],
              ➡ [3.3, 3.5],
            [4.4, 3.7], [5.2, 2.8], [6.1, 2.8], [6.5, 2.4],
              ➡ [7.4, 1.5]
          ]$
(%i2)   wxplot2d([discrete, xydata], [style, points])$
```

---

3 We'll see in Worksheet 3.1 how this option can also be used to choose the size and color of points.

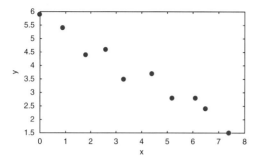

3.1.5.3 Plotting Data in Matrices

If the data is in a matrix with x values in column 1 and y values in column 2, you will have to convert it into a nested list before it can be used by plot2d. This is easily done using the args function:

```
(%i1)   data: matrix([0.0, 5.9], [0.9, 5.4], [1.8, 4.4], [2.6,
          ➥ 4.6], [3.3, 3.5],
(%i2)                   [4.4, 3.7], [5.2, 2.8], [6.1, 2.8], [6.5,
                          ➥ 2.4], [7.4, 1.5])$
(%i3)   wxplot2d([discrete, args(data)], [style, points])$
```

This produces a plot identical to the last one.

3.1.5.4 Plotting Data with Units

Unfortunately, Maxima's plotting functions don't accept x and y data with attached units. You must divide out the units, or use the qty function to strip off the units before plotting:

```
(%i1)   load(physical_constants)$
(%i2)   xdata: [0`m, 1`m, 2`m];
(%i3)   ydata: xdata^2;
```

$$(\%o2)\quad [0`m, 1`m, 2`m]$$
$$(\%o3)\quad [0`m^2, 1`m^2, 4`m^2]$$

Plotting x and y with units gives us an error message:

```
(%i1)   plot2d([discrete,xdata,ydata])$
```

```
Warning: none of the points have numerical values. plot2d: noth-
ing to plot.
```

Strip the units off with the qty function:

```
(%i1)   plot2d([discrete,qty(xdata),qty(ydata)])$
```

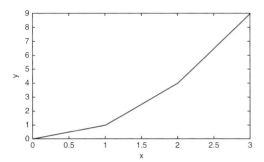

You can also divide out the units (which gives the same graph as above):

```
(%i1)  plot2d([discrete,xdata/(1'm), ydata/(1'm^2)])$
```

3.1.5.5 Plotting Functions and Data Together

To plot both functions and discrete data on the same graph, list both the function and the data in the curve list. The following example plots both discrete data and the function $y = x^2$ on the same graph:

```
(%i1)  data: [ [1,1],[2,4],[3,9],[4,16],[5,25] ]$
(%i2)  plot2d([[discrete, data],x^2], [x,1,5],
                [style, points, lines]
              )$
```

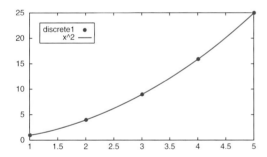

 Worksheet 3.1: Plotting data

In this worksheet, we'll plot experimental data by itself or in combination with theoretical curves over specific ranges of x and y.

3.1.6 Adding Text Labels to Graphs

A graph should be self-explanatory. It must have a title and axis labels that specify units. Multiple curves and points of interest should also be clearly labeled.

Maxima has several commands for annotating graphs. The *xlabel*, *ylabel*, and *title* options label axes and title graphs. The *legend* and *label* options can be used to add text to legends and points. For example,

```
(%i1)  p0 : 1/(1+exp(-1/x));
(%i2)  p1 : exp(-1/x)/(1+exp(-1/x));
(%i3)  plot2d([p0, p1], [x, 0.001, 10],
          [xlabel, "kT/ε"],
          [ylabel, "fraction of population"],
          [title, "Populations of a two-level system as a
            ➡ function of temperature"],
          [legend, "level 0", "level 1"],
          [label, ["As T→ ∞, the populations become equal",
            ➡ 3.5, 0.5]]
(%i4)  )$
```

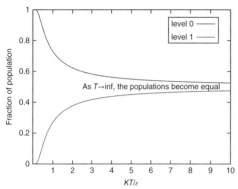

Table 3.1 shows the format for each of these options.

3.1.7 Plotting Rapidly Rising Functions

Rapidly rising functions can produce misleading or useless graphs. We will sometimes be forced to manually change the range on the y axis or use a logarithmic scale for y.

Trigonometric and exponential functions can contain singularities over the plotting range. Consider a plot of $\tan(x)$ with $-2 \le x \le +2$:

```
(%i1)  plot2d(tan(x), [x,-2,2])$
```

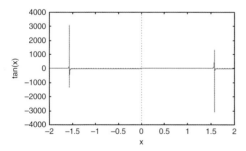

The tangent function is infinite at $-\pi/2$ and $\pi/2$. The function appears to be zero between those points. It isn't. It also isn't clear that the function falls to $-\infty$ and rises to $+\infty$ on either side of the points. We can get a clearer view by limiting the range of the y axis:

```
(%i1)  plot2d(tan(x), [x,-2,2], [y, -20, +20])$
```

Table 3.1 Options to control the display of a `plot2d` plot.

Option	Result
`[axes, false]`	Suppresses plotting of both axes.
`[axes, x]`	Shows only the x axis.
`[axes, y]`	Shows only the y axis.
`[box, false]`	Suppresses drawing a bounding box around the entire plot.
`[color, color_1, color_2, ...]`	Sets curve colors. Possible color names are `red`, `green`, `blue`, `magenta`, `cyan`, `yellow`, `orange`, `violet`, `brown`, `gray`, `black`, `white`, or a hexadecimal RGB code (a string starting with "#" followed by six hexadecimal digits, two each for the red, green, and blue components). If you omit the color option, Maxima chooses the color of each curve on a plot with multiple curves in that order. You can provide multiple colors for plots with multiple curves.
`[grid2d, true]`	Draws a 2D grid in the xy plane, with dotted lines corresponding to tics on the x and y axes.
`[label, [text, x, y], ...]`	Places a text label (or several labels) at the indicated x, y coordinates. For example, `[label, ["cutoff point", 0.5, −10]]` prints a label "cutoff point" at x = 0.5 and y = −10.
`[legend, false]`	Supresses printing a legend on plots with multiple curves.
`[legend, legend1, legend2, ...]`	Specifies the legend for each successive curve on a plot with multiple curves. By default, the names or formulas of the the different curves are used in the legend.
`[nticks, n]`	Sets the number of points computed along a curve. Set n to a higher number for a smoother curve.
`[point_type, symbol_1, symbol_2, ...]`	Sets the symbol used for points. Possible options are `bullet`, `circle`, `plus`, `times`, `asterisk`, `box`, `square`, `triangle`, `delta`, `wedge`, `nabla`, `diamond`, and `lozenge`. For multiple curves, you can list additional point types. For example, `[point_type, bullet, circle]` plots the first curve with filled circles, and the second with open circles.
`[same_xy, true]`	Use the same scale for both the x and y axes.
`[style, style1, style2, ...]`	Specifies each curve style. Possible styles are `points` for isolated points, `lines` for points connected with straight line segments, `linespoints` for both lines and points, `dots` for small dots, and `impulses` for vertical lines. If you omit the style option, Maxima chooses `lines` as the default style.
`[title, text]`	Sets the graph title to *text*.
`[xlabel, text]`	Sets the x axis label to *text*.
`[xtics, initial x, x increment, final x]`	Places tic marks on the x axis at *initial x*, *initial x* + *x increment*, ... *final x*. If *initial x* and *final x* are omitted, they will be chosen automatically.
`[ylabel, text]`	Sets the y axis label to *text*.
`[ytics, initial y, y increment, final y]`	Places tic marks on the y axis at *initial y*, *initial y* + *y increment*, ... *final x*. If *initial y* and *final y* are omitted, they will be chosen automatically.
`[yx_ratio, ratio]`	Sets the ratio between the height and width of the plot. If *ratio* is 2, the y axis twice as long as the x axis; if *ratio* is 1, the plot is square; if *ratio* is 1/2, the x axis will be twice as long as the y axis.

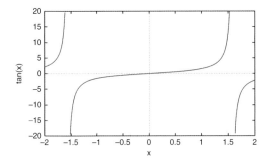

Maxima warns you that some values of the function in this plot are outside of the plotting area. You can also specify a logarithmic scale to show details of the function over a much larger range. For example, plot the function exp(x) between $x = 0$ and $x = 700$:

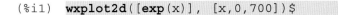

(%i1)   **wxplot2d([exp(x)], [x,0,700])$**

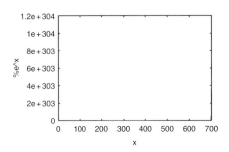

Not much of the function is visible because values of exp(x) on the right side of the range are so much larger than those to the left. Select Plot ⟩ Plot 2d and check the "logscale" box for the y axis,

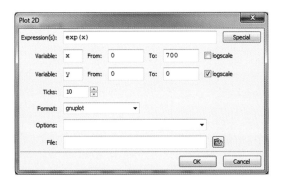

or type *logy* as an option in plot2d directly:

(%i1)   **wxplot2d([exp(x)], [x,0,700], [logy])$**

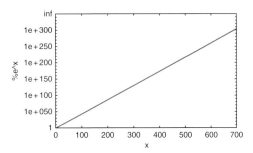

You can also scale the *x* axis logarithmically with the `logx` option.

3.1.7.1 Solving Axis Scaling Problems

Suppose we want to plot the potential energy *E* for a proton and an electron separated by a distance 0.1×10^{-9} m $< r < 1 \times 10^{-9}$ m, using Coulomb's law:

$$E = \frac{q_1 q_2}{4\pi\epsilon_0 r} \qquad \text{Coulomb's law} \tag{3.1}$$

where the charges are $q_1 = 1.60219 \times 10^{-19}$ coulombs, $q_2 = -q_1$, and the permittivity of a vacuum ϵ_0 is 8.85419×10^{-12} farads per meter.

Plotting Coulomb's law directly in SI units gives us an error:

```
(%i1)   q_0 : 1.60219e-19$   /* charge of a proton, in coulombs
          ➡ */
(%i2)   q_1 : -q_0$          /* charge of an electron, in
          ➡ coulombs */
(%i3)   e_0 : 8.85419e-12$   /* permittivity of a vacuum, in
          ➡ farads per meter */
(%i4)   plot2d([q_0*q_1/(4*%pi*e_0*r)], [r,0.1e-9,1e-9])$
```

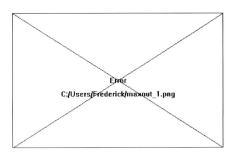

Maxima refuses to plot the graph if the *x* axis would be unreadable! We can fix the *x* axis by dividing the *r* values by 10^{-9}, and multiplying *r* by 10^{-9} in the Coulomb expression. We should change the *x* axis label to show that we've scaled *r*. Since *r* in meters divided by 10^{-9} is just *r* in nanometers, we can add the option [xlabel, "r / nm"] to label the *x* axis. We'll also label the *y* axis with the option [ylabel, "potential energy / J"].

```
(%i5)   plot2d([q_0*q_1/(4*%pi*e_0*(r*1e-9))], [r,0.1,1],
            [xlabel, "r / nm"],
            [ylabel, "potential energy / J"])$
```

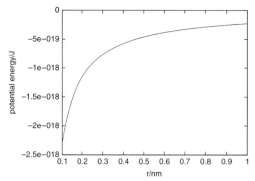

The y values are all around 10^{-19}, so we can make the y axis presentable by dividing all the y values by 10^{-19}. We need to change the y axis label to show that we've scaled it. We do this by adding [ylabel, "potential energy / 10^19 J"] to the plot2d options.

```
(%i6)   plot2d([q_0*q_1/(4*%pi*e_0*(r*1e-9))/1e-19], [r,0.1,1],
           [xlabel, "r / nm"],
           [ylabel, "potential energy / 10^-19 J"])$
```

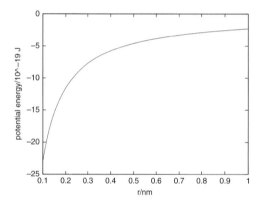

Let's use the *title* and *grid2d* options to add a title and a grid to the Coulomb potential plot.

```
(%i7)   plot2d([q_0*q_1/(4*%pi*e_0*(r*1e-9))/1e-19], [r,0.1,1],
           [xlabel, "r / nm"],
           [ylabel, "potential energy / 10^-19 J"],
           [title, "Coulomb potential between a proton and
               ➦ electron"],
           [grid2d, true])$
```

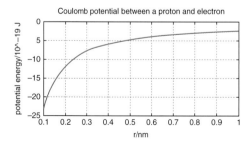

3.1.7.2 Positioning the Legend

By default, the legend for graphs with multiple curves is placed in the top right-hand corner of the graph. This is sometimes a problem when the curves overlap the legend. For example,

```
(%i1)  data: [[1,1], [2,2], [3,3], [4,4], [5,5], [6,6]]$
(%i2)  plot2d([[discrete, data], x], [x, 0, 7], [style, points,
       ➥ lines], [ylabel, "y"], [legend, "data", "y=x"])$
```

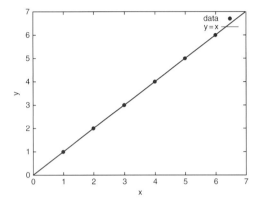

Unfortunately, `plot2d` doesn't have a command for moving the legend, but Gnuplot does. We can use the *gnuplot_preamble* option to send commands directly to Gnuplot. In this case the command is `set key right center`:

```
(%i1)  data: [[1,1], [2,2], [3,3], [4,4], [5,5], [6,6]]$
(%i2)  plot2d([[discrete, data], x], [x, 0, 7], [style, points,
       ➥ lines], [ylabel, "y"], [legend, "data", "y=x"], [
       ➥ gnuplot_preamble, "set key right center"])$
```

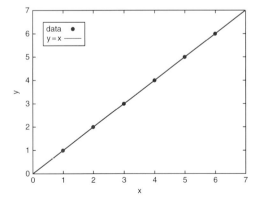

Notice that the Gnuplot command in *gnuplot_preamble* must be placed between double quotes.

Different positions are possible. The first position is horizontal and can be `left`, `center`, or `right`. The second position is vertical and can be `top`, `center`, or `bottom`. You can also place the legend outside the graph with `set key outside`, or below the graph with `set key below`.

Worksheet 3.2: Formatting graphs

In this worksheet, we'll explore Maxima's commands for labeling and formatting graphs, and see how to fix common problems with axis scaling.

3.1.8 Parametric Plots

Problems that focus on the rotational motion of molecules are important in quantum chemistry and spectroscopy. These problems can be more naturally represented in terms of radii and angles than in terms of xy coordinates.

In two dimensions, we can use polar coordinates r and θ, where r is the distance from the origin and θ specifies the angle of the line between the point at r and the origin it makes with the x axis. The polar coordinates are related to the xy coordinates by

$$x = r\cos\theta \qquad y = r\sin\theta \qquad \text{Polar coordinates} \qquad (3.2)$$

Functions of polar coordinates are parametric curves, curves that depend on parameters like r and θ. Maxima provides a simple way to plot parametric curves over a specific range of parameters.

If r is fixed, and θ varies from 0 to 2π, the x and y values calculated from Equation (3.2) lie along a circle with radius r. In xy coordinates, the equation for a circle with radius 1 is $1 = \sqrt{x^2 + y^2}$, but in polar coordinates the equation is simply $r = 1$! Let's verify this by plotting a circle with a radius equal to one.

We can use plot2d with the *parametric* option to specify the x and y coordinates as a function of the angle:

 [parametric, *x as a function of t, y as a function of t, [t, min t, max t], options*]
Specifies a parametric curve when placed in the *curve list* for plot2d. The curve will be plotted for values of x and y computed as the parameter t varies from *min* t to *max* t.

You can type the *parametric* option directly into the curve list for plot2d, or use the Plot ⟩ Plot 2d... dialog to do insert it for you:

1) Select Plot ⟩ Plot 2d... . The *Plot 2D* dialog appears.
2) Click on the **Special** button, and then select Parametric plot . The *Parametric plot* dialog appears.
3) In the x= box, type cos(t).
4) In the y= box, type sin(t).
5) Set the variable t to vary from 0 to 2*%pi.
6) Click **OK**.
7) Set the variable x to vary from -1 to 1.
8) Click **OK**.

The inserted code and the graph will look like this:

```
(%i1)   plot2d([[parametric, cos(t), sin(t), [t, 0, 2*%pi], [
            ➡ nticks, 300]]], [x,-1,1])$
```

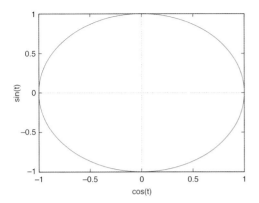

That doesn't look like a circle, because by default the ratio of *y* to *x* in plots is 3/4. To make the ratio one to one, add the option [*same_xy*, true] to the code:

```
(%i1)   plot2d([[parametric, cos(t), sin(t), [t, 0, 2*%pi], [
         ➡ nticks, 300]]], [x,-1,1], [same_xy, true])$
```

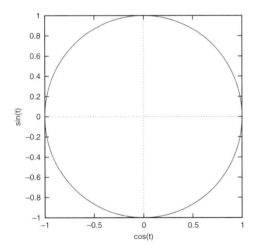

The parameter t corresponds to *θ* when we're using polar coordinates. The range on the parameter t ([t, 0, 2*%pi]) and any curve-specific options must be nested *inside* the *parametric* option. For example, the *Plot 2D* dialog inserted [nticks, 300] inside the *parametric* option; this controls the number of points that Maxima actually computes to draw the curve.

You can have a list of curves for the first argument if you want to plot multiple curves. Here is one branch of an Archimedean spiral plotted with the circle:

```
(%i1)   plot2d([
               [parametric, cos(t), sin(t), [t, 0, 2*%pi]],
               [parametric, t*cos(t)/(2*%pi), t*sin(t)/(2*%pi)
                 ➡ , [t, 0, 2*%pi]]
               ],
               [x,-1,1],
```

```
              [same_xy, true]
       )$
```

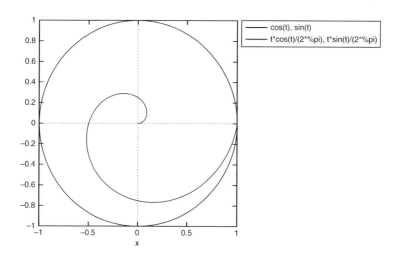

Parametric curves are extremely useful in plotting complex functions. Recall from Chapter 2 that points in the complex plane can be represented by taking the real part of the function as its *x* coordinate, and the imaginary part as the *y* coordinate (see Figure 2.1). For example, consider the following plot of the wavefunction for a particle moving on a circular ring:

```
(%i1)   psi(phi,ml)  :=  exp(%i*ml*phi)/sqrt(2*%pi);
(%i2)   wxplot2d(
           [parametric, realpart(psi(t,1)), imagpart(psi(t,1))
              ➡ , [t, 0, 2*%pi]],
           [same_xy, true]
        )$
```

$$(\%o1) \quad \Psi(\phi, ml) := \frac{\exp(i \cdot ml \cdot \phi)}{\sqrt{2 \cdot \pi}}$$

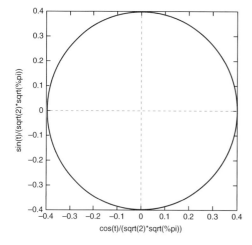

Until now we've plotted x and y as functions of polar coordinates. How can we plot expressions that give one polar coordinate in terms of the other? For example, suppose we want to plot a hyperbolic spiral $r(\theta) = 10/\theta$, with $0 \leq \theta \leq 2\pi$.

First, define one polar coordinate as a function of the other. Then apply Equation (3.2) to write out the x and y functions:

```
(%i1)  r(t) := 10/t$
(%i2)  plot2d([parametric, r(t)*cos(t), r(t)*sin(t), [t,1,10*%
          ➥ pi]]);
```

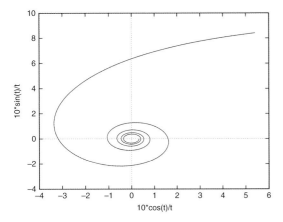

M **Worksheet 3.3: Parametric Plots**

In this worksheet, we'll plot complex functions, wavefunctions for the particle-on-a-ring model, and the angular parts of hydrogenic orbitals in terms of polar coordinates.

3.1.9 Implicit Plots

An implicit equation is one that cannot necessarily be solved to give a single equation whose graph is the complete relationship. For example, the equation for a unit circle

$$x^2 + y^2 = 1 \qquad \text{Implicit equation for a unit circle} \qquad (3.3)$$

can be solved for y as $y = \sqrt{1 - x^2}$ and $y = -\sqrt{1 - x^2}$. These relationships each capture only a semicircle, rather than the full equation. Two values of y are possible for each value of x.

You'll often encounter implicit equations in equilibrium problems, real gas law calculations, chemical kinetics, and quantum chemistry – the solutions of differential equations are often implicit equations. Explicitly solving such equations in terms of their variables may be difficult or impossible, so we may not be able plot them using any of the tools we've seen so far. Maxima can plot implicit equations directly. Its `implicit_plot` function supports all of the options we've already seen for `plot2d`. The function takes the form:

```
implicit_plot( equation list, x range, y range,
    options)
```

The first argument is an equation or list of equations or equation names to be plotted. The second and third arguments give the ranges of the variables we want on the *x* and *y* axes, respectively (typed as usual as [x, *min x*, *max x*], [y, *min y*, *max y*]). You can name the *x* and *y* variables anything you like. You must load(implicit_plot) to load this function.

Let's plot the equation for a unit circle implicitly:

```
(%i1)  load(implicit_plot)$
(%i2)  implicit_plot(x^2+y^2=1, [x, -2, 2], [y, -2, 2], [
    ➡ same_xy, true])$
```

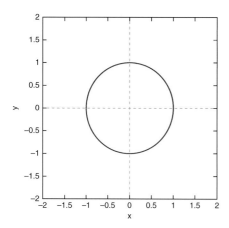

Like plot2d, implicit_plot opens the graph in a separate window that must be closed before you can return to the worksheet. Use wximplicit_plot to embed the plot inline. Unlike plot2d, implicit_plot needs the ranges for *both* the *x* and *y* variables.

Equilibrium laws are usually implicit equations that are approximated to obtain explicit solutions. In Maxima, we can look for graphical solutions to the equations without making such approximations, so we can check the validity of an approximation.

Let's plot the pH of a weak acid solution as a function of the acid's molarity *C*. If we combine the acid dissociation equilibrium law, the dissociation of water, charge and mass balance relations, and the definition of pH, we obtain Equation (3.4):

$$K_a = \frac{10^{-pH}\left(10^{-pH} - \frac{K_w}{10^{-pH}}\right)}{C - 10^{-pH} + \frac{K_w}{10^{-pH}}} \qquad \text{Weak acid dissociation including water equilibria}$$

(3.4)

It's difficult to solve this equation to obtain pH as a function of *C*. The equation is a cubic polynomial in 10^{-pH}; it has three cumbersome solutions (and two of them are imaginary). If we assume that the hydrogen ion concentration 10^{-pH} is much smaller than the concentration *C*, and if we ignore the dissociation of water, we obtain a much simpler relationship:

$$10^{-pH} = \sqrt{K_a C} \qquad \text{Approximate weak acid dissociation}$$

(3.5)

To see how these simplifications affect the predicted pH, we might plot both equations on the same graph, explicitly solving each equation for pH as a function of C. The second equation gives us $pH = -\log \sqrt{K_a C}$, but as we've already seen, the first equation isn't so easily solved. Equation (3.4) is an implicit equation in pH and C. Let's plot it along with its approximation:

```
(%i1)   exact_weak_acid: Ka = ((10^-pH)*(10^-pH-Kw/(10^-pH)))/(C
        ➥   - 10^-pH + Kw/(10^-pH));
(%i2)   approx_weak_acid: (10^-pH) = sqrt(Ka*C);
(%i3)   Ka: 1.75e-5$
(%i4)   Kw: 1.0e-14$
```

$$(\%o1) \quad Ka = \frac{\frac{1}{10^{pH}} - Kw\,10^{pH}}{10^{pH}\left(Kw\,10^{pH} - \frac{1}{10^{pH}} + C\right)} \qquad (\%o2)\frac{1}{10^{pH}} = \sqrt{Ka\,C}$$

```
(%i1)   implicit_plot([exact_weak_acid,approx_weak_acid], [C,
        ➥  0.01, 0.1],[pH,2.8,3.4])$
```

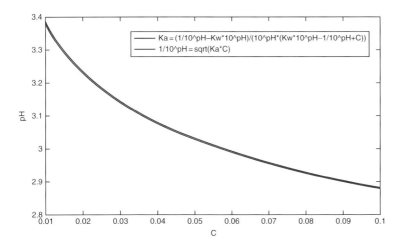

The approximation is good at this concentration range. Using the cursor in `implicit_plot`, the exact solutions can be read directly from the "exact" curve, so there is little need for the approximation. However, the exact equation may have nonphysical solutions that show up on the graph, and a good approximation is useful for picking out the physical solution.

 Worksheet 3.4: Implicit Plots

In this worksheet, we'll find solutions of nonlinear equations like the van der Waals equation and equilibrium mass action laws by reading points from the implicitly plotted equations.

3.1.10   Histograms

Histograms show how data is distributed, that is, how often data points appear in a given dataset. Student grade curves are histograms. The grades are sorted into intervals or *bins* like

"0 to 10," "10 to 20," "20 to 30," and so on. A bar graph is constructed, with one bar per bin. The height of the bar is the number of grades that fall into that bin.

The following example generates 1000 normally distributed random numbers with a mean of zero and a standard deviation of one, using Maxima's `random_normal` function. The list is stored as `data`, and passed to the `histogram` function which builds and displays the histogram. The `stats` package must be loaded first.

```
(%i1)   load(stats)$
(%i2)   data: random_normal(0, 1, 1000)$
(%i3)   histogram(data)$
```

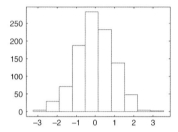

The `histogram` function is built on top of Maxima's `draw` package, so its options have a slightly different format. `plot2d` and friends use lists as options; `histogram` uses options that have the format `option=value` instead of `[option, value]`.

By default, Maxima will sort the data into 10 bins. You can change the number of bins to any number N by adding the `nclasses=N` option. You can also add `xlabel="x axis title"` and `ylabel="y axis title"` options to label the axes, and `title="Graph Title"` to title the plot. For example, using `histogram(data,nclasses=20,xlabel="x", ylabel="frequency")` in the example above gives you a histogram like this:

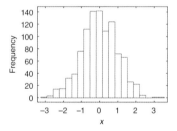

By default, `histogram` plots the absolute frequencies of items in the dataset. You can change this with the *frequency* option, which can be set to `absolute`, `relative`, `percent`, and `density`. The option `frequency=density` makes the total area of the histogram equal to one by dividing all of the absolute frequencies by the number of measurements; we'll use it to plot normalized probability densities in Chapter 8.

Many other options are also available in `histogram`. We'll use some of them in Worksheet 3.5.

 Worksheet 3.5: Plotting histograms

In this worksheet, we'll use the built-in `histogram` function to make plots that show how data is distributed.

3.2 Plotting in Three Dimensions

Many chemically important functions (like atomic and molecular orbitals, and equations of state) are functions of more than one variable. Maxima can interactively plot functions of two variables in three dimensions. The plots can be rotated, zoomed, and realized in different forms so that we can see the function's behavior from different perspectives. Maxima's three-dimensional plotter `plot3d` is similar to `plot2d`; both functions share many of the same options.

The `plot3d` function interactively displays a surface defined as a function of two variables. You can type it directly, or you can select Plot ⟫ Plot 3d... from the menu.

```
plot3d(expression, x range, y range, options)
```

 Plot an expression in two variables x and y in three dimensions. The `x range` and `y range` are required; they are specified as they are in two-dimensional plotting. Many of the `options` from `plot2d` apply; options peculiar to `plot3d` are listed in Table 3.2. Substitute `wxplot3d` for `plot3d` for plots that are placed inline in the worksheet.

You are free to name the two variables anything you'd like. For example, consider the following plot of the function:

$$f(u, v) = sin(u) * exp(-v), \quad 0 \le u \le 10, \quad 0 \le v \le 2$$

```
(%i1)   plot3d(sin(u)*exp(-v), [u, 0, 10], [v, 0, 2])$
```

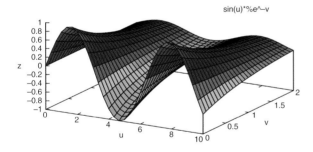

You can use your mouse or trackball to adjust the plot view. Drag the mouse over the plot to rotate it; a mouse wheel can be used to pan, zoom, or stretch the plot. Use the mouse wheel to pan the view from side to side.

3.2.1 Plotting Functions of x, y, and z

In `plot2d`, we plotted functions of a single variable as $y = f(x)$. In `plot3d`, we can plot functions of two variables as $z = f(x, y)$. But, many chemical functions – such as atomic orbitals – are functions of *three* variables. Consider the unnormalized $2p_z$ hydrogenic orbital:

```
(%i1)   pz(x,y,z):=z*exp(-sqrt(x^2+y^2+z^2)/2)$
```

Table 3.2 **Special options for** `plot3d`.

`[azimuth, 30]`	Sets the angle to rotate the xy plane around the z axis, in degrees.
`[box, true]`	Draws a box around the plot.
`[color, blue, green]`	Colors the top of the surface blue, and the underside of the surface green. If only one color is given and no altitude shading is being used, this option will set the color of the mesh lines.
`[colorbox, true]`	Shows a color scale to the right of the plot when altitude shading is being used.
`[elevation, 60]`	Sets the angle to rotate the z axis around the x axis. Setting elevation to zero views the surface down the z axis.
`[grid, 30, 30]`	Sets the number of grid points in the x and y directions to be calculated.
`[mesh_lines_color, black]`	When altitude shading is being used, this option sets the color of the mesh lines; setting it to `false` hides the mesh lines.
`[nticks, 29]`	The number of points calculated when plotting a parametric surface.
`[palette, false]`	Turn off altitude shading.
`[palette, [hue, 0.2, 1, 1, 0.8]]`	Sets the palette for altitude shading. The first three numbers are the hue, saturation, and value (lightness) of the lowest z value; they must be between zero and one. The last number gives the increase in hue added to the initial value for the highest value of z. You can also use `saturation` or `value` in place of `hue`.
`[same_xyz, true]`	Use identical scales for the x, y, and z axes.
`[transform_xy, spherical_to_xyz]`	Transforms spherical coordinates to xyz coordinates.
`[transform_xy, polar_to_xy]`	Transforms cylindrical coordinates to xyz coordinates.
`[zlabel, "text"]`	Labels the z axis.

Many of the options listed previously for `plot2d` also work in `plot3d`. Any of the numbers or colors listed in the examples below are defaults; you can change them if you like.

$$\psi_{2p_z}(x, y, z) = z \exp\left(-\frac{\sqrt{x^2 + y^2 + z^2}}{2}\right) \qquad \text{The 2p}_z \text{ hydrogenic orbital} \qquad (3.6)$$

To see the function as it is, we'd have to plot it in *four* dimensions: x, y, z, and ψ_{2p_z}! The best we can do with three-dimensional plots is look at the wavefunction's amplitude in the xy, xz, and yz planes.

In the xy plane, $z = 0$ everywhere. This is the nodal plane through the nucleus.

```
(%i2)  plot3d(pz(x,y,0), [x,-5,5], [y,-5,5], [zlabel, "2p_z"])$
```

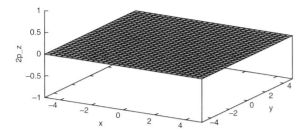

In the *xz* plane, $y = 0$:

```
(%i3)   plot3d(pz(x,0,z),  [x,-5,5],  [z,-5,5],  [ylabel,"z"],[
     ➥ zlabel, "2p_z"])$
```

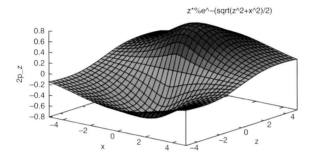

In the *xz* and *yz* planes there is a positive "lobe" on one side of the nucleus (at 0,0), and a negative lobe on the other.

3.2.2   Plotting Multiple Surfaces

To plot multiple surfaces, use this variation of `plot3d`:

```
plot3d([
        [surface_1, xrange_1, yrange_1],
        [surface_2, xrange_2, yrange_2],
            ...
], global_xrange, global_yrange, options)
```

Plot a list of multiple surfaces on the same plot. Each surface is specified by a three-item list containing an expression that depends on *x* and *y* followed by *x* and *y* ranges that pertain to that specific expression. The ranges for each surface can be different from the global *x* and *y* ranges for the complete plot.

For example, to plot the amplitude of the 1s and $2\mathrm{p}_z$ orbital in the *xz* plane,

```
(%i4)   one_s(x,y,z)  :=  exp(-sqrt(x^2+y^2+z^2));
(%i5)   wxplot3d(
            [
                [pz(x,0,z),  [x,-5,5],[z,-5,5]],
                [one_s(x,0,z),  [x,-5,5],[z,-5,5]]
            ],
```

```
        [ylabel, "z"],
        [zlabel, "psi"]
    )$
```

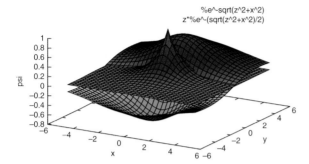

Notice that global x and y ranges are optional, but the x and y ranges for each surface must be specified.

It is possible to change the palette for altitude coloring. You can change the colors for the separate surfaces, too. We'll see how in Worksheet 3.6.

3.2.3   Plotting in Spherical Coordinates

Atomic orbitals are often written in terms of spherical coordinates, a three-dimensional extension of the polar coordinates we saw in Section 3.1.8. The spherical coordinates are r, the distance of a point to the origin; θ, the angle between the line from point to origin with the z axis; and ϕ, the angle between the line from the point's projection in the xy plane and the x axis (see Figure 3.3).

Equation (3.7) relates Cartesian and spherical coordinates.

$$
\begin{aligned}
x &= r\sin\theta\cos\phi \\
y &= r\sin\theta\sin\phi \\
z &= r\cos\theta
\end{aligned}
\qquad
\begin{array}{l}
\text{Transforming spherical coordinates} \\
\text{to Cartesian coordinates}
\end{array}
\tag{3.7}
$$

The x, y, and z coordinates can range from $-\infty$ to $+\infty$. Spherical coordinates cover the same space with the following ranges:

$$
\begin{aligned}
0 &\le r \le \infty \\
0 &\le \theta \le \pi \\
0 &\le \phi \le 2\pi
\end{aligned}
\qquad
\text{Ranges of the spherical coordinates}
\tag{3.8}
$$

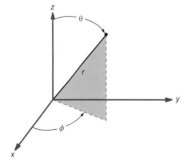

Figure 3.3 The spherical coordinates r, θ, and ϕ are related to the Cartesian coordinates x, y, and z.

Let's use Maxima to look at the shape of the $3d_{z^2}$ orbital, which has an angular component $3\cos^2\theta - 1$.

To plot functions in spherical coordinates using `plot3d`, add the `[transform_xy, spherical_to_xyz]` option. This automatically applies Equation (3.7) to transform the function into xyz coordinates.[4]

```
(%i1)  plot3d(3*cos(theta)^2-1, [theta, 0, %pi], [phi, 0, 2*%pi
       ➥ ],
              [transform_xy, spherical_to_xyz])$
```

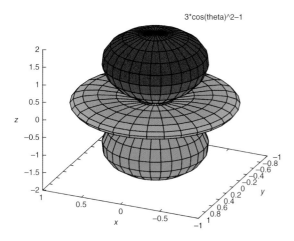

3.2.4 Plotting in Cylindrical Coordinates

For systems with a clearly defined axis (like a electron moving around a fused ring system, or a rotating water molecule) it is more natural to use cylindrical coordinates r, ϕ, and z, where r is the distance of a point from the z axis, ϕ is the azimuthal angle (the angle the projection of the point on the xy plane makes with the x axis), and z is the altitude above the xy plane. See Figure 3.4.

Equation (3.9) relates Cartesian and cylindrical coordinates.

$$\begin{aligned} x &= r\cos\phi \\ y &= r\sin\phi \\ z &= z \end{aligned} \qquad \text{Transforming cylindrical coordinates to Cartesian coordinates} \qquad (3.9)$$

Figure 3.4 The cylindrical coordinates r, ϕ, and z are related to the Cartesian coordinates x, y, and z.

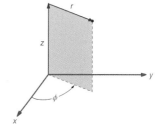

---

4 Other built-in transformations are available, or you can define your own. A project in Worksheet 3.6 will do this using Maxima's `make_transform` function.

The `transform_xy` option can be set to `polar_xy` to plot functions written in cylindrical coordinates. For example, let's make a three-dimensional plot of the particle-on-a-ring wavefunction [17]:

$$\psi_m(\phi) = \frac{1}{\sqrt{2\pi}} \exp(im\phi) \qquad \text{Wavefunction for a particle on a ring} \qquad (3.10)$$

for $r \approx 1$, with $m = 2$:

```
(%i1)  psi(phi,m) := 1/sqrt(2*%pi)*exp(%i*m*phi);
(%i2)  plot3d(realpart(psi(phi,m)), [r, 0.999,1], [phi, 0, 2*%
       ➡ pi], [transform_xy, polar_to_xy]), m=2;
```

$$(\%o1) \quad \psi(\phi, m) := \frac{1}{\sqrt{2\pi}} \exp(i\, m\, \phi)$$

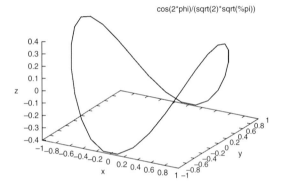

The wavefunction is complex, so we've used the `realpart` function in the expression to be plotted. The ranges of the r and ϕ parameters are set in the second and third arguments. We use the `[transform_xy, polar_to_xy]` option to transform cylindrical coordinates to Cartesian coordinates automatically. Finally, we set the quantum number m using an evaluation environment.

3.2.5   Parametric Surface Plots

It's also possible to do two-parameter plots in three dimensions in a more general way, with the following form of `plot3d`:

```
plot3d([x expression, y expression, z expression],
       s range, t range, options)
```
Plot x, y, and z expressions that are functions of two parameters s and t.

You can choose the names of the coordinates to be anything you like.

Let's plot the boundary surface for the d_{xy} orbital parametrically using the spherical coordinates ϕ and θ as parameters. The coordinate r will be the amplitude of the d_{xy} orbital, which is a sum of `spherical_harmonic` functions:

```
(%i1)  l : 2$
(%i2)  m : 2$
```

```
(%i3)  r(theta,phi) := (spherical_harmonic(l,m,theta,phi) +
       ➥ spherical_harmonic(l,-m,theta,phi))/sqrt(2);
```

$$(\%o1) \quad \mathrm{r}\,(\theta,\phi) := \frac{\mathrm{Y}_l^m\,(\theta,\phi) + \mathrm{Y}_l^{-m}\,(\theta,\phi)}{\sqrt{2}}$$

The X, Y, and Z functions explicitly transform spherical coordinates into Cartesian coordinates using Equation (3.7).

```
(%i4)  X(theta,phi) := r(theta,phi) * sin(theta) * cos(phi);
(%i5)  Y(theta,phi) := r(theta,phi) *sin(theta) * sin(phi);
(%i6)  Z(theta,phi) := r(theta,phi) * cos(theta);
```

$$(\%o2) \quad \mathrm{X}\,(\theta,\phi) := \mathrm{r}\,(\theta,\phi)\,\sin(\theta)\,\cos(\phi)$$
$$(\%o3) \quad \mathrm{Y}\,(\theta,\phi) := \mathrm{r}\,(\theta,\phi)\,\sin(\theta)\,\sin(\phi)$$
$$(\%o4) \quad \mathrm{Z}\,(\theta,\phi) := \mathrm{r}\,(\theta,\phi)\,\cos(\theta)$$

Now draw a parametric plot:

```
(%i7)  plot3d([X(theta,phi), Y(theta, phi), Z(theta, phi)], [
       ➥ theta, 0, 2*%pi], [phi, 0, 2*%pi], [grid,
       ➥ 100,100])$
```

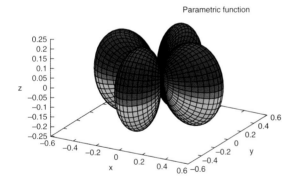

We've used the *grid* option to compute a 100 × 100 grid for better resolution. By default, plot3d computes a 30 × 30 grid.

We saw in the previous section that Maxima can automatically do this transformation for us with the *transform_xy* option. The following code produces a plot identical to the one above (except that it will not be labeled with "Parametric Plot"):

```
(%i1)  l : 2$
(%i2)  m : 2$
(%i3)  r(theta,phi) := (spherical_harmonic(l,m,theta,phi) +
       ➥ spherical_harmonic(l,-m,theta,phi))/sqrt(2);
(%i4)  plot3d(r(theta,phi), [theta, 0, 2*%pi], [phi, 0, 2*%pi],
       ➥ [grid, 100,100], [transform_xy, spherical_to_xyz
       ➥ ]);
```

Use `transform_xy` for cylindrical or spherical coordinate transformations. You should only define X, Y, and Z functions for `plot3d` if you are transforming into some other coordinate system. In Worksheet 3.6, we'll take this approach to plot a function in elliptical coordinates.

3.2.6 Plotting Discrete Three-Dimensional Data

3D scatter plots of experimental data are common in spectroscopy and in studies of quantitative structure–activity relationships. Three-dimensional plots can reveal important structure and clustering in the data that a flat 2D plot might miss, but they can be difficult to read. The ability to view the plots from multiple perspectives is essential [18].

Maxima's `plot3d` function can't plot discrete 3D data directly, but the more powerful `draw3d` function can. The syntax of the `draw3d` function and its arguments is quite different from `plot3d`.

Suppose we want to plot x, y, z points stored in a matrix P. Each row of P is a point; the three columns list the x, y, and z coordinates, respectively.

```
(%i1)   P: matrix([1,1,2],[2,4,4],[3,9,6],[4,16,8],
              [5,25,10],[6,36,12],[7,49,14])$
```

Unlike `plot3d`, `draw3d` can accept matrix data. The call to plot the points is `draw3d(points(P))$`, where P is a data matrix.[5]

```
(%i2)   draw3d(points(P))$
```

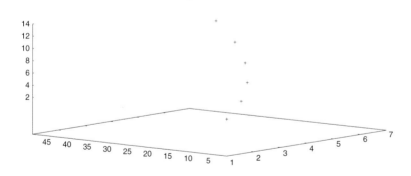

Separate arrays could also be used for the x, y, and z data:

```
(%i3)   X: [1,2,3,4,5,6,7]$
(%i4)   Y: [1,4,9,16,25,36,49]$
(%i5)   Z: [2,4,6,8,10,12,14]$
(%i6)   draw3d(points(X,Y,Z))$
```

---

5 In older versions of Maxima, `draw3d` must be loaded from the `draw` package before it can be used.

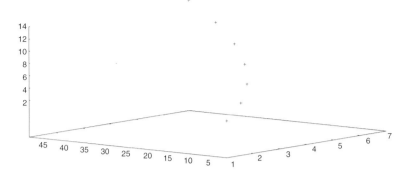

draw3d has many options for formatting the plot. For example, to draw lines between the points and change the point symbols to circles,

```
(%i7)    draw3d(points_joined = true,
         point_type = circle,
         points(P))$
```

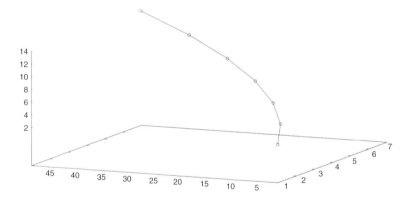

Some additional options for draw3d are listed in Table 3.3. You'll find many more in the Maxima manual. Experiment!

3.2.7 Contour Plotting

Surface plots for complicated functions can be difficult to interpret. A contour plot can provide a clearer and more detailed view of the surface.

There are several ways to do contour plots in Maxima. The simplest way is to use the contour_plot function, which takes the same arguments as the plot3d function:

contour_plot(*surface expression, x range, y range, options*)

Plot an expression in two variables as a series of contours that have constant values of the expression. The *options* are the same as those available in plot3d.

Table 3.3 Examples of options for `draw3d` in Maxima.

`axis_3d = false`	Hides the axes.
`color = red`	Sets the pen color to red.
`colorbox = false`	Hides the color key when altitude color mapping is being used.
`enhanced3d = true`	The surface color will change with altitude.
`explicit(f(x,y), x, -5, 5, y, -10, 10)`	Plots the surface $f(x, y)$ with $-5 \le x \le 5$ and $-10 \le y \le 10$.
`key = "text"`	Sets the legend for the next plotted surface.
`label(["text",1,2,3])`	Places a label "text" in 3D space at coordinates (1,2,3).
`line_width = 2`	Sets the pen width to 2.
`logx = true`	Uses a log scale for the x axis.
`palette = gray`	Changes the palette used for altitude color mapping to grayscale.
`parametric(u(t),v(t),w(t), t,0,2)`	Plots a parametric 3D graph with $x = u(t)$, $y = v(t)$, $z = w(t)$, and $0 \le t \le 2$.
`points(X,Y,Z)`	Draws isolated points from separate arrays containing the x, y, and z coordinates.
`points_joined=true`	Draws line segments between points.
`point_size = 3`	Sets the point size to 3.
`points_type = filled_circle`	Draws filled circles for points.
`proportional_axes = xyz`	Makes the axes proportional to their relative lengths.
`surface_hide=true`	Makes a wiremesh surface opaque.
`title = "text"`	Sets the plot title to "text."
`xaxis = true`	Shows the x axis (true by default).
`xaxis_color = blue`	Colors the x axis blue.
`xaxis_width = 2`	Sets the width of the x axis to 2.
`xlabel = "text"`	Labels the x axis with "text."
`xtics = false`	Hides tics on the x axis.
`user_preamble = "set size ratio -1"`	Sends the `set size ratio -1` command to GnuplotGnuplot.
`wired_surface = true`	Draws a wire mesh on a surface.

You can change any of the numbers in the examples, and y or z can be substituted for x.

For example, let's draw a contour plot of the $2p_z$ orbital in the yz plane:

```
(%i1)  pz(x,y,z):=z*exp(-sqrt(x^2+y^2+z^2)/2)$
(%i2)  contour_plot(pz(0,y,z),[y,-10,10],[z,-10,10],
         [gnuplot_preamble, "set cntrparam levels incremental
            ➡ -0.4, 0.1, 0.4;"])$
```

Table 3.4 Gnuplot commands for controlling contour plots.

`set cntrparam levels 7;`	Draws seven evenly spaced contours between the minimum and maximum values of *z*.
`set cntrparam levels incremental -0.4, 0.1, 0.4;`	Draw evenly spaced contours, with the lowest contour at −0.4, the highest contour at 0.4, and the intermediate contours spaced by 0.1 between those limits.
`set cntrparam levels discrete -0.2, -0.5, 0.0, 0.2, 0.5;`	Draw unevenly spaced contours. In this case, draw five contour lines where the function is equal to −0.2, −0.5, 0.0, 0.2, and 0.5.

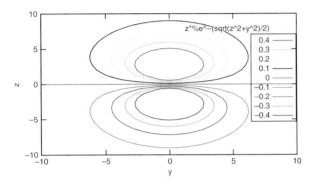

Unfortunately, the `contour_plot` does not have options for controlling the number and spacing of contour lines. You must pass commands directly to Gnuplot, using the *gnuplot_preamble* option. This option has the general form `[gnuplot_preamble, "commands"]`, where the commands between the double quotes are separated by semicolons. Some useful Gnuplot commands for controlling contour plots are given in Table 3.4.

It's also possible to draw contours on the plane underneath a surface plot using the Gnuplot commands shown in the following example:

```
(%i3)   plot3d(pz(0,y,z),[y,-10,10],[z,-10,10],
            [gnuplot_preamble, "
                set contour base;
                set cntrparam levels incremental -0.4, 0.1, 0.4;
                set nokey;
            "])$
```

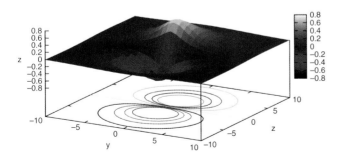

The Gnuplot command `set contour base;` draws the contour graph in the plane below the surface; `set nokey` supresses the display of the contour line key.

Setting the elevation of a surface plot to zero draws a color contour map. The Gnuplot command `set view map;` accomplishes this:

```
(%i4)  plot3d(pz(0,y,z),[y,-10,10],[z,-10,10],
           [gnuplot_preamble,"set view map;"]
       )$
```

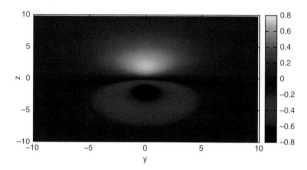

To make a polar contour plot in the *xy* plane, use the *polar_to_xy* option for *transform_xy*. For example,

```
(%i1)  pz(r,theta) := r*cos(theta)*exp(-r/2)$
(%i2)  contour_plot(pz(r,theta), [r,0,10], [theta, 0, 2*%pi], [
          ➥ transform_xy, polar_to_xy],
          [gnuplot_preamble, "set cntrparam levels 10;"])$
```

The `[transform_xy, polar_to_xy]` option causes the first coordinate to be interpreted as distance from the *z* axis, and the second coordinate to be taken as an azimuthal (polar) angle.

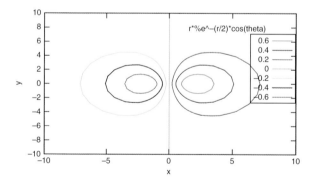

Ⓜ **Worksheet 3.6: Plotting in Three Dimensions**

In this worksheet, we'll plot functions and data in Cartesian, cylindrical, and spherical coordinates, and as contour maps.

4

Programming Maxima

First solve the problem. Then write the code.

– John Johnson

In this chapter, we'll see how Maxima can be used to write programs that perform iterative calculations and conditionally execute commands.

Section 4.1 focuses on how to control the execution of expressions using Maxima's noun and verb forms. In Section 4.2, we'll write multiline functions with local variables using `block` and `ev` statements for performing complex calculations that must be repeated many times. In Section 4.3, we'll use logical variables and the `if` statement to conditionally evaluate commands and groups of commands. Section 4.4 introduces recursive functions, which have many applications in quantum chemistry. In Section 4.5, we'll see how to use Maxima's context system to make assumptions, restrict domains, and declare properties for variables and operations in a session. Finally, in Section 4.6, we'll see how to use the `for` statement to write loops that repeat a calculation until a logical condition is met.

4.1 Nouns and Verbs

Sometimes we want to display or manipulate an expression before we let Maxima actually evaluate it. For example, suppose you want to display the rate law for a second-order reaction as

$$\frac{dx}{dt} = k(a - x)(b - x)$$

In Section 1.2.3, we saw that the `diff` function could be used to display a derivative. But, just typing the derivative in the equation evaluates it:

```
(%i1)   diff(x,t) = k*(a-x)*(b-x);
```

```
(%o1)   0 = k (a − x) (b − x)
```

The solution is to make the `diff` function a noun, that is, a symbol that isn't executed. To do this, type a single quote ' before the function name:

```
(%i2)   'diff(x,t) = k*(a-x)*(b-x);
```

```
(%o2)   d/dt x = k (a − x) (b − x)
```

Symbolic Mathematics for Chemists: A Guide for Maxima Users, First Edition. Fred Senese.
© 2019 John Wiley & Sons Ltd. Published 2019 by John Wiley & Sons Ltd.
Companion website: http://booksupport.wiley.com

The arguments will still be evaluated. For example, the value of *a* is inserted into the quoted expression `'diff(a,x)`:

```
(%i1)    a : sqrt(x)+x$
(%i2)    'diff(a, x);
```

$$(\%o2) \quad \frac{d}{dx}(2x)$$

To prevent an entire expression from being executed, put parentheses around the expression after the single quote.

```
(%i3)    '(diff(a, x));
```

$$(\%o3) \quad \frac{d}{dx}(a)$$

The single quote stops values from being substituted for variables and functions, but it won't keep Maxima from simplifying the expression.

```
(%i4)    '(a+a);
```

$$(\%o4) \quad 2a$$

The single quote also won't stop purely numerical calculations with built-in functions like `sqrt`, `sin`, and `exp`:

```
(%i5)    'sqrt(3.0);
```

$$(\%o5) \quad 1.732050807568877$$

By default, variables and functions that appear in function definitions are nouns. They will *not* be replaced by their current values at the time the function is defined. For example, suppose we want to define a function that converts grams to moles, using a variable `molar_mass` that is defined outside the function body:

```
(%i1)    molar_mass: 18.05$
(%i2)    grams_to_moles(x)  := x/molar_mass$
```

The current value of `molar_mass` (18.05) will be used when the function is evaluated:

```
(%i3)    grams_to_moles(36.10);
```

$$(\%o3) \quad 2.0$$

If we change the `molar_mass` value and call the function again, the new value will be used in the calculation:

```
(%i4)    molar_mass : 36.10$
(%i5)    grams_to_moles(36.10);
```

$$(\%o5) \quad 1.0$$

If you want the values of variables and functions in a function definition to be inserted when the function is defined, you can use the quote–quote operator " in the function definition.[1] Consider the difference between the following functions f and g:

```
(%i1)    a: 10$
(%i2)    f(x)  :=  a+x;
(%i3)    g(x)  :=  ''a + x;
```

$$(\%o2) \quad f(x) := a + x$$
$$(\%o2) \quad g(x) := 10 + x$$

By default, function names are verbs, that is, they are evaluated in expressions. You can make a function name a noun by using the `declare` function.

```
(%i1)    f(x)  :=  3*cos(x)  -  1;
(%i2)    f(%pi);
```

$$(\%o1) \quad f(x) := 3\cos(x) - 1$$
$$(\%o2) \quad -4$$

```
(%i3)    declare(f, noun)$
(%i4)    f(%pi);
```

$$(\%o4) \quad f(\pi)$$

Use `remove` to remove a function's noun status.

```
(%i5)    remove(f, noun)$
(%i6)    f(%pi);
```

$$(\%o6) \quad -4$$

To evaluate all the nouns in an expression, use the *nouns* switch:

```
(%i1)    'integrate(x^2*cos(x),  x);
(%i2)    %, nouns;
```

$$(\%o1) \quad \int x^2 \cos(x)\, dx$$
$$(\%o2) \quad (x^2 - 2)\,\sin(x) + 2x\cos(x)$$

(M) **Worksheet 4.0: Using nouns and verbs**

In this worksheet, we'll use Maxima's noun and verb forms to control the display and execution of expressions.

---

1 You can also use the `define` function to do this, as we saw in Section 1.1.8.

4.2 Writing Multiline Functions

In Section 1.1.8, we saw how to build functions to define a single-line calculation that may need to be used many times. To define more complicated functions that perform commands in a definite sequence, use an evaluation environment (ev) or a `block` statement as the body of the function. The syntax of the `block` statement is

```
block(
    [local1, local2, ...],
    command1,
    command2,
    ⋮
    lastcommand
);
```

A block statement evaluates a series of commands from left to right and returns the value of `lastcommand`. Each command in the block is separated by a comma, *not* a semicolon or dollar sign.

The variables `local1`, `local2`, ... listed at the beginning of the block are defined *only within the block*. They are temporary variables that are automatically killed after the block executes.

Let's define the `pH_buffer` function we defined in Section 1.1.8 again, this time using a block statement. We'll compute [A⁻] and [HA] as local variables A and HA, and then use the Henderson–Hasselbalch equation (Equation 1.2) to compute the pH directly:

```
(%i1)   log10(x) := log(x)/log(10)$
(%i2)   pH_buffer(Ka,Ca,Cb,Va,Vb) :=
            block(
                [A, HA],
                A : Vb*Cb/(Va+Vb),
                HA : (Va*Ca-Vb*Cb)/(Va+Vb),
                -log10(Ka) + log10(A/HA)
            )$
(%i3)   pH_buffer(1.75e-5, 1, 1, 0.025, 0.0125), numer;
(%o3)   4.756961951313706
```

The local variables A and HA aren't defined outside the block:

```
(%i4)   values;
```

```
(%o4)   []
```

Block statements can contain other block statements. Evaluation environments (Section 1.3) can also be nested inside of block statements and function definitions. In this case, though, you'll need to enclose the environment in an ev (...) function. In fact, we could have coded the function above entirely as an evaluation environment:

```
(%i1)   log10(x) := log(x)/log(10)$
(%i2)   pH_buffer(Ka,Ca,Cb,Va,Vb) :=
```

```
    ev(
        -log10(Ka) + log10(A/HA),
        A=Vb*Cb/(Va+Vb),
        HA=(Va*Ca-Vb*Cb)/(Va+Vb)
    )$
```
(%i3) pH_buffer(1.75e-5, 1, 1, 0.025, 0.0125), **numer**;

(%o3) 4.756961951313706

Notice two differences between `block` and `ev` environments:

- The order of statements is different. `block` returns the result of the last command, while `ev` returns the result of the *first* command.
- There is no need to list local variables in an `ev` environment; all of the variables you set in the environment persist only inside the `ev`. In `block` statements, a list of local variable names has to be provided.

Command history is also different within a `block`. The system variable for the value of the previous statement within a `block` is `%%`, *not* `%`. This is demonstrated in the following example.

```
(%i1)  degree_of_dissociation(Kp, P) := block(
            [P_A, P_A2, alpha, n0, n_A2, n_2A,n_tot],
            n_A2 : (1-alpha)*n0,
            n_2A : 2*alpha*n0,
            n_tot : n_A2 + n_2A,
            Kp = P_A^2/P_A2,
            subst(n_2A/n_tot*P, P_A, %%),
            subst(n_A2/n_tot*P, P_A2, %%),
            solve(%%,alpha),
            last(%%),
            rhs(%%)
        )$
(%i2)  degree_of_dissociation(0.05, 1);
```

(%o2) $\dfrac{1}{9}$

The function computes the degree of dissociation α for the ideal gas reaction $A_2 \rightleftharpoons 2A$. If there are initially n0 moles of A_2 and α is the fraction of A_2 that dissociates, we have

	A_2	$\rightleftharpoons$	$2A$
Initial moles:	n0		0
Change:	$-\alpha$ n0		$+2\alpha$ n0
Moles at equilibrium:	$(1-\alpha)$ n0		2α n0

with the moles at equilibrium stored in n_A2 and n_2A. The total number of moles is n_tot. The equilibrium law is Kp = P_A^2 / P_A2, with the partial pressure P_A and P_A2 computed as n_2A/n_tot*P and n_A2/n_tot*P, respectively. We solve the equation for α, select the last (positive) solution, and return the right-hand side rhs of the equation.

Finally, we test the function by computing α with `Kp = 0.05` and `P = 1`.

 Worksheet 4.1: Writing Multiline Functions

In this worksheet, we'll write multiline functions to perform common equilibrium, kinetics, gas law, and quantum mechanical calculations.

4.3 Decision Making

All of the functions and sequences of Maxima commands we have written so far execute from beginning to end in a linear fashion, one statement after another. We sometimes need a function or a program to make decisions or repeat a sequence of commands until a task has been completed.

The `if` statement is the fundamental command for decision making in Maxima. It can take several forms:

```
if logical expression then command;
```

If the `logical expression` is true, `command` is executed; otherwise it is skipped.

```
if logical expression then command1 else command2;
```

If `logical expression` is true, `command1` is executed; otherwise `command2` is executed.

```
if logical expression 1 then command1
elseif logical expression 2 then command2
    ⋮
else commandN
```

The *i*-th command is executed if the *i*-th logical expression is true and all preceding conditions are false; `commandN` is executed only if all of the conditions are false.

The commands can be any Maxima commands, including `block` statements or other `if` commands. The conditions must be logical expressions which evaluate to `true` or `false`.

We can use `if` statements to define piecewise functions. For example, the following function describes an artificial *crater potential* that can be used to simply model radioactive decay processes [19]. The potential energy of a particle is V_0 if it is inside the nucleus of a heavy atom, V_1 if it is within a distance a of the nucleus, and V_2 if it is at a distance greater than a from the nucleus. The distance x is set to zero at the surface of the nucleus.

```
(%i1)   crater_potential(x) := block(
            if (x < 0) then V0
            elseif (x > a) then V2
            else V1
        )$
(%i2)   plot2d(crater_potential(x), [x, -1, 2], [y, 0, 11], [ylabel
           ➥ ,"crater potential"]), a=1, V0=5, V1=10, V2=2$
```

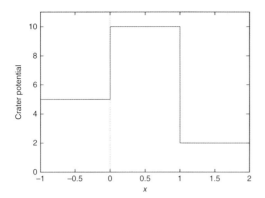

4.4 Recursive Functions

By default, the last expression in a block is the value the statement returns. You can change this using the `return` statement:

> `return(`*expression*`)`
> Terminates a block or loop and returns the value of *expression*.

This is convenient for defining recursive functions – functions that call themselves. For example, suppose we wanted to define a function that computes the Laguerre polynomials $L_n(x)$, which are encountered in the quantum mechanical description of the hydrogen atom and the three-dimensional harmonic oscillator. The Laguerre polynomials can be defined recursively by

$$
\begin{aligned}
L_0(x) &= 1, \\
L_1(x) &= 1 - x, \\
L_n(x) &= \frac{(2n - 1 - x)L_{n-1}(x) + (1 - n)L_{n-2}(x)}{n}
\end{aligned}
$$

The Laguerre polynomials   (4.1)

Recursion relations like this can be used directly in defining the function in Maxima:

```
(%i1)   L[n](x)  := block(
            if n=0 then return(1),
            if n=1 then return(1-x),
            ((2*n-x-1)*L[n-1](x)+(1-n)*L[n-2](x))/n,
            ratsimp(%%)
        )$
```

Notice that we've defined a *subscripted function*; the subscript n on the function is also a function argument. The special cases $n = 0$ or $n = 1$ stop the recursion. Without them, the function would call itself until memory was exhausted!

Let's check the function with $n = 6$ against Maxima's built-in Laguerre polynomial function, `laguerre`:

```
(%i2)   L[6](x), expand;
(%i3)   laguerre(6,x);
```

(%o2) $\quad \dfrac{x^6}{720} - \dfrac{x^5}{20} + \dfrac{5\,x^4}{8} - \dfrac{10\,x^3}{3} + \dfrac{15\,x^2}{2} - 6\,x + 1$

(%o3) $\quad \dfrac{x^6}{720} - \dfrac{x^5}{20} + \dfrac{5\,x^4}{8} - \dfrac{10\,x^3}{3} + \dfrac{15\,x^2}{2} - 6\,x + 1$

 Worksheet 4.2: Decision Making and Recursive Functions

In this worksheet, we'll use `if` statements and logical expressions to plot titration curves. We'll also write recursive functions to compute orthogonal polynomials used in quantum chemistry, and look at *kernel density estimation* for plotting noisy discrete data.

4.5 Contexts

Maxima calculations are performed over the entire field of complex numbers. While this makes results completely general, it sometimes prevents the computation of an integral or the solution of an equation when no completely general solution exists.

Restricting the domain of the problem before the calculation can help. For example, we can tell Maxima that x is a nonnegative integer, or that a is a positive constant. Maxima remembers these facts in further computations.

Facts are added using the `assume` function.

     assume(*fact1, fact2, ...*)
Teaches Maxima one or more facts (logical expressions) about a calculation.

(%i4)   **assume**(a>b, b>c);

(%o3) $[a > b, b > c]$

You can test deductions made from facts with the `is` function.

     is(*logical expression*)
Returns `true`, `false`, or `unknown` for a logical expression, considering known facts.

Maxima remembers these facts in further computations. A collection of known facts is called a context. Maxima can make limited deductions from a context. Use the `is` function to see if a logical expression involving known facts is true or false:

(%i5)   **is**(a-c > 0);

(%o4)   *true*

The `facts` function lists all of the facts you've entered.

(%i6)   **facts**();

(%o5) $[a > b, b > c]$

To list all facts that you've entered about a particular variable, provide the variable name:

(%i7)   **facts**(a);

(%o6)   [*a* > *b*]

The fact can be any logical expression, but symbolic equality (like a=d) won't be accepted. You must use algebraic equality (equal(a,d)).

(%i8)   **assume**(**equal**(a,d));
(%i9)   **is**(d > c);

(%o7)   [*equal*(*a*, *d*)]
(%o8)   *true*

The forget function deletes facts; for example, typing forget(a>b) removes that fact from the list.

Maxima recognizes mathematical properties of function and variables called features. We can assign features to a variable or function name using the declare function.

declare(*noun*, *feature*)
Associates a *feature* with a noun. For example, declare(R,constant) declares R to be a constant. A list of properties is given in Table 4.1. *noun* can be a list of nouns and *feature* can be a list of features.

The declare function expects its first argument to be a noun. If you want to declare a verb, use the nounify function on it. For example, to declare the product function as multiplicative, type declare(nounify(product), multiplicative).

You can see what features are available for declare by looking at the *features* system variable:

(%i1)   **features**;

(%o1)   [*integer, noninteger, even, odd, rational, irrational, real, imaginary, complex, analytic, increasing, decreasing, oddfun, evenfun, posfun, constant, commutative, lassociative, rassociative, symmetric, antisymmetric, integervalued*]

To see all properties associated with a variable or function, use the properties function:

(%i1)   **properties**(**sum**);

(%o1)   [*"specialevaluationform", "conjugatefunction", outative, noun, rule*]

To test a variable or function for a specific feature, use the featurep predicate:

featurep(*name*, *feature*)
Returns true if the variable or function name is declared as *feature*.

Table 4.1 Features that can be associated with a variable or function name using `declare(f, feature)`.

Feature	Effect
additive	Treat f as an additive function, for example, `f(x + y + z)` will be replaced by `f(x) + f(y) + f(z)`.
antisymmetric	Treat f as an antisymmetric function, for example, `f(y,x)` is equal to `−f(x,y)`.
complex	Treat f as a complex variable or function.
constant	Treat f as a symbolic constant; it will be factored out of expressions in the calculation of derivatives and integrals.
decreasing	Treat f as a decreasing function (with a negative first derivative).
even	Treat f as an even integer.
imaginary	Treat f as a multiple of i, the square root of minus one.
increasing	Treat f as an increasing function (has a positive first derivative).
integer	Treat f as an integer.
integervalued	Treat f as an integer-valued function.
irrational	Treat f as an irrational variable or function.
linear	Treat f as a linear function, for example, `f(a*(x+y)` will be replaced with `a*(f(x) + f(y))`.
mainvar	Treat f as a main variable. During simplification, expressions are arranged so that the main variable comes last.
multiplicative	Treat f as a multiplicative function, for example, `f(x*y*z)` will be replaced with `f(x)*f(y)*f(z)`.
noninteger	Treat f as a noninteger.
nonscalar	Treat f as a nonscalar: a vector, list, or matrix.
noun	Treat f as a noun (a quantity that will not be evaluated).
odd	Treat f as an odd integer.
outative	Treat f as a function which scales as its arguments scale, for example, `f(a*x)` simplifies to `a*f(x)`, where a is a constant.
posfun	Treat f as a positive function.
rational	Treat f as a rational variable or function.
real	Treat f as a real variable or function.
scalar	Treat f as a scalar: not a vector, list, or matrix.
symmetric	Treat f as a symmetric or commutative function, for example, `f(y,x)` is equal to `f(x,y)`.

For example,

(%i1) **featurep(sum, linear);**

(%o1) *false*

The sum function doesn't have the linear feature, as we can also see by trying this:

(%i2) **sum(a*x[i]+b*y[i], i, 1, N);**

(%o2) $\displaystyle\sum_{i=1}^{N} b\,y_i + a\,x_i$

The sum doesn't simplify by default. Let's declare `sum` to be linear, and see how this affects simplification:

```
(%i3)    declare(sum,linear);
(%i4)    sum(a*x[i]+b*y[i], i, 1, N);
```

$$(\%\text{o}3)\quad \textit{done}$$

$$(\%\text{o}4)\quad b\left(\sum_{i=1}^{N} y_i\right) + a\sum_{i=1}^{N} x_i$$

The `remove` function undeclares a property:

```
(%i5)    remove(sum,linear);
(%i6)    sum(a*x[i]+b*y[i], i, 1, N);
```

$$(\%\text{o}5)\quad \textit{done}$$

$$(\%\text{o}6)\quad \sum_{i=1}^{N} b\,y_i + a\,x_i$$

The `featurep` function can be used with `sublist` (Section 2.3.6) to choose items from a list that have certain features. For example, suppose we want to pick out only real solutions from a list of solutions for an equation. We can define a function that tests to see whether an item x has the `real` feature, and then use it as a predicate for `sublist`. For example, to list only the real solutions of $x^3 + 2x^2 + 4x - 9 = 0$,

```
(%i1)    realp(x) := featurep(rhs(x),real)$
(%i2)    soln: solve(x^3+2*x^2+4*x-9 = 0,x), numer;
(%i3)    sublist(soln, realp);
```

$(\%\text{o}2)\quad [x = -0.39803656228642\,(0.86602540378443\,i - 0.5)$
$+\, 2.233184016520717\,(-0.86602540378443\,i - 0.5) - 0.66666666666666,$
$x = 2.233184016520717\,(0.86602540378443\,i - 0.5)$
$-\, 0.39803656228642\,(-0.86602540378443\,i - 0.5) - 0.66666666666666,$
$x = 1.168480787567629]$
$(\%\text{o}3)\quad [x = 1.168480787567629]$

Notice that the `pick` function must look at the right-hand side of x, since each item in the list is actually an equation.

 Worksheet 4.3: Using Contexts

In this worksheet, we'll set up and use contexts with the `assume` and `declare` functions. We will also see two ways to pick real solutions for an equation from a list of possible solutions.

4.6 Iteration

We often need to repeat a calculation until some desired result is obtained. For example, we may want to perform the calculation for each element of a list, or each term in a sum, or we may want to repetitively refine an approximate solution to a problem. Repetitive calculations of this kind are called iterations. The programming construct that performs iterations is called a loop.

Maxima has two mechanisms for iteration:

1) If we need all the iteration results, we can store the input data in a list, and then use Maxima's map function to apply the calculation to each item. For example, if a function $f(x)$ is to be evaluated on inputs a, b, and c,

```
(%i1)   map(f, [a,b,c]);
(%o1)   [f (a), f (b), f (c)]
```

2) If we're only interested in the last iteration, loop constructs are more convenient. Loops repeat a calculation a fixed number of times or until a specific stopping criterion is satisfied. In Maxima, loops are implemented with the for statement. "For" loops can take several different forms, as we'll see in the following sections.

4.6.1 Indexed Loops

Loop iterations must have a condition for stopping – otherwise, they may run forever! One way to determine when a loop should terminate is to use an index variable i to count the number of times a calculation is performed. The iteration stops when i reaches a certain limit.

for (*i* : *i1* thru *limit*) do *expression*;
Calculate *expression* repetitively for each value of *i*, where *i* = *i1*, *i1*+1, ..., *limit*. The *expression* can be a block or ev environment.

For example, let's demonstrate that when $|x| < 1$, we can expand $(1 - x)^{-1} = 1 + x + x^2 + x^3 + \cdots$ using a for statement.[2]

```
(%i1)   total : 0$
(%i2)   for i:0 thru 15 do
            total : total + x^i;
(%i3)   total;
(%i4)   total, x=0.1;
(%i5)   1/(1-x), x=0.1;

(%o2)   done
(%o3)   x^15 + x^14 + x^13 + x^12 + x^11 + x^10 + x^9 + x^8 + x^7 + x^6 + x^5 + x^4 + x^3 + x^2 + x + 1
(%o4)   1.111111111111111
(%o5)   1.111111111111111
```

$$x^{15} + x^{14} + x^{13} + x^{12} + x^{11} + x^{10} + x^9 + x^8 + x^7 + x^6 + x^5 + x^4 + x^3 + x^2 + x + 1$$

The for statement is a "loop" that iteratively adds a term x^i to total for each value of i. Evaluating the sum with $x = 0.1$ gives essentially the same result as computing $1/(1 - x)$.

---

2 We'll see in Section 5.1 that Maxima provides a sum function that makes this calculation even simpler. The code total: 0; for i:0 thru 15 do total : total + x^i; total; is equivalent to total : sum(x^i, i, 0, 15).

Notice that the loop itself returns only the value "done"; it does not display any other result. We must save results within the loop, and use or display them after the loop is done.

By default, every iteration of the loop increments the index variable i by one. Sometimes we want to use a different increment. For example, suppose we want to compute the sum $2\left(x + \frac{x^3}{3} + \frac{x^5}{5} + \cdots\right)$, which approximates $\ln\frac{1+x}{1-x}$ when $|x| < 1$. The most convenient values of i would be 1, 3, 5, ..., so we would like i to start at 1 and be incremented by 2 every time the loop executes. We can use the following variant of the `for` statement:

for (*i* : *i1* step *increment* thru *limit*) do *expression*;
Calculate *expression* repetitively for each value of *i*, where *i* = *i1*, *i1+increment*, ..., *limit*. Negative increments are allowed. The *expression* can be a `block` or `ev` environment.

```
(%i1)   total : 0$
(%i2)   for i : 1 thru 15 do
            total : total + x^i/i;
(%i3)   2*total;
(%i4)   %, x=0.1;
(%i5)   log((1+x)/(1-x)), x=0.1;
```

(%o2)   *done*

(%o3) $2\left(\dfrac{x^{15}}{15} + \dfrac{x^{13}}{13} + \dfrac{x^{11}}{11} + \dfrac{x^9}{9} + \dfrac{x^7}{7} + \dfrac{x^5}{5} + \dfrac{x^3}{3} + x\right)$

(%o4)   0.20067069546215

(%o5)   0.20067069546215

To see the value of a variable for each iteration, use the `display` or `disp` functions.

display(*x*, *y*, ...)
Shows equations that give the values of expressions *x*, *y*, ...

disp(*x*, *y*, ...)
Shows the values of expressions *x*, *y*, ...

For example, this loop prints the values of the first 10 numbers in the Fibonacci sequence (each number in the sequence is the sum of the previous two numbers)[3]

```
(%i1)   for i:0 thru 10 do disp(fib(i));
```

0
1
1
2
3
5
8
13

___
3 The `fib(i)` function returns the ith Fibonacci number.

21
34
55
(%o1)   *done*

The output from `display` or `disp` isn't saved as the output from the `for` command, so we can't do any further calculations with these numbers. To save values from each iteration, build a list. For example,

```
(%i1)   results : []$
(%i2)   for i:0 thru 10 do
            results : endcons(fib(i),results)$
(%i3)   results;
```

(%o3) $[0, 1, 1, 2, 3, 5, 8, 13, 21, 34, 55]$

We start with an empty list `results`, and use the `endcons` function[4] to append the calculated Fibonacci number `fib(i)` onto the end of the list for each turn of the loop.[5]

4.6.2 Conditional Loops

Sometimes we'd like the loop to continue *while* a condition exists, or perhaps *unless* a condition exists. These variants of the `for` statement can be used:

```
while condition do expression;
```
Calculate *expression* repetitively while *condition* is true. The *expression* can be a `block` or `ev` environment.

For example, let's use a `while` loop to estimate the unit round-off error in default arithmetic in Maxima. Recall from Section 2.1 that the unit round-off error is the smallest number ϵ such that computing $\epsilon + 1$ gives a result of 1. In this case, the condition for continuing the loop is $1 + \epsilon/2 > 1$:

```
(%i1)   epsilon: 1.0$
(%i2)   while (1.0+epsilon/2 > 1.0) do epsilon : epsilon/2;
(%i3)   epsilon;
```

(%o3) $2.220446049250313 \; 10^{-16}$

Since Maxima uses 64-bit floating point numbers, of which 52 bits are used to store the significant digits of the number [20], we expect the unit round-off error to be 2^{-52}. It is.

The `unless` loop is similar to the `while` loop:

```
unless condition do expression;
```
Calculate *expression* repetitively unless *condition* is true. The *expression* can be a `block` or `ev` environment.

---

4 Lists and list-building functions were covered in Section 2.3.
5 We could have built this list more simply using `makelist`. The code above is equivalent to `results : makelist(fib(i), i, 0, 10)`.

Let's repeat the calculation of the unit round-off error with an `unless` loop:

```
(%i1)   epsilon : 1.0$
(%i2)   unless (1.0+epsilon/2 = 1.0) do epsilon : epsilon/2;
(%i3)   epsilon;
```

$$(\%o3) \quad 2.220446049250313 \; 10^{-16}$$

We keep halving `epsilon` *unless* that would give us a value that vanishes when added to one. In the `while` loop, we kept halving `epsilon` *while* that gave us a value that didn't vanish when added to one.

Conditional loops can solve difficult equations numerically by computing a series of successive approximations $y_1, y_2, \ldots, y_n$, where we stop the loop when the relative difference between successive approximations falls below a certain threshold `tol`:

$$\left| \frac{y_n - y_{n-1}}{y_n} \right| \leq \text{tol} \qquad \text{Convergence criterion} \qquad (4.2)$$

We'll use this approach to solve equilibrium problems (and equations that can't be solved analytically with `solve`) in Worksheet 4.4.

4.6.3 Looping Over Lists

Another form of the `for` loop can process elements in a list:

for *variable* in *list* do *expression*;
Assign each element in *list* successively to *variable*, and calculate *expression*. The *expression* can be a `block` or `ev` environment.

For example, suppose we wanted to find which item in a list has the maximum absolute deviation from the mean:

```
(%i1)   x_data: [1.3, 2.5, 8.7, 0.3, 5.9, 3.6]$
(%i2)   x_mean : mean(x_data)$
(%i3)   max_deviation: -1000$
(%i4)   for item in x_data do
            if (max_deviation < abs(item - x_mean)) then
                [outlier, max_deviation] : [item, abs(item -
                  ➥ x_mean)]$
(%i5)   printf(true, "The mean is ~5f. ", x_mean)$
(%i6)   printf(true, "The point ~f has the maximum absolute
            ➥ deviation ~3f from the mean.", outlier,
            ➥ max_deviation)$
```

The mean is 3.717. The point 8.7 has the maximum absolute deviation 5.0 from the mean.

Here, we've used the `mean` function to compute the mean of the data. The loop successively assigns each element of `x_data` to the variable `item` and updates the values of `outlier` and `max_deviation` simultaneously using a list assignment if the maximum deviation is less than the absolute deviation for `item`. The loop terminates when all of the elements of `x_data` have been processed. The formatted printing function `printf` (Section 2.5) summarizes the results.

You can also use this form of the `for` loop to apply a series of functions to the same data:

```
(%i1)    for f in [sin, cos, tan, sec] do
(%i2)         display(f, f(%pi/6));
```

$$f = sin$$
$$f\left(\frac{\pi}{6}\right) = \frac{1}{2}$$
$$f = cos$$
$$f\left(\frac{\pi}{6}\right) = \frac{\sqrt{3}}{2}$$
$$f = tan$$
$$f\left(\frac{\pi}{6}\right) = \frac{1}{\sqrt{3}}$$
$$f = sec$$
$$f\left(\frac{\pi}{6}\right) = \frac{2}{\sqrt{3}}$$

4.6.4 Nested Loops

Nested loops can be used to build nested lists with more than one index. For example, suppose we want to build a list of x, y, z points with $z = F(x, y)$, with $xmin \leq x \leq xmax$ and $ymin \leq y \leq ymax$. We want the separation between adjacent x and y values to be `deltax` and `deltay`, respectively.

```
(%i1)    xmin : 0$
(%i2)    xmax : 9$
(%i3)    deltax : 3$
(%i4)    ymin : -5$
(%i5)    ymax : 5$
(%i6)    deltay : 2$
(%i7)    F(x,y) := x^2 + y^2 + x*y$

(%i9)    points: []$
(%i10)   for x : xmin step deltax thru xmax do
                 for y : ymin step deltay thru ymax do
                     points: cons([x, y, F(x,y)], points)$
(%i11)   points;
```

$$(\%o10) \quad \begin{array}{l} [[9,5,151],\ [9,3,117],\ [9,1,91],\ [9,-1,73],\ [9,-3,63],\ [9,-5,61],\ [6,5,91], \\ [6,3,63],\ [6,1,43],\ [6,-1,31],\ [6,-3,27],\ [6,-5,31],\ [3,5,49],\ [3,3,27], \\ [3,1,13],\ [3,-1,7],\ [3,-3,9],\ [3,-5,19],\ [0,5,25],\ [0,3,9],\ [0,1,1],\ [0,-1,1], \\ [0,-3,9],\ [0,-5,25]] \end{array}$$

  Worksheet 4.4: Iterative Calculations

In this worksheet, we'll use program loops to perform iterative calculations.

5

Algebra

Grandad taught me that the alien signs and symbols of algebraic equations were not just marks on paper. They were not flat. They were three-dimensional, and you could approach them from different directions, look at them from different ways, stand them on their heads. You could take them apart and put them back together in a variety of shapes, like Legos. I stopped being scared of them.

– Mal Peet [21]

Algebra allows many chemical problems to be expressed in symbolic form. It encodes essential relationships between variables, but we often need to rewrite, reorganize, and simplify the code before we can solve the problem.

Maxima won't simplify equations or expressions on its own; you have to request it. In this chapter, we'll see how to use Maxima to rewrite, factor, expand, and extract pieces of expressions and equations. We'll also use it solve equations and systems of equations, and connect discrete data points with lines and polynomials.

In Sections 5.1 and 5.2, we'll use Maxima to represent and manipulate sums, series, and products of terms. Sections 5.3 and 5.4 focus on simplifying, rewriting, and solving equations and systems of equations. Finally, in Section 5.5 we'll estimate the values of a variable between its previously measured values using interpolating lines or polynomials.

5.1 Series

A series is a sum of terms that have a common form. For example, all of the terms in a power series have the form:

$$a_1 + a_2x + a_3x^2 + \cdots = \sum_{n=1}^{\infty} a_i x^{i-1} \qquad \text{A power series} \qquad (5.1)$$

Power series also provide a way to represent numerical solutions for many important equations in chemistry, as we'll see in Chapter 11. To understand the properties of these solutions, we'll need to be able to rewrite, evaluate, and simplify power series.

Power series make it possible to approximate functions. For example, the exponential function e^x can be written as

$$e^x = 1 + \frac{x}{1!} + \frac{x^2}{2!} + \frac{x^3}{3!} + \cdots = \sum_{i=0}^{\infty} \frac{x^i}{i!} \qquad \text{Power series expansion for } e^x \qquad (5.2)$$

This series converges to a single value (e^x) as the number of terms approaches infinity.[1]

---

1 Other series may not converge to a single value; they are called *divergent series*. We'll see how to derive power series for a given expression and test power series for convergence in Section 6.6.

Symbolic Mathematics for Chemists: A Guide for Maxima Users, First Edition. Fred Senese.
© 2019 John Wiley & Sons Ltd. Published 2019 by John Wiley & Sons Ltd.
Companion website: http://booksupport.wiley.com

The `sum` function provides a convenient way to write both infinite and finite series in Maxima:

sum(*expression*, i, i_1, i_N)
Add values of *expression* as the index *i* varies from *i_1* to *i_N*.

You must have $i\_N > i\_1$, or the sum returns a zero value.

Selecting [Calculus] ⟩ [Calculate Sum...] brings up a dialog for the `sum` function. For the sum in Equation (5.2),

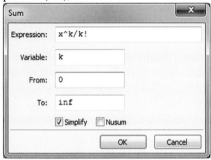

which inserts the code

```
(%i1)    sum(x^k/k!, k, 0, inf);
```

$$(\%o1) \quad \sum_{k=0}^{\infty} \frac{x^k}{k!}$$

The bounds of the sum can be expressions. For infinite sums you can use `minf` or `inf` as bounds.

5.1.1 Simplifying Sums

Checking the `Simplify` box in the dialog applies additional rules that can simplify the sum. For example, $\sum_{k=0}^{\infty} x^k$ is recognized as $1/(1-x)$, when $|x| < 1$:

```
(%i2)    sum(x^k, k, 0, inf), simpsum;
```

Is $|x| - 1$ positive, negative or zero? `negative`;

$$(\%o2) \quad \frac{1}{1-x}$$

The switch *simpsum* can simplify many power series, when simpler forms exist. This is helpful in statistical mechanics, where we often need to write a power series as a simple expression. For example, consider a molecule with an infinite number of equally spaced, nondegenerate energy levels spaced by ϵ. The molecule's partition function q can be written as

```
(%i1)    q : sum(exp(-beta*i*epsilon), i, 0, inf);
```

$$(\%o1) \quad \sum_{i=0}^{\infty} e^{-\beta \epsilon i}$$

where β is $1/kT$ and $i\epsilon$ is the energy of the ith level. We'd like to write q as a simple expression rather than an infinite sum:[2]

```
(%i1)   q, simpsum
```

Is $e^{\beta\epsilon} - 1$ positive, negative or zero? pos;

$$(\%o2) \quad \frac{1}{1 - e^{-\beta\epsilon}}$$

The *simpsum* switch knows some identities for simplifying sums, but when it fails, try the much more powerful `simplify_sum` function:

> `simplify_sum(expression)`
> Tries to simplify all sums that appear in *expression* to closed form expressions. You must `load(simplify_sum)` before using this function.

For example, the following sums are simplified by `simplify_sum` but not by *simpsum*:

```
(%i1)   load(simplify_sum)$
(%i2)   sum(k*x^(k-1), k, 1, N), simpsum;
(%i3)   simplify_sum(%);
```

$$(\%o2) \quad \sum_{k=1}^{N} k\, x^{k-1}$$

$$(\%o3) \quad \frac{x^{N}\,(xN - N - 1)}{(x-1)^2} + \frac{1}{(x-1)^2}$$

```
(%i4)   sum(binomial(n+k,k)/2^k,k,1,n), simpsum;
(%i5)   simplify_sum(%);
```

$$(\%o4) \quad \sum_{k=1}^{n} \frac{\binom{n+k}{k}}{2^k}$$

$$(\%o5) \quad \frac{x^{N}\,(xN - N - 1)}{(x-1)^2} + \frac{1}{(x-1)^2}$$

Sums of log terms can be simplified using the `logcontract` function. For example,

```
(%i1)   ((b-a)*log(c-d)+(a-c)*log(b-d)+(c-b)*log(a-d)+(a-b)*log
    ➥ (c)+(log(b)-log(a))*c-a*log(b)+(log(a))*b);
(%i2)   logcontract(%);
```

$$(\%o1) \quad (b-a)\log(c-d) + (a-c)\log(b-d) + (c-b)\log(a-d) + (a-b)\log(c) + (\log(b) - \log(a))\,c - a\log(b) + \log(a)\,b$$

$$(\%o2) \quad a\log\left(\frac{c\,(d-b)}{b\,(d-c)}\right) + c\log\left(\frac{b\,(d-a)}{a\,(d-b)}\right) + b\log\left(\frac{a\,(d-c)}{c\,(d-a)}\right)$$

---

2 In Section 6.6, we'll use the `powerseries` function to perform the reverse operation (writing a power series from a simple expression).

5.1.2 Reindexing and Combining Sums

The changevar function can simplify and combine sums by shifting the summation index:

changevar(*expression, substitution, new_index, old_index*)
Change sum or product indices in *expression* from *old_index* to *new_index* as specified by *substitution*.

For example, we can replace the index i in the following sum with j, where $j = i - 1$:

```
(%i1)  sum(a[i]*x^(i-1), i, 0, inf) + beta*sum(b[i]*x^i, i, 0,
       ➡ inf);
(%i2)  changevar(%, j=i-1, j, i);
```

$$(\%o1)\quad \beta\left(\sum_{i=0}^{\infty} b_i\,x^i\right) + \sum_{i=0}^{\infty} a_i\,x^{i-1}$$

$$(\%o2)\quad \beta\left(\sum_{j=-1}^{\infty} b_{j+1}\,x^{j+1}\right) + \sum_{j=-1}^{\infty} a_{j+1}\,x^j$$

We'll encounter this particular sum again in Section 11.2, when writing power series solutions for ordinary differential equations. We'll need to collect the coefficients of like powers and combine the sums. This can be done by shifting the index of the sum involving the a_i coefficients by one (without changing the other sum) and then rewrite the result as a single sum.

First, we need to target only one of the sums with changevar. We can do this by substitution, but there is a simpler way. The bashindices function assigns a unique index to every sum and product in an expression:

```
(%i1)  sum(a[i]*x^(i-1), i, 0, inf) + beta*sum(b[i]*x^i, i, 0,
       ➡ inf);
(%i2)  bashindices(%);
```

$$(\%o1)\quad \beta\left(\sum_{i=0}^{\infty} b_i\,x^i\right) + \sum_{i=0}^{\infty} a_i\,x^{i-1}$$

$$(\%o2)\quad \beta\left(\sum_{j2=0}^{\infty} b_{j2}\,x^{j2}\right) + \sum_{j1=0}^{\infty} a_{j1}\,x^{j1-1}$$

Now we can shift the index of the sum over j1 by 1:

```
(%i3)  changevar(%, j2=j1-1, j2, j1);
```

$$(\%o3)\quad \left(\sum_{j2=-1}^{\infty} a_{j2+1}\,x^{j2}\right) + \beta\sum_{j2=0}^{\infty} b_{j2}\,x^{j2}$$

The sums must run over the same index before they can be combined. The `niceindices` function renames each index to *i*. If *i* already appears in the expression being summed, it will try to choose successive names from the list *j, k, l, m, n*.

(%i4)   **niceindices(%);**

$$(\%o4) \quad \left(\sum_{i=-1}^{\infty} a_{i+1} x^i \right) + \beta \sum_{i=0}^{\infty} b_i x^i$$

Move the constant β into the sum. The `intosum` function moves all factors into a sum:

(%i5)   **intosum(%);**

$$(\%o5) \quad \left(\sum_{i=-1}^{\infty} a_{i+1} x^i \right) + \sum_{i=0}^{\infty} \beta \, b_i x^i$$

Finally, combine the two sums using `sumcontract`:

(%i6)   **sumcontract(%);**

$$(\%o6) \quad \left(\sum_{i=0}^{\infty} a_{i+1} x^i + \beta \, b_i x^i \right) + \frac{a_0}{x}$$

5.1.3   Applying Functions to Sums and Products

Sums and products are stored internally as lists, so the map function introduced in Section 2.3.3 is also useful for applying functions to each term in a sum or product. For example,

(%i1)   **map(sin, a+b+c);**
(%i2)   **map(sqrt, a*b*c);**

(%o1) $\sin(c) + \sin(b) + \sin(a)$
(%o2) $\sqrt{a} \, \sqrt{b} \, \sqrt{c}$

Because sums, products, and function argument lists are stored as lists, all of the functions for editing lists introduced in Section 2.3.4 work on them as well:

(%i1)   **L : sum(a[i], i, 1, 10);**
(%i2)   **rest(L, 5);**
(%i3)   **last(L);**
(%i4)   **first(L);**
(%i5)   **delete(a[5], L);**

(%o1) $a_{10} + a_9 + a_8 + a_7 + a_6 + a_5 + a_4 + a_3 + a_2 + a_1$
(%o2) $a_5 + a_4 + a_3 + a_2 + a_1$
(%o3) a_1
(%o4) a_{10}
(%o5) $a_{10} + a_9 + a_8 + a_7 + a_6 + a_4 + a_3 + a_2 + a_1$

Worksheet 5.0: Using Series

In this worksheet, we'll look at several common series encountered in statistical mechanics, gaseous equations of state, and quantum theory, and we'll use Maxima can be used to represent, simplify, and compute the series.

5.2 Products

In statistical mechanics, we connect the microscopic properties of individual molecules with the properties of a sample containing a large number of molecules. The essential link between the microscopic and macroscopic properties is establised using *partition functions*, which tell us how energy is partitioned among molecules or among independent modes of motion. For example, if we want to know how energy is partitioned among translational, rotational, and vibrational modes, we can write the total partition function f as a product of the partition function f_i for each independent mode:

$$f = f_1 f_2 f_3 \cdots f_{3N} = \prod_{i=1}^{3N} f_i \qquad \text{Partition function for independent modes } i \text{ in a system of molecules with } N \text{ atoms} \qquad (5.3)$$

In Maxima we can write products like this with the `product` function:

   `product(expression, i, i_1, i_N)`
Multiply values of `expression` as the index i varies from i_1 to i_N.

Selecting [Calculus] ⟩ [Calculate Product...] brings up a dialog for the `product` function. For example, to write the product in Equation (5.3),

which inserts the code

(%i1)   `f = `**`product`**`(f[i],i,1,3*N);`

$$(\%o1) \quad f = \prod_{i=1}^{3N} f_i$$

For the vibrational part of the partition function for a nonlinear molecule, each of $3N - 6$ vibrational modes with frequency v_i has $f_i = (1 - \exp(-hv_i/kT)^{-1}$. For a molecule with three atoms, we could write

```
(%i1)   f[vib] = product(f[i],i,1,3*N-6),
(%i2)          f[i] = (1-exp(-h*nu[i]/(k*T)))^-1,
(%i3)          N=3;
```

$$(\%o1) \quad f_{vib} = \frac{1}{\left(1 - e^{-\frac{v_1 h}{kT}}\right)\left(1 - e^{-\frac{v_2 h}{kT}}\right)\left(1 - e^{-\frac{v_3 h}{kT}}\right)}$$

The *simpproduct* switch can do some rudimentary simplifications on products. For example,

```
(%i1)   product(k,k,1,N);
(%i2)   %, simpproduct;
```

$$(\%o1) \quad \prod_{k=1}^{N} k$$

$(\%o2) \quad N!$

Products of roots can be simplified to roots of products using the `rootscontract` function. For example,

```
(%i1)   sqrt(x)*y^(5/2)*z^(1/2);
(%i2)   rootscontract(%);
```

$(\%o1) \quad \sqrt{x}\, y^{\frac{5}{2}}\, \sqrt{z}$

$(\%o2) \quad \sqrt{x\, y^5\, z}$

Products and quotients of factorials can be combined with the `factcomb` function:

```
(%i1)   k!/(k*(k-1));
(%i2)   factcomb(%);
```

$$(\%o1) \quad \frac{k!}{(k-1)\, k}$$

$(\%o2) \quad (k-2)!$

Logarithms of products can be simplified to sums using the *logexpand*=all switch:

```
(%i1)   log(product(a[i],i,1,n));
(%i2)   %, logexpand=all;
```

$$(\%o1) \quad \log\left(\prod_{i=1}^{n} a_i\right)$$

$$(\%o2) \quad \sum_{i=1}^{n} \log(a_i)$$

5.3 Equations

Equations are expressions that use the equality operator =. They can be stored in variables like any other expression:

```
(%i1)   vdW_equation:  (P+n^2*a/V^2)*(V-n*b)  =  n*R*T;
```

$$(\%o1) \quad \left(\frac{a\,n^2}{V^2} + P \right)(V - b\,n) = n\,R\,T$$

 Remember the difference between an *assignment* and an *equation*. An assignment stores a value in a variable. Typing a:1 will assign a value of 1 to the variable *a*. An equation $a = 1$ is simply a statement that *a* is equal to 1. *An equation will not change the actual value of a.*

5.3.1 Simplifying Equations

To simplify an equation or expression, use the ratsimp function.

 ratsimp(*expression, main variable*)
Simplifies *expression* and all of its sub-expressions by writing it as a ratio of two polynomials in the *main variable*. If you omit the (optional) main variable, ratsimp chooses it for you.

For example,

```
(%i2)   ratsimp(vdW_equation);
```

$$(\%o4) \quad \frac{P\,V^3 - b\,n\,P\,V^2 + a\,n^2\,V - a\,b\,n^3}{V^2} = n\,R\,T$$

Each side of an equation is simplified separately. Terms are not moved from one side of the equation to the other. In this case, *V* was chosen as the main variable. Choosing *P* as the main variable gives a slightly different result:

```
(%i3)   ratsimp(vdW_equation, P);
```

$$(\%o5) \quad \frac{P\left(V^3 - b\,n\,V^2\right) + a\,n^2\,V - a\,b\,n^3}{V^2} = n\,R\,T$$

Notice that after simplification, the coefficients of the main variable are polynomials in the other variables.

When you want to simplify individual terms in a sum or product, or expressions embedded in the argument of a function, apply ratsimp using map (see Section 2.3.3):

```
(%i1)   x:   a/(a^3+a)+(b^3+b)/b;
(%i2)   ratsimp(x);
(%i3)   map(ratsimp, x);
```

$$(\%o1) \quad \frac{b^3 + b}{b} + \frac{a}{a^3 + a}$$

$$(\%o2) \quad \frac{\left(a^2 + 1\right) b^2 + a^2 + 2}{a^2 + 1}$$

$$(\%o3) \quad b^2 + \frac{1}{a^2 + 1} + 1$$

5.3.2 Simplifying Trigonometric and Exponential Functions

In Section 1.2.2, we saw that the `radcan` and `trigsimp` functions could simplify expressions that involve exponentials and trigonometric functions. They apply to equations as well. Another useful simplification function is `trigreduce`, which combines products and powers of trigonometric functions:

```
(%i1)   (sqrt(sin(theta)^2 + cos(theta)^2) + 2*sin(theta)*cos(
        ➥ theta) - 2*sin(theta)^2)^2-1;
(%i2)   trigsimp(%);
(%i3)   trigreduce(%);
(%i4)   ratsimp(%);
```

$$(\%o1) \quad \left(\sqrt{\sin(\theta)^2 + \cos(\theta)^2} - 2 \cdot \sin(\theta)^2 + 2 \cdot \cos(\theta) \cdot \sin(\theta)\right)^2 - 1$$

$$(\%o2) \quad \left(8 \cdot \cos(\theta)^3 - 4 \cdot \cos(\theta)\right) \cdot \sin(\theta)$$

$$(\%o3) \quad 2 \cdot \left(\frac{\sin(4 \cdot \theta)}{2} - \frac{\sin(2 \cdot \theta)}{2}\right) + \sin(2 \cdot \theta)$$

$$(\%o4) \quad \sin(4 \cdot \theta)$$

Complex exponentials, sines, and cosines are related through Euler's formula (Equation (2.4)). Maxima's `demoivre` function applies this equation to rewrite complex exponential expressions in terms of trigonometric functions[3]:

```
(%i1)   demoivre(exp(%i*theta));
(%i2)   demoivre(A*exp(%i*(k*x-omega*t)));
```

$$(\%o1) \quad \%i \sin(\theta) + \cos(\theta)$$

$$(\%o2) \quad (\%i \sin(k\,x - \omega\,t) + \cos(k\,x - \omega\,t))\,A$$

The `exponentialize` function reverses this operation. It converts circular and hyperbolic functions into complex exponentials:

```
(%i1)   exponentialize(%i*sin(theta)+cos(theta)), ratsimp;
(%i2)   exponentialize(A*sin(k*x-omega*t));
```

$$(\%o1) \quad e^{\%i\,\theta}$$

$$(\%o2) \quad \frac{\%i\left(e^{\%i\,(k\,x - \omega\,t)} - e^{-\%i\,(k\,x - \omega\,t)}\right) A}{2}$$

---

3 The function name refers to de Moivre's theorem,

$$e^{in\theta} = \cos(n\theta) + i \sin(n\theta),$$

which can be deduced from Euler's formula [15].

Maxima applies Euler's formula when it extracts the real and imaginary parts of exp($i\theta$) using the functions `realpart` and `imagpart`:

```
(%i1)  realpart(exp(%i*theta));
(%i2)  imagpart(exp(%i*theta));
```

```
(%o1)  cos (θ)
(%o2)  sin (θ)
```

 Worksheet 5.1: Simplifying Equations

In this worksheet, we'll simplify equations and convert expressions between exponential and trigonometric forms. The project will apply some of what we learn to understand the interference pattern obtained in the classic double-slit experiment.

5.3.3 Extracting Expressions From an Equation

We often need to extract pieces of equations in derivations and proofs. To isolate the right or left side of an equation, use `rhs` and `lhs`. For example,

```
(%i1)  vdW_equation: (P+n^2*a/V^2)*(V-n*b) = n*R*T$
(%i2)  lhs(vdW_equation);
(%i3)  rhs(vdW_equation);
```

$$(\%o2) \quad \left(P + \frac{a\,n^2}{V^2} \right) (V - b\,n)$$

$$(\%o3) \quad nRT$$

Similarly, extract the numerator and denominator of a fraction using the `num` and `denom` functions. For example,

```
(%i4)  fraction: ratsimp(lhs(vdW_equation));
(%i5)  num(fraction);
(%i6)  denom(fraction);
```

$$(\%o4) \quad \frac{P \cdot V^3 - b \cdot n \cdot P \cdot V^2 + a \cdot n^2 \cdot V - a \cdot b \cdot n^3}{V^2}$$

$$(\%o5) \quad P \cdot V^3 - b \cdot n \cdot P \cdot V^2 + a \cdot n^2 \cdot V - a \cdot b \cdot n^3$$

$$(\%o6) \quad V^2$$

Let's use these functions to rewrite the van der Waals equation as a polynomial in V, with zero for the right-hand side. We'll multiply both sides of the equation by the denominator on the left-hand side, and then subtract the right-hand side from both sides of the equation. The simplified result will be stored as `V_polynomial`.

```
(%i7)  ratsimp(vdW_equation);
(%i8)  denom(lhs(%)) * %;
(%i9)  V_polynomial: % - rhs(%), ratsimp;
```

(%o7) $\dfrac{P\,V^3 - b\,n\,P\,V^2 + a\,n^2\,V - a\,b\,n^3}{V^2} = n\,R\,T$

(%o8) $P\,V^3 - b\,n\,P\,V^2 + a\,n^2\,V - a\,b\,n^3 = n\,R\,T\,V^2$

(%o9) $P\,V^3 + (-n\,R\,T - b\,n\,P)\,V^2 + a\,n^2\,V - a\,b\,n^3 = 0$

Notice how arithmetic on an equation works. Operations on an equation are applied to both sides.

We can extract coefficients in a polynomial with the `coeff` function:

`coeff(`*expression*`, x, n)`
Extracts the coefficient of x^n in *expression*.

For example, the following command extracts the coefficient of V^2 on the left-hand side of `V_polynomial`, and saves it as "C":

(%i10) **C: coeff(lhs**(V_polynomial)**,V,2);**

(%o10) $-n\,R\,T - b\,n\,P$

The `part` function allows even finer control over which pieces of an expression or equation you can extract:

`part(`*expression*`, i)`
`part(`*expression*`, i, j, ...)`
Extract part *i* of *expression*. The second form returns part *j* of part *i*. Additional indices take parts of the previously extracted part.

Let's apply `part` to the following equation, which we'll encounter in the series solution of a differential equation in Chapter 11.

(%i1) eqn: (**sum**(((k^2+(2*i+3)*k+i^2+3*i+2)*a[k+i+2]+(-2*k-2*i
       ➥ +2*alpha)*a[k+i])*x^(k+i),i,0,**inf**))+(k^2+k)*a[k
       ➥ +1]*x^(k-1)+(k^2-k)*a[k]*x^(k-2)=0;

(%o1) $\left(\displaystyle\sum_{i=0}^{\infty} \left(\left(k^2 + (2\,i+3)\,k + i^2 + 3\,i + 2\right) a_{k+i+2} + (-2\,k - 2\,i + 2\,\alpha)\,a_{k+i} \right) x^{k+i} \right) +$
$\left(k^2 + k\right) a_{k+1}\, x^{k-1} + \left(k^2 - k\right) a_k\, x^{k-2} = 0$

Since the expression in this equation must apply for any value of *x*, the coefficients of each power of *x* must be zero. We'd like to set up an equation that sets the coefficient of x^{k+i} to zero. This can be done with a series of calls to `part`:

(%i2) **part(eqn,1);**
(%i3) **part(%, 1);**
(%i4) **part(%, 1);**
(%i5) **part(%, 1);**
(%i6) **%=0;**

(%o2) $\left(\sum_{i=0}^{\infty} \left(\left(k^2 + (2i+3)k + i^2 + 3i + 2\right) a_{k+i+2} + (-2k - 2i + 2\alpha) a_{k+i}\right) x^{k+i}\right) +$

$\left(k^2 + k\right) a_{k+1} x^{k-1} + \left(k^2 - k\right) a_k x^{k-2}$

(%o3) $\sum_{i=0}^{\infty} \left(\left(k^2 + (2i+3)k + i^2 + 3i + 2\right) a_{k+i+2} + (-2k - 2i + 2\alpha) a_{k+i}\right) x^{k+i}$

(%o4) $\left(\left(k^2 + (2i+3)k + i^2 + 3i + 2\right) a_{k+i+2} + (-2k - 2i + 2\alpha) a_{k+i}\right) x^{k+i}$

(%o5) $\left(k^2 + (2i+3)k + i^2 + 3i + 2\right) a_{k+i+2} + (-2k - 2i + 2\alpha) a_{k+i}$

(%o6) $\left(k^2 + (2i+3)k + i^2 + 3i + 2\right) a_{k+i+2} + (-2k - 2i + 2\alpha) a_{k+i} = 0$

...or we could have extracted the coefficient and set it to zero in a single line:

(%i7) **part(eqn,1,1,1,1)** = 0;

(%o8) $(k^2 + (2i+3)k + i^2 + 3i + 2)a_{k+i+2} + (-2k - 2i + 2\alpha)a_{k+i} = 0$

The system variable *piece* holds the last expression extracted using part:

(%i8) **piece;**

! (%o7) $\left(k^2 + (2i+3)k + i^2 + 3i + 2\right) a_{k+i+2} + (-2k - 2i + 2\alpha) a_{k+i}$

The part function's ability to pick out parts of expressions is powerful, but it will take some practice to master. Type example(part) for more examples of applying part to expressions, integrals, and equations.

To pick out all terms that contain a particular variable, use the partition function:

partition(*expression*, x)
Returns a list; the first item is the parts of the *expression* that don't contain x; the second item is the parts that *do* contain x.

For example, the Gibbs free energy for a reaction of the form $R + \nu H^+(aq) \rightarrow P$ can be written as

$$\Delta G = \Delta G^\circ + RT \ln \frac{a_P}{a_R} - \nu RT \ln a_{H^+} \qquad (5.4)$$

where a_P, a_R, and a_{H^+} are the activities of P, R, and H^+, respectively. To partition the right-hand side of the equation into pH-independent and pH-dependent parts, we can write

(%i1) eqn: DeltaG = DeltaGo + R*T***log**(aP/aR) - nu*R*T***log**(aH);
(%i2) **partition(rhs**(eqn), aH);

(%o1) $DeltaG = \log\left(\frac{aP}{aR}\right) RT - \nu \log(aH) RT + DeltaGo$

(%o2) $[\log\left(\frac{aP}{aR}\right) RT + DeltaGo, -\nu \log(aH) RT]$

The first list item contains terms that don't depend on aH; the second item is the aH-dependent term.

We'll use the `partition` function to separate series expansions into classical and quantum mechanical terms in Section 6.7. In Section 11.6, `partition` will help separate variables in solving partial differential equations.

> **Worksheet 5.2: Manipulating equations**
>
> In this worksheet, we'll use Maxima to build and rearrange equations in proofs and derivations. We'll also extract terms and expressions from equations.

5.3.4 Expanding Expressions

Suppose we wanted Maxima to display the van der Waals equation with the terms in the parentheses multiplied out. We can use the `expand` function (which is inserted automatically if you select $\boxed{\text{Simplify}} \gg \boxed{\text{Expand Expression}}$ from the menu or the $\boxed{\text{Expand}}$ function from the *General Math* pane):

 expand (*expression*)
Multiply out products of sums in *expression*, split numerators into separate terms, and distribute all multiplications over additions.

```
(%i3)   expand(vdW_equation);
```

$$(\%o10) \quad P\,V + \frac{a\,n^2}{V} - \frac{a\,b\,n^3}{V^2} - b\,n\,P = n\,R\,T$$

Expanding an equation expands the right-hand and left-hand sides separately; it will not move terms from one side of the equation to the other.

The `expand` function can be applied as a switch in evaluation environments. In the following example, the evaluation environment in line 2 is the same as typing `expand(expr)`:

```
(%i1)   expr: (x+1)^5 + (y-1)^3+(z+2)^-2;
(%i2)   expr, expand;
```

$$(\%o1) \quad \frac{1}{(z+2)^2} + \left(y-1\right)^3 + (x+1)^5$$

$$(\%o2) \quad \frac{1}{z^2+4\,z+4} + y^3 - 3\,y^2 + 3\,y + x^5 + 5\,x^4 + 10\,x^3 + 10\,x^2 + 5\,x$$

The `expandwrt` function provides finer control over which parts of the expression to expand.

 expandwrt (*expression*, *x1*, ... *xn*)
Expand *expression* with respect to the variables or expressions *x1*, ..., *xn*.

For example, to expand only the *x* term,

```
(%i3)   expandwrt(expr, x)
```

$$(\%o3) \quad \frac{1}{(z+2)^2} + \left(y-1\right)^3 + x^5 + 5\,x^4 + 10\,x^3 + 10\,x^2 + 5\,x + 1$$

Unlike `expand`, the `expandwrt` function won't expand denominators by default. You can set the switch *expandwrt_denom* to true to expand denominators. For example,

(%i1)   `f : (x+a)*(x-a)/((b-x)*(c-x));`

(%o1) $\dfrac{(x-a)\,(x+a)}{(b-x)\,(c-x)}$

(%i2)   `expandwrt(f,x);`

(%o2) $\dfrac{x^2}{(b-x)\,(c-x)} - \dfrac{a^2}{(b-x)\,(c-x)}$

(%i3)   `expandwrt(f,x), expandwrt_denom: true;`

(%o3) $\dfrac{x^2}{x^2 - cx - bx + bc} - \dfrac{a^2}{x^2 - cx - bx + bc}$

To force the expansion to place all terms over a common denominator, use `ratexpand` (rational expansion) instead of `expand` or `expandwrt`.

`ratexpand(`*expression*`)`
Expand *expression* into a sum of terms which all have a common denominator.

Compare the results from `expand` with those of `ratexpand`:

(%i1)   `expr: (x+1)^5 + (y-1)^3+(z+2)^-2;`

(%o1) $\dfrac{1}{(z+2)^2} + \left(y-1\right)^3 + (x+1)^5$

(%i2)   `expand(expr);`

(%o2) $\dfrac{1}{z^2 + 4z + 4} + y^3 - 3y^2 + 3y + x^5 + 5x^4 + 10x^3 + 10x^2 + 5x$

(%i3)   `ratexpand(expr);`

(%o3) $\dfrac{y^3 z^2}{z^2+4z+4} - \dfrac{3y^2 z^2}{z^2+4z+4} + \dfrac{3yz^2}{z^2+4z+4} + \dfrac{x^5 z^2}{z^2+4z+4} + \dfrac{5x^4 z^2}{z^2+4z+4} + \dfrac{10x^3 z^2}{z^2+4z+4} +$

$\dfrac{10x^2 z^2}{z^2+4z+4} + \dfrac{5xz^2}{z^2+4z+4} + \dfrac{4y^3 z}{z^2+4z+4} - \dfrac{12y^2 z}{z^2+4z+4} + \dfrac{12yz}{z^2+4z+4} + \dfrac{4x^5 z}{z^2+4z+4} +$

$\dfrac{20x^4 z}{z^2+4z+4} + \dfrac{40x^3 z}{z^2+4z+4} + \dfrac{40x^2 z}{z^2+4z+4} + \dfrac{20xz}{z^2+4z+4} + \dfrac{4y^3}{z^2+4z+4} - \dfrac{12y^2}{z^2+4z+4} +$

$\dfrac{12y}{z^2+4z+4} + \dfrac{4x^5}{z^2+4z+4} + \dfrac{20x^4}{z^2+4z+4} + \dfrac{40x^3}{z^2+4z+4} + \dfrac{40x^2}{z^2+4z+4} + \dfrac{20x}{z^2+4z+4} +$

$\dfrac{1}{z^2+4z+4}$

Ratios of polynomials that are difficult to integrate or differentiate can often be simplified by splitting them up into a sum of "partial fraction" terms. The `partfrac` function can do this:

 `partfrac(expression, x)`
Expand *expression* into fractions with powers of *x* in the denominator.

For example,

```
(%i1)    (x^3-3*x^2-25*x-21)/(x^3-2*x^2-56*x+192);
(%i2)    partfrac(%,x);
```

$$(\%o1) \quad \frac{x^3 - 3\,x^2 - 25\,x - 21}{x^3 - 2\,x^2 - 56\,x + 192}$$

$$(\%o2) \quad -\frac{25}{8\,(x+8)} + \frac{35}{8\,(x-4)} - \frac{9}{4\,(x-6)} + 1$$

Partial fractions are useful for detecting points where an expression becomes infinite. For example, let's solve the van der Waals equation $(P + a/V^2)(V - b) = RT$ for P, and then write the solution as an expansion in terms of V, which appears in the denominator:

```
(%i1)    solve((P+a/V^2)*(V-b)=R*T, P);
(%i2)    first(%);
(%i3)    partfrac(%, V);
```

$$(\%o1) \quad [P = \frac{R\,T\,V^2 - a\,V + a\,b}{V^3 - b\,V^2}]$$

$$(\%o2) \quad P = \frac{R\,T\,V^2 - a\,V + a\,b}{V^3 - b\,V^2}$$

$$(\%o3) \quad P = \frac{R\,T}{V - b} - \frac{a}{V^2}$$

Writing the result in this form shows us that the pressure will become infinite as V approaches either b or zero.

Trigonometric functions can be expanded with the `trigexpand` function. This is especially useful when we want to eliminate sums of angles or multiple angles from an expression. For example,

```
(%i1)    sin(3*x)/sin(x);
(%i2)    trigexpand(%);
(%i3)    trigsimp(%);
```

$$(\%o1) \quad \frac{\sin(3\,x)}{\sin(x)}$$

$$(\%o2) \quad \frac{3\cos(x)^2\sin(x) - \sin(x)^3}{\sin(x)}$$

$$(\%o3) \quad 4\cos(x)^2 - 1$$

If a result still contains sums or multiples of angles, you can apply `trigexpand` again:

```
(%i4)    cos(2*x+y)$
```

```
(%i5)    trigexpand(%);
(%i6)    trigexpand(%);
(%i7)    trigsimp(%);
```

$$(\%o2) \quad \cos(2x)\cos(y) - \sin(2x)\sin(y)$$
$$(\%o3) \quad \left(\cos(x)^2 - \sin(x)^2\right)\cos(y) - 2\cos(x)\sin(x)\sin(y)$$
$$(\%o4) \quad \left(2\cos(x)^2 - 1\right)\cos(y) - 2\cos(x)\sin(x)\sin(y)$$

5.3.5   Factoring Expressions

The `factor` function is the inverse of `expand`. It rewrites an expression as a product of terms.

factor(*expression*)
Factors *expression* into a product of irreducible integers or expressions.

For example,

```
(%i1)    factor(1000);
```

$$(\%o1) \quad 2^3\, 5^3$$

```
(%i2)    factor(y^4-9*y^2+4*y+12);
```

$$(\%o2) \quad (y-2)^2\,(y+1)\,(y+3)$$

To factor individual terms in a sum or product, or expressions embedded in the argument of a function, use `map(factor,expression)` (see Section 2.3.3):

```
(%i1)    sin(y^3+3*y^2+3*y+1);
(%i2)    factor(%);
(%i3)    map(factor,%);
```

$$(\%o1) \quad \sin\left(y^3 + 3y^2 + 3y + 1\right)$$
$$(\%o2) \quad \sin\left(y^3 + 3y^2 + 3y + 1\right)$$
$$(\%o3) \quad \sin\left((y+1)^3\right)$$

The system variable `dontfactor` can be set to a list of variables that should not be factored. For example,

```
(%i1)    expr: x^2*y+4*x*y+4*y-3*x^2-12*x-12;
(%i2)    factor(expr);
(%i3)    factor(expr), dontfactor : [x];
```

$$(\%o1) \quad x^2 y + 4xy + 4y - 3x^2 - 12x - 12$$
$$(\%o2) \quad (x+2)^2\,(y-3)$$
$$(\%o3) \quad \left(x^2 + 4x + 4\right)\,(y-3)$$

The `factorout` function can be used to factor out expressions in particular variables:

factorout(*expression*, var_1, var_2, ...)
Rewrite *expression* as a product of factors that include the variables var_1, var_2, ...

For example,

```
(%i1)   expr: 6*x*y^2+3*y^2+6*x^2*y+x*y-y-2*x^2-x;
(%i2)   factor(expr);
(%i3)   factorout(expr, x);
(%i4)   factorout(expr, y);
```

$$(\%o1) \quad 6\,x\,y^2 + 3\,y^2 + 6\,x^2\,y + x\,y - y - 2\,x^2 - x$$
$$(\%o2) \quad (2\,x + 1)\,(y + x)\,(3\,y - 1)$$
$$(\%o3) \quad 3\,(2\,x + 1)\,y^2 + (2\,x + 1)\,(3\,x - 1)\,y - 2\,x^2 - x$$
$$(\%o4) \quad 3\,y^2 + x\,(2\,y + 1)\,(3\,y - 1) + 2\,x^2\,(3\,y - 1) - y$$

 Worksheet 5.3: Expanding and factoring expressions

In this worksheet we'll rewrite expressions and equations by selectively factoring and expanding them.

5.3.6   Substitution

Substitutions are indispensible in simplifying and deriving equations.

We've already seen how Maxima's `subst` function (inserted with ⎡Subst...⎤ on the *General Math* pane, or ⎡Simplify⎤ ≫ ⎡Substitute...⎤ from the menu) does purely syntactical substitution (Section 1.2.2). This is useful when we want to make a substitution *after* evaluation of an expression. For example,

```
(%i1)   subst(1, x, diff(x^3, x));
```

```
(%o1)   3
```

The `subst` function will only do a substitution when the target appears explicitly in the expression. For example, substituting x for $\sqrt{y}$ in $y^{\frac{3}{2}}$ won't work:

```
(%i1)   subst(x, sqrt(y), y^(3/2);
```

$$(\%o1) \quad y^{\frac{3}{2}}$$

We've also seen that the evaluation environment can be used to perform substitutions in parallel `before` evaluation of an expression. For example,

```
(%i1)   n*R*T/V, n=1, R=0.08206, T=273.15, V=22.4;
```

```
(%o1)   3
```

Evaluation environments shouldn't be used for substitutions that must be done *after* evaluation of the expression. For example, the substitution we did with `subst` won't work with an evaluation environment:

```
(%i1)   diff(x^3, x), x=1;
```

diff: second argument must be a variable; found 1.
– an error. To debug this try: debugmode(true);

Like `subst`, substitutions in evaluation environments are syntactical, not mathematical. The target expression must appear explicitly:

```
(%i1)   y^(3/2), sqrt(y)=x;
```

$$(\%o1) \quad y^{\frac{3}{2}}$$

```
(%i2)   2*sqrt(y), sqrt(y)=x;
```

$$(\%o1) \quad 2x$$

When the substitution needs to be mathematical rather than syntactic, use `ratsubst` rather than `subst` or an evaluation environment:

`ratsubst(a, b, expression)`
Substitute a for b in `expression`. The substitution occurs *after* the expression is evaluated.

```
(%i1)   ratsubst(x, sqrt(y), y);
```

$$(\%o1) \quad x^3$$

Suppose we want to write the equilibrium constant K for an ionic dissociation $AB \rightleftharpoons A^+ + B^-$ in terms of experimentally measurable quantities. The equilibrium law is

```
(%i1)   eqn: K =   a[A\+]*a[B\-]/a[AB];
```

$$(\%o1) \quad K = \frac{a_{A+}\, a_{B-}}{a_{AB}}$$

... in terms of the activities for each species, where the + and − are included in the variable names by escaping them with a backslash. Maxima will interpret them as characters rather than addition and subtraction.

Next, we replace each activity a_i with its activity coefficient γ_i times its molality m_i. We can do this manually, with a series of substitutions:

```
(%i2)   ratsubst(m[A\+]*%gamma[A\+], a[A\+], eqn);
(%i3)   ratsubst(m[B\-]*%gamma[B\-], a[B\-], %);
(%i4)   ratsubst(m[AB]*%gamma[AB], a[AB], %);
```

$$(\%o2) \quad K = \frac{\gamma_{A+}\, m_{A+}\, a_{B-}}{a_{AB}}$$

$$(\%o3) \quad K = \frac{\gamma_{A+}\, m_{A+}\, \gamma_{B-}\, m_{B-}}{a_{AB}}$$

$$(\%o4) \quad K = \frac{\gamma_{A+}\, m_{A+}\, \gamma_{B-}\, m_{B-}}{\gamma_{AB}\, m_{AB}}$$

Using `ratsubst` becomes tedious when we have many similar substitutions to do. Maxima provides a way to define general simplification rules like "replace a_i with $\gamma_i m_i$, where i is anything." Defining a new rule requires two steps:

1) Declare i as a wildcard with the `matchdeclare` function:

   `matchdeclare(symbol, predicate)`
   Treat *symbol* as a wildcard when *predicate* is true.

   Try to use a predicate that is as specific as possible. However, using "all" or "true" as the predicate lets us apply the rule to expressions as well as variables and values.

2) Tell the simplifier about the new rule with the `tellsimpafter` function:

   `tellsimpafter(pattern, replacement)`
   Replace a *pattern* with *replacement*, after first applying all of the built-in rules for simplifying expressions. The *pattern* contains wildcards previously declared by `matchdeclare`; any wildcards in *replacement* are assigned values matched in the actual expression.

After defining the rule, Maxima applies it automatically in all further simplifications:

```
(%i5)   matchdeclare(_i, true)$
(%i6)   tellsimpafter(a[_i], %gamma[_i]*m[_i])$
(%i7)   a[Cl\-]*a[Na\+];
```

$$(\%o7) \quad \gamma_{Cl-} \cdot m_{Cl-} \cdot \gamma_{Na+} \cdot m_{Na+}$$

To delete all of the rules you have defined, type `clear_rules()`.

Now we want to substitute the mean activity coefficient γ_m into the expression, with $\gamma_m^2 = \gamma_{A+}\gamma_{B-}$, and the total molality m with $m_{A+} = m_{B-} = m\alpha$ and $m_{AB} = m(1-\alpha)$. We can also assume that the activity coefficient of the undissociated molecule γ_{AB} is 1:

```
(%i8)    ratsubst(%gamma[m]^2,   %gamma[A\+]*%gamma[B\-], eqn2)$
(%i9)    ratsubst(m*alpha, m[A\+], %)$
(%i10)   ratsubst(m*alpha, m[B\-], %)$
(%i11)   ratsubst(m*(1-alpha), m[AB], %)$
(%i12)   eqn3 : %, %gamma[AB] = 1;
```

$$(\%o12) \quad K = -\frac{\alpha^2\, m\, \gamma_m^2}{\alpha - 1}$$

For the first ionization of H_2SO_3 in water, with a total molality of 1.0273 m and experimentally determined values $\alpha = 0.1340$ and $\gamma_m = 0.783$, we have [22]

```
(%i13)   eqn3, alpha=0.1340, %gamma[m]=0.783, m=1.0273, numer;
```

$$(\%o2) \quad K = 0.01305907803244$$

To make several substitutions in parallel before evaluating an expression, use an evaluation environment.

To make purely syntactical substitutions after evaluating an expression, use subst.

To make a mathematical substitution, use ratsubst.[4]

 Worksheet 5.4: Substitution and Simplification

In this worksheet, we'll make serial and recursive substitutions on expressions. We'll also see how to define our own rules for simplifying expressions.

5.3.7 Solving an Equation Symbolically

In Section 1.2.2, we used the solve function to solve simple algebraic equations. In this section, we'll see how the solutions provided by solve can be extracted and used in subsequent calculations.

Recall that the solve function has two arguments: an equation (or list of equations) to be solved, and a variable (or list of variables) to be solved for. For example, let's solve the van der Waals equation for pressure and save the solution as solution:

```
(%i1)   vdW_equation: (P + n^2*a/V^2)*(V-n*b) = n*R*T;
(%i2)   solution: solve(vdW_equation, P);
```

$$(\%o1) \quad \left(\frac{a\,n^2}{V^2} + P \right)(V - b\,n) = n\,R\,T$$

$$(\%o2) \quad [P = \frac{n\,R\,T\,V^2 - a\,n^2\,V + a\,b\,n^3}{V^3 - b\,n\,V^2}]$$

Maxima returns a list of solutions; in this case, the list contains only one item. The list format makes it easy to assign the solutions to variables in evaluation environments. For example, let's assign the solution's value of P to P_vdw:

```
(%i3)   P_vdw : P, solution;
```

$$(\%o3) \quad \frac{n\,R\,T\,V^2 - a\,n^2\,V + a\,b\,n^3}{V^3 - b\,n\,V^2}$$

The evaluation environment evaluates P using the solution, and assigns the result to Pvdw.

The subst function can also be used to substitute a solution into an expression. For example, to substitute the solution for P into the equation $Z = PV/RT$, type

```
(%i4)   subst(solution, Z = P*V/(R*T));
```

$$(\%o4) \quad Z = \frac{V\left(n\,R\,T\,V^2 - a\,n^2\,V + a\,b\,n^3\right)}{R\,T\left(V^3 - b\,n\,V^2\right)}$$

---

4 To apply a mathematical substitution repeatedly to an expression, use fullratsubst, as demonstrated in Worksheet 5.4. The fullratsubst function is found in the lrats package.

5.3.7.1 Handling Multiple Solutions

An equation will often have more than one solution. For example, in solving the equilibrium expression

$$K_a = \frac{[\text{H}^+][\text{A}^-]}{[\text{HA}]} \approx \frac{x^2}{C_{HA} - x} \qquad \text{Equilibrium law for acid dissociation} \qquad (5.5)$$

for *x*, we obtain two solutions:

```
(%i1)   K[a]= x^2/(C[HA] - x);
(%i2)   solution: solve(%, x);
```

$$(\%o1) \quad K_a = \frac{x^2}{C_{HA} - x}$$

$$(\%o2) \quad [x = -\frac{\sqrt{4\,K_a\,C_{HA} + K_a^2} + K_a}{2}, x = \frac{\sqrt{4\,K_a\,C_{HA} + K_a^2} - K_a}{2}]$$

Only the positive solution has physical meaning. We can pick the solution we want using subscripts or Maxima's ordinal functions first, second, third, fourth, fifth, sixth, seventh, eighth, ninth, tenth, or last (Section 2.3.2).

```
(%i3)   solution[2];
(%i4)   second(solution);
```

$$(\%o3) \quad x = \frac{\sqrt{4\,K_a\,C_{HA} + K_a^2} - K_a}{2}$$

$$(\%o4) \quad x = \frac{\sqrt{4\,K_a\,C_{HA} + K_a^2} - K_a}{2}$$

The sublist function (Section 2.3.6) can be used to filter longer lists of solutions. For example, suppose we want to find the molar volume of CO_2 gas at 298 K and 1 atm using the van der Waals equation, we get several nonphysical complex solutions:

```
(%i1)   (P+a/Vm^2)*(Vm-b)  = R*T;
(%i2)   solution: solve(%, Vm), P=1, R=0.082059, T=298, a=3.640,
    ➥   b=0.04267, float;
```

$$(\%o1) \quad (Vm - b)\left(P + \frac{a}{Vm^2}\right) = R\,T$$

$(\%o2) \quad [Vm = 8.074353906564303\,(0.8660254037844386\,i - 0.5)$
$+ 8.107237761890614\,(-0.8660254037844386\,i - 0.5)$
$+ 8.165417333333332, Vm = 8.107237761890614\,(0.8660254037844386\,i - 0.5)$
$+ 8.074353906564303\,(-0.8660254037844386\,i - 0.5)$
$+ 8.165417333333332, Vm = 24.34700900178825]$

We're only interested in the real solution. We can pick it out using sublist and the featurep predicate (Section 4.5).

```
(%i3)   realp(x)  := featurep(x, real)$
(%i4)   sublist(solution, realp);
```

$(\%o4) \quad [Vm = 24.34700900178825]$

5.3.8 Solving an Equation Numerically

Some equations have solutions that cannot be easily written symbolically. For example, consider the equation $x^2 = 2^x$:

(%i1)   **solve(x^2=2^x, x);**

(%o1) $[x = -2^{\frac{x}{2}}, x = 2^{\frac{x}{2}}]$

The solve function failed to find a symbolic solution; x's appear on the right-hand side of each "solution." But the equation can be solved numerically using the find_root function:

find_root(*equation*, x, a, b)
Find a solution of *equation* for x on the interval $[a,b]$.

To find an interval $[a, b]$, rewrite the equation in the form $f(x) = 0$ and plot $f(x)$. The function will cross the x axis at one or more points if there is a real solution; these points are called the *roots* of $f(x)$.

(%i2)   **plot2d(x^2-2^x, [x, -5, 5])\$**

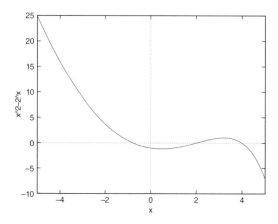

There are three roots: one between $x = -2$ and $x = 0$, one at $x = 2$, and one at $x = 4$. A suitable interval will be one where the function changes sign on either side of the root. For the negative root,

(%i3)   find_root(x^2=2^x, x, -2, 0);

(%o3) -0.76666469596212

(M) **Worksheet 5.5: Solving Equations**

In this worksheet, we'll obtain the equilibrium concentrations of reactants and products in a single reaction both symbolically and numerically.

5.4 Systems of Equations

We often need to solve several equations simultaneously in chemistry, in more than one variable. This process can be extremely tedious to do by hand, but in Maxima all we need to do is be able to list the equations, list the unknowns, and pass the lists to the equation solver. For example,

```
(%i1)   eqn1: x+y+z = 1$
(%i2)   eqn2: x+2*y = 4$
(%i3)   eqn3: y+z = 8$
(%i4)   solve([eqn1,eqn2,eqn3], [x,y,z]);
```

$$(\%o4) \quad [[x = -7, y = \frac{11}{2}, z = \frac{5}{2}]]$$

5.4.1 Eliminating Variables

One straightforward approach to solving a system of equations is to successively solve each equation for one variable, and then substitute the solution into the remaining equations to eliminate the variable. The process can continue until a single equation remains in one variable. The last equation can then be solved by applying `solve`.

Maxima provides the `eliminate` function to automate this process:

`eliminate(equation list, variable list)`
Eliminates each variable in *variable list* from the equations in the *equation list*. Each variable eliminates one equation; if the number of variables is equal to the number of equations, a single expression equal to the last variable is displayed.

For example, let's take the equation for kinetic energy E of a particle with mass m and velocity v, and combine it with the de Broglie equation for the wavelength λ of a quantum mechanical particle, and a third equation that gives the restriction on wavelength for a standing wave along a one-dimensional wire of length L:

```
(%i1)   eqns: [
            E=(1/2)*m*v^2,        /* kinetic energy of a particle
              ➡ */
            %lambda=h/(m*v),      /* de Broglie wavelength  */
            n*%lambda/2 = L       /* standing wave wavelength
              ➡ restriction */
        ];
(%i2)   eliminate(eqns, [%lambda, v, E]);
```

$$(\%o1) \quad [E = \frac{m\,v^2}{2}, \lambda = \frac{h}{m\,v}, \frac{n\,\lambda}{2} = L]$$

$$(\%o2) \quad [\frac{h^2\,n^2}{8\,m\,L^2}]$$

This last expression is equal to the last variable in the elimination list, E. It gives the energy of a quantum mechanical particle trapped in a one-dimensional wire of length L.

The `eliminate` function is useful in solving multiple equilibrium problems. For example, let's derive and solve an equation that computes the hydrogen ion concentration for a weak acid

HA in water in terms of its acid dissociation constant K_a and its formal concentration C_a. The relevant equations are

$$K_a = \frac{[H^+][A^-]}{[HA]} \qquad \text{Law of mass action} \tag{5.6}$$

$$C_a = [A^-] + [HA] \qquad \text{Mass balance} \tag{5.7}$$

$$[H^+] = [A^-] + [OH^-] \qquad \text{Charge balance} \tag{5.8}$$

$$K_w = [H^+][OH^-] \qquad \text{Ion product of water} \tag{5.9}$$

```
(%i1)   eqn1: Ka = H*A/HA$
(%i2)   eqn2: Ca = HA+A$
(%i3)   eqn3: H = A + OH$
(%i4)   eqn4: Kw = H*OH$
(%i5)   eliminate([eqn1,eqn2,eqn3,eqn4], [A, HA, OH]);
```

$$(\%o5) \quad [H\left(H^3 + Ka\,H^2 + (-Kw - Ca\,Ka)\,H - Ka\,Kw\right)]$$

The roots of the final expression give the hydrogen ion concentration. For example, choosing specific values of C_a, K_a, and K_w,

```
(%i6)   f: first(%), Ka=1.77e-4,Ca=2.00e-4,Kw=1.01e-14$
(%i7)   plot2d(f, [H, 0, 0.0002], [ylabel, "f"])$
```

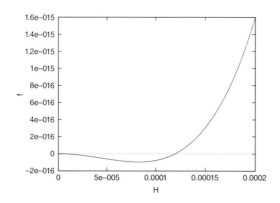

There is a trivial solution at $H = 0$, and another between $H = 0.0001$ and $H = 0.00015$, which is

```
(%i8)   find_root(f, H, 0.0001, 0.00015);
```

$$(\%o2) \quad 1.1942372416198077\,10^{-4}$$

5.4.2 Solving Systems of Equations Without Elimination

Chemical equilibria often involve many variables (the concentrations of reactants and products) with many constraints (laws of mass action, stoichiometric equations, mass balance equations, and charge balance equations). We often need to find all of the concentrations subject to those constraints, without eliminating variables.

For example, consider the formation of $Ag(NH_3)_2^+$ complexes in solutions made by mixing Ag^+ in aqueous NH_3 solutions. Complexes like this associate and dissociate in a stepwise fashion. The equilibria are

$$K_{f1} = \frac{[Ag(NH_3)^+]}{[Ag^+][NH_3]}$$

$$K_{f2} = \frac{[Ag(NH_3)_2^+]}{[Ag(NH_3)^+][NH_3]}$$

We know the total silver concentration C_{Ag}, which is

$$C_{Ag} = [Ag^+] + [Ag(NH_3)^+] + [Ag(NH_3)_2^+]$$

We also know the total ammonia concentration C_{NH_3}, which is

$$C_{NH_3} = [NH_3] + [Ag(NH_3)^+] + 2[Ag(NH_3)_2^+]$$

Given that $K_{f1} = 2.5 \times 10^3$ and $K_{f2} = 1.0 \times 10^4$, let's compute the equilibrium concentrations $[Ag^+]$, $[Ag(NH_3)^+]$, $[Ag(NH_3)_2^+]$, and $[NH_3]$ for a solution that is 0.010 M Ag^+ in 0.020 M NH_3.

First, set the constants and write out the equations:

```
(%i1)   K_f1 : 2.5e3$
(%i2)   K_f2 : 1e4$
(%i3)   CAg : 0.010$
(%i4)   CNH3 : 0.020$

(%i6)   eqn1 : K_f1=AgNH3p/(Agp*NH3);
(%i7)   eqn2 : K_f2=AgNH32p/(AgNH3p*NH3);
(%i8)   eqn3 : CAg=Agp + AgNH3p + AgNH32p;
(%i9)   eqn4 : CNH3 = NH3 + AgNH3p + 2*AgNH32p;
```

$$(\%o5) \quad 2500.0 = \frac{AgNH3p}{Agp\,NH3}$$

$$(\%o6) \quad 10000.0 = \frac{AgNH32p}{AgNH3p\,NH3}$$

$$(\%o7) \quad 0.01 = Agp + AgNH3p + AgNH32p$$

$$(\%o8) \quad 0.02 = NH3 + AgNH3p + 2\,AgNH32p$$

Now solve the four equations simultaneously in the concentrations for each species.

```
(%i10)   results: solve([eqn1,eqn2,eqn3,eqn4],[Agp,AgNH3p,AgNH32p
    ➡ ,NH3]);
```

$(\%o9)$ $[[Agp = 2.4172\ 10^{-4}, AgNH3p = 7.3829\ 10^{-4}, AgNH32p = 0.009, NH3 = 0.0012]$, $[Agp = 6.2939\ 10^{-4}\,i + 2.1247\ 10^{-4}, AgNH3p = -7.9181\ 10^{-4}\,i - 0.001, AgNH32p = 1.6242\ 10^{-4}\,i + 0.01, NH3 = 4.6697\ 10^{-4}\,i - 6.6086\ 10^{-4}]$,

$[Agp = 2.1247\ 10^{-4} - 6.2939\ 10^{-4}\ i, AgNH3p = 7.9181\ 10^{-4}\ i - 0.001, AgNH32p = 0.01 - 1.6242\ 10^{-4}\ i, NH3 = -4.6697\ 10^{-4}\ i - 6.6086\ 10^{-4}]]$

We obtain a list of solutions. Only the first solution is real; the rest are mathematically correct (but imaginary) solutions, and we can ignore them.[5]

```
(%i1)   first(%);
```

(%o10) $[Agp = 2.4172\ 10^{-4}, AgNH3p = 7.3829\ 10^{-4}, AgNH32p = 0.009, NH3 = 0.0012]$

Ⓜ **Worksheet 5.6: Solving Systems of Equations**

In this worksheet, we'll solve systems of equations using elimination, recursive substitution, and direct solution using numerical or symbolic approaches. We'll also construct fractional composition diagrams that show the abundance of various species in a solution where multiple equilibria occur.

5.5 Interpolation

Interpolation is the process of estimating the value of a variable within the range of its previously measured values. For example, we may want to use data points $(x_0, y_0), (x_1, y_1), \dots, (x_n, y_n)$ to find a function $y = f(x)$ that passes through all of the points $(y_i = f(x_i))$. The function can be used for data reduction, for locating errors in the dataset, for "smoothing" noisy data, for numerical integration of experimental data (Section 6.5.6), and in the numerical solution of differential equations (Section 11.5).

Lagrange interpolation finds a polynomial $f(x)$ of degree n that passes through $n + 1$ data points (x_i, y_i):

$$f(x) = \sum_{i=0}^{n} nL_i(x)y_i, \text{ where } L_i(x) = \prod_{j \neq i} \frac{x - x_j}{x_i - x_j} \qquad \text{Lagrange interpolation} \qquad (5.10)$$

A pair of points gives a line; three points give a parabola, and so on. Maxima can compute the interpolation formula for a dataset with the `lagrange` function, found in the `interpol` package:

`lagrange(data)`
Returns the Lagrange interpolation expression in x for `data`, which can be a two-column matrix, a list of points, or a list of y values (in which case x is assumed to be [1, 2, 3, ...]). You must `load(interpol)` before using this function.

Let's build a dataset by sampling the function $\cos(4x)e^{-x^2}$ on the interval $[-2, 2]$. The dataset is a list of points:

```
(%i1)   true_f(x) := cos(4*x)*exp(-x^2)$
(%i2)   data : makelist([i, true_f(i)], i, -2, 2, .5), numer;
```

---

5 See Section 4.5 for a way to use the `sublist` function to display only the real solutions in a list.

(%o2)   [[−2, −0.0026649260775371], [−1.5, 0.10120120366029],
[−1.0, −0.24046204996858], [−0.5, −0.3240954821756], [0.0, 1.0],
[0.5, −0.3240954821756], [1.0, −0.24046204996858], [1.5, 0.10120120366029],
[2.0, −0.0026649260775371]]

Now use the Lagrange interpolation formula to define a function $f(x)$:

```
(%i3)   load(interpol)$
(%i4)   f(x) := ''(lagrange(data))$
```

This function definition is equivalent to Equation 5.10. Remember that function definitions aren't executed, so we must use the quote–quote operator " (two single quote marks) to force `lagrange(data)` to execute and correctly insert the function definition.

 You must use parentheses to execute the entire expression. In this case, `f(x) := "(lagrange(data))` correctly defines the function; `f(x) := "lagrange(data)` won't work.

The function correctly returns the y at any point in `data`:

```
(%i5)   f(0);
(%i6)   true_f(0);
```

```
(%o5)   1.0
(%o6)   1
```

Values between data points are interpolated, and don't match up with the true function as accurately:

```
(%i7)   f(0.25);
(%i8)   true_f(0.25);
```

```
(%o7)   0.56567444222918
(%o8)   0.50756704400077
```

Plotting the Lagrange interpolation formula along with the original function and the sampled points reveals that the interpolating polynomial passes through all the data points, but it oscillates between them.

```
(%i9)   plot2d([[discrete, data], true_f(x), f(x)], [x, -2.1,
          ➡ 2.1],
               [style, points, lines, lines],
               [legend, "data", "true function", "lagrange"],
               [color, "#6666ff", "#6666ff", black]
        )$
```

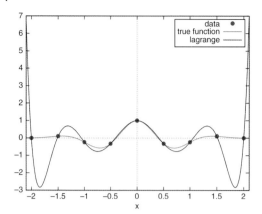

The oscillation increases on the edges of the x range, and the interpolating polynomial becomes completely inaccurate outside the range of the data. The problem becomes especially severe with high-degree polynomials. It's wise to limit `lagrange` to small groups of perhaps three or four points around the region you'd like to interpolate, rather than applying it to an entire dataset.

5.5.1 Piecewise Linear Interpolation

In piecewise linear interpolation, lines are drawn between each adjacent pair of points in the data. Use the `linearinterpol` function:

`linearinterpol(`*data*`)`
Returns a formula for piecewise linear interpolation in x for *data*, which can be a two-column matrix, a list of points, or a list of y values (in which case x is assumed to be [1, 2, 3, …]). You must `load(interpol)` before using this function.

Using the same data as before,

```
(%i10)  f(x) := ''(linearinterpol(data))$
(%i11)  plot2d([[discrete, data], true_f(x), f(x)], [x, -3, 3],
                [style, points, lines, lines],
                [legend, "data", "true function", "piecewise
                    ➡ linear"],
                [color, "#6666ff", "#6666ff", black]
        )$
```

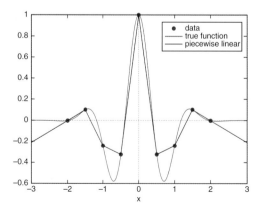

The linear interpolation function is often used in numerical integrations using the *trapezoid rule*, as we'll see in Section 6.5.6.

5.5.2 Spline Interpolation

We've seen that Lagrange interpolation may behave poorly when applied to more than a few points, producing an oscillating polynomial. Piecewise linear functions avoid oscillation but show sharp cusps at each of the points, making them almost useless for applications like numerical differentiation.

Spline interpolation avoids both of these problems by fitting smoothly varying piecewise polynomials (called *splines*). Usually splines are chosen to be cubic polynomials, fit to groups of four data points.

cspline(*data, options*)
Returns a formula for cubic spline interpolation in x for *data*, which can be a two-column matrix, a list of points, or a list of y values (in which case x is assumed to be [1, 2, 3, ...]). The options 'd1=*number* and 'dn=*number* can be used to specify the first derivative at the first and last points, respectively. You must load(interpol) before using this function.

Applying the method to the same data as before, we have

```
(%i12)   f(x) := ''(cspline(data))$
(%i13)   plot2d([[discrete, data], true_f(x), f(x)], [x, -3, 3],
                [style, points, lines, lines],
                [legend, "data", "true function", "cubic spline"
                  ➥ ],
                [color, "#6666ff", "#6666ff", black]
         )$
```

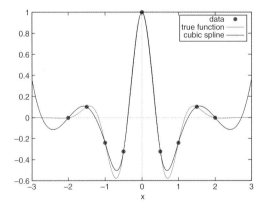

If the derivatives at the endpoints are unknown, a "natural" spline is chosen so that the endpoints are points of inflection (the interpolating function changes from concave up to concave down on passing through these points). You can change this if you know the derivatives at those points. For example,

```
(%i14)   f2(x) := ''(cspline(data, 'd1=0.06, 'dn=-0.06))$
```

```
(%i15)  plot2d([[discrete, data], true_f(x), f(x), f2(x)], [x,
        ➡ -2.1, 2.1],
(%i16)       [style, points, lines, lines, lines],
(%i17)       [legend, "data", "true function", "cubic spline"
             ➡ , "cubic spline w/ derivs"],
(%i18)       [color, "#6666ff", "#6666ff", black, red]
(%i19)  );
```

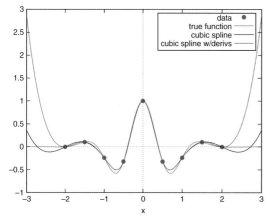

Setting the derivatives yields somewhat more accurate interpolation within the intervals on the ends of the range (but in this case it makes the spline even less accurate outside the range of the *x* data).

Ⓜ **Worksheet 5.7: Interpolation**

In this worksheet, we'll compute Lagrange, piecewise linear and spline functions to interpolate discrete data. We'll also see how to use interpolating functions to estimate derivatives from discrete data.

6

Differentiation, Integration, and Minimization

> *Ask yourself whether our language is complete–whether it was so before the symbolism of chemistry and the notation of the infinitesimal calculus were incorporated in it; for these are, so to speak, suburbs of our language.*
>
> – Ludwig Wittgenstein [23]

Calculus is the native language of physical chemistry. Thermodynamics, kinetics, statistical mechanics, and quantum chemistry rely on the concept of infinitesimal change, and it is essential in solving problems and in analyzing data in many other branches of chemistry as well.

In this chapter, we'll use Maxima to perform the basic operations of calculus, including computation of limits, differential expansions, derivatives, and integrals. We'll also see how to minimize and maximize functions, find points of inflection, and compute power series and Taylor series expansions for functions.

6.1 Limits

Sometimes we can't directly compute a function's value at a point. For example, we can't compute $\sin(x)/x$ at $x = 0$ directly, because division by zero is undefined. But, we can find the value by studying what happens as *x approaches* zero, as shown in Figure 6.1.

We can't claim that $\sin(0)/0 = 1$, but we can say that the *limit* of $\sin(x)/x$ as x approaches zero is one. This is written symbolically as

$$\lim_{x \to 0} \frac{\sin(x)}{x} = 1 \qquad \text{Limit notation} \tag{6.1}$$

Similarly, we can't evaluate an expression like $1/\infty$ because infinity is not a number. But we can say that $\lim_{x \to \infty} \frac{1}{x} = 0$, because as the denominator x gets larger, the fraction $1/x$ approaches zero.[1]

In chemistry, limits are encountered when studying the kinetics of reactions and the behavior of materials under extreme conditions. For example, real gases obey the ideal gas law as the pressure of the gas approaches zero; activity coefficients are given by the Debye–Hückel law as concentration approaches zero; gaseous reactions can exhibit different orders at high- and low-pressure limits.

---

1 This isn't just a matter of notation. It's wrong to use ∞ as an arbitrarily large number in calculations. For example, $\lim_{x \to \infty} \left(1 + \frac{1}{x}\right)^x = e$. But if you try to substitute ∞ for x in $\left(1 + \frac{1}{x}\right)^x$, you get $\left(1 + \frac{1}{\infty}\right)^\infty = (1 + 0)^\infty = 1$, which is incorrect. If you try to compute this in Maxima with `(1+1/x)^x, x=inf`, the result is $\left(\frac{1}{\infty} + 1\right)^\infty$, which Maxima correctly refuses to simplify.

Symbolic Mathematics for Chemists: A Guide for Maxima Users, First Edition. Fred Senese.
© 2019 John Wiley & Sons Ltd. Published 2019 by John Wiley & Sons Ltd.
Companion website: http://booksupport.wiley.com

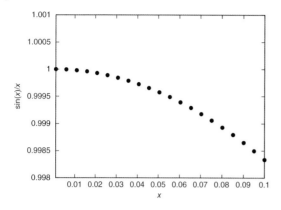

Figure 6.1 The value of $\sin(x)/x$ approaches one as x approaches zero.

The `limit` function can compute limits by approaching the limit point from one or both sides:

`limit(expression, x, a)`
`limit(expression, x, a, direction)`
Compute the limit of `expression` when the variable x approaches a value a. Optionally, the limit can be computed when x approaches a from a particular direction; the `direction` can be specified as `plus` or `minus`.

Let's compute the limit of C_V for an atomic solid as temperature approaches absolute zero, and as temperature approaches infinity. Einstein predicted that the constant-volume heat capacity C_V of a lattice of N atoms vibrating around their lattice points with frequency v is

$$C_V = 3Nk\frac{x^2 e^x}{(e^x - 1)^2}, \quad x = \frac{hv}{kT} \qquad \text{Einstein model for the heat capacity of an atomic solid}$$

(6.2)

```
(%i1)    assume(k>0, h>0, nu>0)$
(%i2)    C[v] : 3*N*k*x^2*exp(x)/(exp(x)-1)^2, x = h*nu/(k*T);
(%i3)    limit(C[v], T, 0);
(%i4)    limit(C[v], T, inf);
```

$$(\%o2) \qquad \frac{3\,h^2\,v^2\,R\,e^{\frac{hv}{kT}}}{k^2\,T^2\left(e^{\frac{hv}{kT}} - 1\right)^2}$$

$(\%o3) \quad 0$

$(\%o4) \quad 3\,kN$

If you omit the `assume` statement in line 1, Maxima will ask you whether or not k, h, and v are positive, negative, or zero. Line 3 shows that C_V approaches zero as T approaches zero; line 4 shows that as the temperature approaches positive infinity, C_V approaches $3kN$ (or $3R$).

Like many other Maxima functions, when `limit` doesn't know how to take a limit, it will simply echo your input back:

```
(%i1)    limit(f(x)/f(x 2), x, inf);
```

$$(\%o1) \qquad \lim_{x \to \infty} \frac{f(x)}{f(x^2)}$$

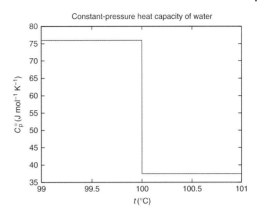

Figure 6.2 The limit of the heat capacity of water as the temperature approaches the boiling point doesn't exist; different limits are obtained by approaching the point from different directions.

6.1.1 Limits for Discontinuous Functions

Limits don't exist at points that are breaks or jumps in the value of a function. For example, consider finding the constant-pressure heat capacity of H_2O at its boiling point in Figure 6.2. From the negative direction, the limit is $\lim_{t \to 100^-} C_p^\circ = 75.327\ \mathrm{J\ mol^{-1}\ K^{-1}}$. Approaching from the positive direction gives $\lim_{t \to 100^+} C_p^\circ = 37.47\ \mathrm{J\ mol^{-1}\ K^{-1}}$. The limits from the left and from the right give C_p° for water and steam at 100 °C, respectively.

You'll often find discontinuous functions in routine chemical calculations. For example, the hydrogen ion concentration $[H^+]$ can be approximately computed for a monoprotic weak acid with formal concentration C and acid dissociation constant K_a by

$$K_a = \frac{[H^+]^2}{C - [H^+]} \qquad \text{Approximate } [H^+] \text{ for a weak acid} \qquad (6.3)$$

The expression on the right-hand side has a singularity at $C = [H^+]$:

```
(%i1)  f(C,H) := H^2/(C-H)$
(%i2)  plot2d(f(C,H*scaleH),
              [H,0.99*C/scaleH,1.01*C/scaleH],
              [xlabel, "[H+] / 10^-4 mol/L"],
              [ylabel, "[H+]^2/(C-[H+])"]
       ),C=1.5e-4, scaleH=1e-4$
```

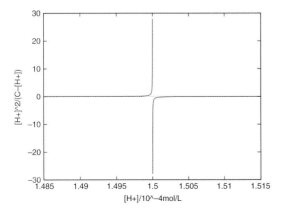

Taking the limit of `f(C,H)` as H approaches C without specifying the direction, we have

```
(%i3)   limit(f(C,H),C,H);
```

```
(%o3)   infinity
```

In Maxima, `infinity` means that the limit of the *absolute value* of the expression is ∞, but the limit of the expression itself is neither $+\infty$ or $-\infty$. You might obtain a limit of `infinity` for an oscillating function (like $\lim_{x\to\infty}(-a)^x$) or a function where the limit is different depending on the direction of approach, as is the case with `f(C,H)`. Taking the limit of `f(C,H)` from the positive direction gives $-\infty$; the limit from the negative direction is $+\infty$.

```
(%i4)   assume(C>0)$
(%i5)   limit(f(C,H),H,C,plus);
(%i6)   limit(f(C,H),C,H,minus);
```

```
(%o5)   −∞
(%o6)   ∞
```

When the function is undefined at the limit point, the result is `und`. For example, the built-in `unit_step` function is discontinuous at zero. The limit at that point is undefined – but directional limits can still be computed:

```
(%i1)   plot2d(unit_step(x),  [x,-2,2], [y,-0.5,1.5])$
(%i2)   limit(unit_step(x),  x,  0);
(%i3)   limit(unit_step(x),  x,  0,  minus);
(%i4)   limit(unit_step(x),  x,  0,  plus);
```

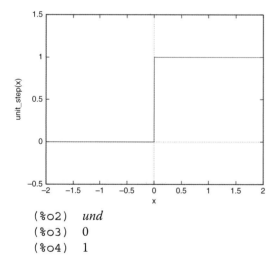

```
(%o2)   und
(%o3)   0
(%o4)   1
```

6.1.2   Limits for Indefinite Functions

Another possible result is `ind` (indefinite, but bounded). For example,

```
(%i1)   limit(cos(1/x),  x,  0);
```

```
(%o1)   ind
```

The function oscillates wildly as x approaches zero:

```
(%i2)   plot2d(cos(1/x), [x, -0.05, 0.05], [y,-1.1,1.1])$
```

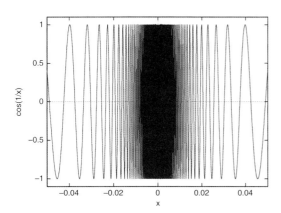

The function is bounded (between -1 and 1) but becomes indefinite around $x = 0$.

 Worksheet 6.0: Limits

In this worksheet, we'll use limits to solve problems in theoretical chemistry and thermodynamics.

6.2   Differentials

The symbol dx means "an arbitrarily small change in x." Infinitesimal differences like dx are called *differentials*.

Differentials are useful in gauging the effect of small perturbations on calculated results. For example, consider the problem of estimating the cross-sectional area of a capillary tube given a measured radius of the tube. The area A is just πr^2. If there is an infinitesimal error dr in the radius measurement, what will be the error dA created in A? The area with an erroneous radius $r + dr$ can be written as

$$A + dA = \pi(r + dr)^2 = \pi r^2 + 2\pi r dr + (dr)^2 \tag{6.4}$$

The $(dr)^2$ term is the product of two very small numbers, so it can be neglected. Subtracting A from both sides of Equation (6.4) gives dA as

$$dA = 2\pi r dr \tag{6.5}$$

In Maxima, we can derive the differential form using the `diff` function:

`diff(expression)`
Compute the total differential of *expression*. If *expression* is a list, `diff` is applied to each item, creating a list of total differentials.

```
(%i1)   A : %pi*r^2;
(%i2)   diff(A);
```

```
(%o1)   π r²
(%o2)   2 π r del (r)
```

Maxima displays the differential dr as `del(r)`.

What happens when the expression has more than one variable? The total differential will have a term for each variable:

$$dF(x_1, x_2, \cdots) = \frac{\partial F}{\partial x_1} dx_1 + \frac{\partial F}{\partial x_2} dx_2 + \cdots \qquad \text{Total differential expansion of F} \qquad (6.6)$$

For example, let's write the total differential for the pressure of an ideal gas:

```
(%i1)   P : n*R*T/V$
(%i2)   diff(P);
```

$$(\%o2) \quad -\frac{n R T \, \mathrm{del}\,(V)}{V^2} + \frac{n R \, \mathrm{del}\,(T)}{V} + \frac{n T \, \mathrm{del}\,(R)}{V} + \frac{R T \, \mathrm{del}\,(n)}{V}$$

Each symbol has a corresponding term in the total differential. The `diff` function assumes that every symbol is a variable, so there is a term in the expansion for R, which is a constant. You must declare constants *before* computing the expansion.[2] For example, the total differential dP for a van der Waals gas is

```
(%i1)   declare([R,a,b], constant)$
(%i2)   P : n*R*T/(V-b) + n^2*a/V^2$
(%i3)   diff(P);
```

$$(\%o3) \quad \left(-\frac{R n T}{(V-b)^2} - \frac{2 a n^2}{V^3} \right) \mathrm{del}\,(V) + \frac{R n \, \mathrm{del}\,(T)}{V-b} + \left(\frac{R T}{V-b} + \frac{2 a n}{V^2} \right) \mathrm{del}\,(n)$$

R, a, and b have been declared as constants so that terms in `del(R)`, `del(a)`, and `del(b)` don't appear in the total differential.

 Worksheet 6.1: Differentials

In this worksheet, we'll write differential expansions for calculations based on measurements, and use them to compute error estimates for the calculated results. We'll also write differential expansions for several common thermodynamic functions.

6.3   Derivatives

The derivative dy/dx gives the instantaneous rate of change of y with respect to x, defined as

$$\frac{dy}{dx} = \lim_{\Delta x \to 0} \frac{\Delta y}{\Delta x} = \lim_{\Delta x \to 0} \frac{f(x + \Delta x) - f(x)}{\Delta x} \qquad \text{The derivative of } y = f(x) \text{ with respect to } x \qquad (6.7)$$

---

2 See Section 4.5 for more about the `declare` function.

The change in y must be the result of the change in x, that is, y *must depend on x.* The derivative dy/dx is not simply a ratio of two differentials!

We can apply Equation (6.7) directly to find derivatives. For example, we can compute the derivative of $\sin(x)$ by

```
(%i1)   limit((sin(x+Delta_x) - sin(x))/Delta_x, Delta_x, 0);
```

```
(%o1)   cos(x)
```

... but there is an easier way. When we know y as an explicit function of x, we can compute derivatives with another variation of the `diff` function:

    `diff(y, x)`
    Compute the derivative of y with respect to x.

For example, `diff(sin(x),x)` yields `cos(x)`.

The derivative at any point along a curve corresponds to the slope of a line tangent to the curve. For example,

```
(%i1)   f(x) := (x+1)*(x-4)$
(%i2)   x0 : 1$
(%i3)   y0 : f(x0)$
(%i4)   m : diff(f(x), x)$
(%i5)   m : m, x=x0$
(%i6)   plot2d([f(x), m*(x-x0)+y0], [x, 0, 2], [legend, "f(x)=(x
        ➥ +1)(x-4)", "tangent line at x=1"])$
```

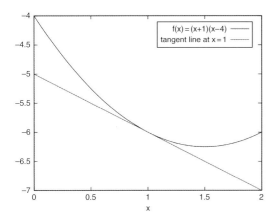

The equation for the tangent line to a function $y = f(x)$ at a point (x_0, y_0) is $y = \left(dy/dx\right)_{x=x_0} (x - x_0) + y_0$. Notice that we must compute the derivative with x as an unassigned variable (in line 4) *before* substituting $x = x_0$ in line 5.

    A single evaluation environment like `diff(f(x),x), x=1` will fail because the substitution is made before the derivative is computed – and `diff(f(1),1)` would be taking the derivative of a numerical value with respect to another numerical value.

We often want to include derivatives in expressions without needing to define y. This can be done by suppressing the evaluation of diff with the single-quote operator ('). For example, to type a first-order rate law,

(%i1)   `diff(A,t) = -k*A;`

(%o1) $\dfrac{d}{dt}A = -kA$

Typing diff(A,t) rather than 'diff(A,t) gives $0 = -kA$, which is incorrect.

 Simply typing diff(y,x) without defining y as a function or expression involving x gives zero, because as far as Maxima knows, y doesn't depend on x.

By default, derivatives are displayed in the Leibniz notation (dy/dx). You can display derivatives as subscripts instead (y_x) by setting the derivabbrev system variable.

(%i1)   **derivabbrev : true$**
(%i2)   **diff(y,x);**
(%i3)   **derivabbrev: false$**
(%i4)   **diff(y,x);**

(%o2) y_x
(%o4) $\dfrac{d}{dx}y$

6.3.1   Explicit Partial and Total Derivatives

Let's compute the derivative dV/dT for an ideal gas. The volume depends on moles n and pressure P as well as the temperature T.

(%i1)   `V : n*R*T/P;`
(%i2)   **diff(V,T);**

(%o1) $\dfrac{nRT}{P}$

(%o2) $\dfrac{nR}{P}$

Maxima has actually computed a partial derivative $(\partial V/\partial T)_P$ by assuming that P is held constant. This is equal to the derivative dV/dT as long as P doesn't change with T.

But, if P *does* change with T, we must tell Maxima about it. There are three ways to do this:

- Explicitly define P as a function of T or assign an expression that depends on T to P.
- Type P(T) instead of P in the expression for V.
- Declare P as a function of T using the depends function:

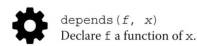 depends(*f*, *x*)
Declare f a function of x.

To keep our code simple and general, let's take the third approach:

```
(%i3)    depends(P,T)$
(%i4)    diff(V,T);
```

$$(\%o4) \quad \frac{nR}{P} - \frac{n\left(\frac{d}{dT}P\right)RT}{P^2}$$

In line 3, the depends function tells Maxima that P depends on T. The relationship between P and T is unknown, so Maxima leaves dP/dT in the result for line 4.

Maxima can recognize chains of dependencies when it computes derivatives. For example,

```
(%i1)    y : w*x + z^2*x^2$
(%i2)    depends(z,x)$
(%i3)    depends(w,z)$
(%i4)    diff(y,x);
```

$$(\%o4) \quad 2x^2 z\left(\frac{d}{dx}z\right) + \left(\frac{d}{dz}w\right)x\left(\frac{d}{dx}z\right) + 2xz^2 + w$$

The system variable dependencies gives us a list of defined dependencies.

```
(%i5)    dependencies;
```

$$(\%o5) \quad [w(z), z(x)]$$

To remove a dependency, use the remove function:

```
(%i6)    remove(w,dependency)$
(%i7)    dependencies;
```

$$(\%o7) \quad [z(x)]$$

```
(%i8)    diff(y,x);
```

$$(\%o8) \quad 2x^2 z\left(\frac{d}{dx}z\right) + 2xz^2 + w$$

6.3.2 Derivatives Evaluated at a Specific Point

We sometimes want to type or substitute derivatives evaluated at a single point in expressions. For example, to type the expression $\left.\frac{d\psi(x)}{dx}\right|_{x=0}$, use the at function[3]:

```
(%i1)    at('diff(psi(x),x), x=0);
```

$$(\%o1) \quad \left.\frac{d}{dx} \cdot \psi(x)\right|_{x=0}$$

---

3 Maxima often produces such expressions when solving differential equations with the desolve method (Section 11.4).

We can use the nouns switch to evaluate the derivative. For example, if $\psi(x) = \sqrt{2/L} \sin(n\pi x/L)$,

```
(%i1)   at('diff(psi(x),x), x=0), psi(x) := sqrt(2/L)*sin(n*%pi*
     ➡ x/L), nouns;
```

$$(\%o1) \qquad \frac{\sqrt{2} \cdot \pi \cdot n}{L^{\frac{3}{2}}}$$

6.3.3 Higher-Order Derivatives

The `diff` function can also compute second third and higher-order derivatives:

     `diff(expression, x, n)`
     Compute the *n*th derivative of `expression` with respect to the variable *x*.

For example, the first, second, and third derivatives of $\sin(x)$ are

```
(%i1)   diff(sin(x), x, 1);
(%i2)   diff(sin(x), x, 2);
(%i3)   diff(sin(x), x, 3);
```

```
(%o1)   cos(x)
(%o2)   −sin(x)
(%o3)   −cos(x)
```

The conditions for locating the critical point of a gas involve the computation of second derivatives. We must have $(\partial P/\partial V)_T = 0$ and $(\partial^2 P/\partial V^2)_T = 0$ at the critical point. Let's solve these conditions simultaneously to find P, V, and T at the critical point of a van der Waals gas:

```
(%i1)   P_vdw : R*T/(V-b) - a/V^2$
(%i2)   eqn1: P = P_vdw;
(%i3)   eqn2: diff(P_vdw, V)=0;
(%i4)   eqn3: diff(P_vdw, V, 2)=0;
```

$$(\%o2) \qquad P = \frac{RT}{V-b} - \frac{a}{V^2}$$

$$(\%o3) \qquad \frac{2a}{V^3} - \frac{RT}{(V-b)^2} = 0$$

$$(\%o4) \qquad \frac{2RT}{(V-b)^3} - \frac{6a}{V^4} = 0$$

Solving the three equations simultaneously, we have

```
(%i5)   solve([eqn1, eqn2, eqn3], [P, V, T]);
```

$$(\%o5) \qquad [[P = \%r1, V = b, T = 0], [P = \frac{a}{27\,b^2}, V = 3\,b, T = \frac{8\,a}{27\,b\,R}]]$$

We get two solutions. The first applies at absolute zero. The `%r1` means that the pressure is arbitrary for that solution. The second solution is the one we want:

```
(%i6)   second(%);
```

$$(\%o6) \quad [P = \frac{a}{27\,b^2}, V = 3\,b, T = \frac{8\,a}{27\,b\,R}]$$

6.3.4   Mixed Derivatives

Differentiating with respect to more than one variable yields a mixed derivative:

$$\frac{\partial}{\partial x}\left(\frac{\partial f}{\partial y}\right) = \frac{\partial^2 f}{\partial x \partial y} \qquad \text{Notation for mixed derivatives} \tag{6.8}$$

In chemical thermodynamics we often use mixed derivatives to derive relationships from the differentials of state functions.[4]

State functions have many convenient mathematical properties. They can be integrated between two points without knowledge of the path that connects the two points, since they depend only on the coordinates of a particular point. If $F(x, y)$ is a state function, differentiating in any order gives the same result:

$$\frac{\partial^2 F}{\partial x \partial y} = \frac{\partial^2 F}{\partial y \partial x} \qquad \text{Test for a state function } F \tag{6.9}$$

To compute mixed derivatives like $\frac{\partial^2 P}{\partial V \partial T}$, nest one `diff` call inside another, or use this form of `diff`:

```
diff(expression, x, nx, y, ny, ...)
```
Compute a mixed partial derivative of `expression` that is an *nx*th derivative with respect to variable *x* and an *ny*th derivative with respect to variable *y*. Additional variable/n pairs can be added. Each variable name *must* be followed by the order of the derivative in that variable.

Let's verify that the pressure of a van der Waals gas is a state function.

```
(%i1)   P : n*R*T/(V-b) - n^2*a/V^2$
(%i2)   diff(P,T,1,V,1);
(%i3)   diff(P,V,1,T,1);
```

$$(\%o2) \quad -\frac{n\,R}{(V-b)^2}$$

$$(\%o3) \quad -\frac{n\,R}{(V-b)^2}$$

Differentiating in either order gives the same result, so P is a state function.

---

4 A state function is a function that depends only on the current state of the system, and not on its history. The differential of a state function is called an *exact differential*.

Worksheet 6.2: Explicit Differentiation

In this worksheet, we'll compute and plot explicit first and second derivatives of single variable and multivariate functions. We'll also locate critical points for several model gases, and use them to rewrite the equations of state in reduced form.

6.3.5 Assigning Partial Derivatives

Partial derivatives describe how one variable changes with another when other variables are held constant. Partial derivatives are indispensable in thermodynamics, where abstract quantities like energy and entropy are connected with experimentally measurable properties like pressure, volume, and temperature. For example, the internal energy U is a function of volume V and temperature T. Its total differential is

$$dU = \left(\frac{\partial U}{\partial V}\right)_T dV + \left(\frac{\partial U}{\partial T}\right)_V dT \quad \text{Total differential of } U \tag{6.10}$$

The derivative $(\partial U/\partial V)_T$ has dimensions of pressure, and is often referred to as the internal pressure π_T. The derivative $(\partial U/\partial T)_V$ is the constant-volume heat capacity, C_V.

These partial derivatives can be assigned values using the `gradef` (gradient define) function.

```
gradef(y, x, expression)
```
Defines $\partial y/\partial x$ as `expression`.[5]

Let's assign a value π_T to $(\partial U/\partial V)_T$. After using `gradef`, Maxima automatically recognizes U as a function of V:

```
(%i1)   gradef(U,V,pi[T])$
(%i2)   'diff(U,V)[T] = diff(U,V);
```

$$(\%o2) \quad \left(\frac{d}{dV}U\right)_T = \pi_T$$

Similarly for $(\partial U/\partial T)_V = C_V$, we have

```
(%i3)   gradef(U,T,C[V])$
(%i4)   'diff(U,T)[V] = diff(U,T);
```

$$(\%o4) \quad \left(\frac{d}{dT}U\right)_V = C_V$$

The total differential expansion is now

```
(%i5)   diff(U);
```

$$(\%o5) \quad \pi_T \, \text{del}\,(V) + C_V \, \text{del}\,(T)$$

---

5 An alternative form of the `gradef` command is `gradef(f(x₁,x₂,x₃,…), y₁,y₂,y₃,…)` which defines $\partial f/\partial x_1 = y_1$, $\partial f/\partial x_2 = y_2$, and so on.

6.3.5.1 Partial Derivatives from Total Differential Expansions

We can use total differential expansions to find other partial derivatives. For example, let's use the total differential expansion for U derived above to find the derivative $(\partial U/\partial T)_P$. We'll divide dU by dT, and then substitute $(\partial V/\partial T)_P$ for `del(V)/del(T)` in the result (since we'll be imposing constant pressure conditions):

```
(%i6)   'diff(U,T)[P] = diff(U)/del(T), expand;
(%i7)   ratsubst( diff(V,T)[P], del(V)/del(T),%);
```

$$(\%o6)\quad \left(\frac{d}{dT}U\right)_P = \frac{\pi_T\,\text{del}(V)}{\text{del}(T)} + C_V$$

$$(\%o7)\quad \left(\frac{d}{dT}U\right)_P = C_V + \left(\frac{d}{dT}V\right)_P \pi_T$$

We want to write this expression in terms of experimentally measurable quantities. The thermal expansion coefficient $\alpha = (1/V)(\partial V/\partial T)_P$ is measurable; substituting it into the equation gives

```
(%i8)   ratsubst(V*alpha, 'diff(V,T)[P], %);
(%i9)   dUdT_P : rhs(%)$
```

$$(\%o8)\quad \left(\frac{d}{dT}U\right)_P = C_V + \alpha\,\pi_T\,V$$

We save the result as `dUdT_P` for later use.

Now let's find the derivative $(\partial U/\partial P)_T$ in terms of measurable quantities. Impose constant-temperature conditions by setting dT to zero in the total differential for U, and then divide by dP.

```
(%i10)  diff(U)$
(%i11)  ratsubst(0, del(T), %)$
(%i12)  'diff(U,P)[T] = %/del(P), expand;
```

$$(\%o12)\quad \left(\frac{d}{dP}U\right)_T = \frac{\pi_T\,\text{del}(V)}{\text{del}(P)}$$

The isothermal compressibility $\kappa_T = -(1/V)(\partial V/\partial P)_T$ is an experimentally measurable quantity, so we can replace $(\partial V/\partial P)_T$ with $-\kappa_T V$:

```
(%i13)  ratsubst(-V*kappa[T], del(V)/del(P), %);
(%i14)  dUdP_T : rhs(%)$
```

$$(\%o13)\quad \left(\frac{d}{dP}U\right)_T = -\kappa_T\,\pi_T\,V$$

where the right-hand side of the equation is saved for later use.

6.3.5.2 Writing Total Differential Expansions in Terms of New Variables

Now write the total differential of U in terms of P and T. Kill the U variable to wipe out previous definitions made with `gradef`, and then define the partial derivatives we've calculated above:

```
(%i15)  kill(U) $
(%i16)  gradef(U,P, dUdP_T) $
(%i17)  gradef(U,T, dUdT_P) $
(%i18)  diff(U);
```

$$(\%o18) \quad \left(C_V + \alpha\,\pi_T\,V\right) \operatorname{del}(T) - \kappa_T\,\pi_T\,V\operatorname{del}(P)$$

Let's use the work we've done to derive a relationship between C_p (the heat capacity at constant pressure) and C_V (the heat capacity at constant volume). C_p is equal to $(\partial H/\partial T)_p$, where the enthalpy H is defined as $U + PV$:

```
(%i19)  depends(V,T) $
(%i20)  H : U + P*V$
(%i21)  C[P] = diff(H,T);
(%i22)  ratsubst(V*alpha, diff(V,T),%);
```

$$(\%o21) \quad C_p = P\left(\frac{d}{dT}\,V\right) + C_V + \alpha\,\pi_T\,V$$

$$(\%o22) \quad C_p = C_V + \left(\alpha\,\pi_T + \alpha\,P\right)\,V$$

where we've used the definition of α to make the substitution $V\alpha = (\partial V/\partial T)_p$.

To make substitutions into higher order or mixed derivatives, it is sometimes necessary to use the *derivsubst* flag. For example, suppose we want to rewrite $\frac{d^2y}{dt^2}$ in terms of x, when $x = \frac{dy}{dt}$:

```
(%i1)  subst (x, 'diff (y, t), 'diff (y, t, 2)) ;  /* doesn't
       ➥ work */
(%i2)  subst (x, 'diff (y, t), 'diff (y, t, 2)) , derivsubst=
       ➥ true;
```

$$(\%o1) \quad \frac{d^2}{dt^2}\,y$$

$$(\%o2) \quad \frac{d}{dt}\,x$$

In Worksheet 7.7, we'll see a more general and powerful procedure for rewriting thermodynamic derivatives in terms of experimental variables.

6.3.6 Implicit Differentiation

Sometimes we would like to find a derivative dy/dx from an equation that relates x to y *without* first solving the equation for y. If the equation can be cast into the form $f(x, y) = 0$, the chain rule for derivatives can be written as

$$\frac{dy}{dx} = -\frac{df(x, y)/dx}{df(x, y)/dy} \qquad \text{Computing a derivative implicitly} \tag{6.11}$$

Let's write a function that implicitly computes dy/dx from an equation,[6] and use it to find dy/dx for a unit circle ($x^2 + y^2 = 1$).

```
(%i1)   implicit_diff(eqn, y, x) := block(
            [f],
            f : lhs(eqn)-rhs(eqn),
            ratsimp(-diff(f,x)/diff(f,y))
        )$
(%i2)   implicit_diff(x^2+y^2=1,y,x);
```

$$(\%o2) \quad -\frac{x}{y}$$

The local variable f holds the function $f(x, y)$, obtained by casting the equation eqn in the form $f(x, y) = 0$. A rational simplification is performed after applying Equation (6.11) to compute dy/dx.

Compare this to explicit differentiation:

```
(%i3)   result: solve(x^2+y^2=1,y);
(%i4)   diff(y,x), result[1];
(%i5)   diff(y,x), result[2];
```

$$(\%o3) \quad [y = -\sqrt{1-x^2}, y = \sqrt{1-x^2}]$$

$$(\%o4) \quad \frac{x}{\sqrt{1-x^2}}$$

$$(\%o5) \quad -\frac{x}{\sqrt{1-x^2}}$$

We get the same result for dy/dx as before, with the substitution $y = \pm\sqrt{1-x^2}$. However, solving the equation explicitly isn't always convenient (or even possible). For example, it's easy to implicitly compute $(\partial V/\partial T)_P$ from the van der Waals equation:

```
(%i6)   vdW_equation: (P + n^2*a/V^2)*(V-n*b) = n*R*T;
(%i7)   implicit_diff(vdW_equation, V, T);
```

$$(\%o6) \quad \left(\frac{a\,n^2}{V^2} + P\right)(V - b\,n) = n\,R\,T$$

$$(\%o7) \quad \frac{n\,R\,V^3}{P\,V^3 - a\,n^2\,V + 2\,a\,b\,n^3}$$

To compute the derivative explicitly, we'd solve the van der Waals equation for V, choose the physically significant solution, and apply the diff function to it:

```
(%i8)   result: solve(vdW_equation, V);
(%i9)   diff(V,T), last(result), ratsimp;
```

---

6 The third-party impdiff package is available to compute implicit derivatives of any degree for multivariable functions. See the Maxima manual for more.

The output isn't shown – try it yourself, and you'll understand why. There are four unwieldy solutions for V, and only the last is real. The derivative is a fearsome expression that is equivalent to the simple implicit derivative we computed above with the solution for V substituted in.

It is also possible to implicitly differentiate an equation with `diff` directly. For example, we can differentiate the definition of enthalpy $H = U + PV$ with respect to P, and solve the result for dU/dP:

```
(%i1)   H(P,T) = U(P,T) + P*V(P,T)$
(%i2)   diff(%,P);
(%i3)   solve(%, diff(U(P,T),P));
```

$$(\%o2) \quad \frac{d}{dP} H(P,T) = P\left(\frac{d}{dP} V(P,T)\right) + \frac{d}{dP} U(P,T) + V(P,T)$$

$$(\%o3) \quad \left[\frac{d}{dP} U(P,T) = -P\left(\frac{d}{dP} V(P,T)\right) + \frac{d}{dP} H(P,T) - V(P,T)\right]$$

 In general it's best to show the dependency of a function f on x and y by typing it as `f(x,y)` in expressions, rather than using the `depends` function to declare dependencies. In the current version of Maxima, `depends` is used in derivatives, but not in integrals or integral transforms.

 Worksheet 6.3: Partial Derivatives

In this worksheet, we'll apply the `gradef`, `depends`, and `diff` functions along with implicit differentiation to derive several thermodynamic relationships commonly encountered in first-semester physical chemistry.

6.4   Maxima, Minima, and Inflection Points

Many chemical problems involve finding critical points (points where the first derivative of a function changes sign). Critical points on molecular potential energy surfaces correspond to stable reactants, products, and intermediates, as well as transition states. The dynamics of chemical reactions can be modeled by equations of motion that connect these points.

Critical points include maxima, minima, and inflection points of a continuous function. The type of critical point can be discerned by looking at the second derivative. At a maximum, the second derivative is positive (the curve is concave up). At a minimum, the second derivative is negative (the curve is concave down). On either side of an inflection point, the second derivative changes sign (the curve is changing from concave up to concave down).

We can locate critical points for a function $f(x, y)$ by simultaneously solving the equations

$$\frac{\partial f(x,y)}{\partial x} = 0$$

$$\frac{\partial f(x,y)}{\partial y} = 0$$

for x and y.

For example, let's find the maxima and minima in the radial wavefunction for the $3p$ orbital,

$$R_{3,1}(\rho) = \frac{1}{\sqrt{486}} \left(\frac{Z}{a_0} \right)^{3/2} \rho(4 - \rho)e^{-\rho/2}, \quad \rho = \left(\frac{2Z}{3a_0} \right) r \qquad \text{Radial wavefunction for the 3p orbital}$$

(6.12)

where a_0 is the Bohr radius, 52.9 pm. It will be helpful to plot the function first:

```
(%i1)   Z : 1$
(%i2)   a0 : 52.9e-12$
(%i3)   R31(r) := block(
                [k,rho],
                k: 1/sqrt(486)*(Z/a0)^(3/2),
                rho : 2*Z/(3*a0)*r,
                k*rho*(4-rho)*exp(-rho/2)
(%i4)   )$
(%i5)   plot2d(R31(r*10^-9)/10^14, [r, 0, 2],
                [xlabel, "r/nm"],
                [ylabel, "R(r)/10^14"]
        );
```

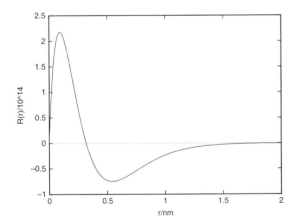

Notice that we've scaled r so that the x axis is in nanometers, and scaled the function so that powers of 10 won't appear on the y axis. Hovering the cursor over the curve shows the maximum and minimum at about $r = 0.093$ nm and $r = 0.54$ nm, respectively. To locate the maximum and minimum precisely, solve $\frac{dR_{3,1}(r)}{dr} = 0$ for r:

```
(%i6)   criticalpoints: solve(diff(R31(r),r)=0, r), numer;
```

```
(%o5)   [r = 9.296430765138981 10^{-11}, r = 5.418356923486106 10^{-10}]
```

These two values of r (in meters) match the values estimated from the graph. But how can we distinguish maxima from minima without a graph?

Critical points correspond to maxima when the sign of the first derivative changes from positive to negative as x increases. The second derivative will be negative (see Figure 6.3). A positive second derivative indicates that the curve was concave-up, and the critical point was a minimum.

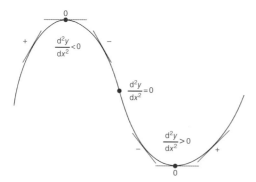

Figure 6.3 The sign of the second derivative d^2/dx^2 can determine whether a critical point is a maximum $(d^2/dx^2 < 0)$, minimum $(d^2/dx^2 > 0)$, or inflection point $(d^2/dx^2 = 0)$.

Let's compute the second derivative for the radial wavefunction at each critical point:

```
(%i7)   second_derivative: diff(R31(r), r, 2), numer$
(%i8)   second_derivative, criticalpoints[1];
(%i9)   second_derivative, criticalpoints[2];
```

```
(%o7)   -2.9481234903172417 10^34
(%o8)   1.7425103985204013 10^33
```

The negative second derivative at $r = 0.09296$ nm indicates a maximum. The point at $r = 0.54186$ nm is a minimum because it has a positive second derivative.

When the second derivative is zero, we have an inflection point where the curvature changes from concave-up to concave-down. The second derivative will be zero at those points, so we can locate them by solving $d^2y/dx^2 = 0$ for x:

```
(%i10)  inflectionpoints: solve(second_derivative=0, r), numer;
```

```
(%o9)   [r = 2.0122353683881923 10^-10, r = 7.509764631611814 10^-10]
```

Let's plot the critical points and inflection points on our graph. We must take care to scale the x and y values consistently with the axis scales we chose in the previous plot; we convert r to nm, and scale the y value by dividing it by 10^{14}.

```
(%i11)  scaledpoint(r) := [r/1e-9, R31(r)/1e14]$
(%i12)  maxpt : scaledpoint(r), criticalpoints[1]$
(%i13)  minpt : scaledpoint(r), criticalpoints[2]$
(%i14)  inflectionpt1 : scaledpoint(r), inflectionpoints[1]$
(%i15)  inflectionpt2 : scaledpoint(r), inflectionpoints[2]$
(%i16)  plot2d([R31(r*10^-9)/10^14,
                [discrete, [maxpt]],
                [discrete, [minpt]],
                [discrete, [inflectionpt1, inflectionpt2]]
            ],
            [r, 0, 2],
            [style, lines, points, points, points],
            [color, blue, black, black, red],
            [point_type, bullet, bullet, circle, diamond],
            [xlabel, "r/nm"],
```

```
    [ylabel, "R(r)/10^14"],
    [legend, "R(r)", "Maximum", "Minimum", "Inflection
        ➥ point"],
    [title, "Critical and inflection points on the
        ➥ hydrogen 3p radial wavefunction"]
) $
```

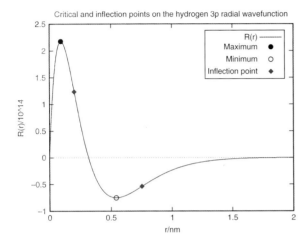

6.4.1 Critical Points of Surfaces

At a critical point (a, b) for a function $f(x, y)$, *both* first derivatives are zero. If the point is a minimum, both second derivatives are greater than zero. If the point is a maximum, both second derivatives are less than zero.

A saddle point is an additional type of critical point that can occur on surfaces. Saddle points are a maximum in one direction and a minimum in the other. The following condition holds if a point on $f(x, y)$ is a saddle point:

$$\left(\frac{\partial^2 f(x, y)}{\partial x^2} \right) \left(\frac{\partial^2 f(x, y)}{\partial y^2} \right) - \left(\frac{\partial^2 f(x, y)}{\partial x \partial y} \right)^2 < 0 \qquad \text{Condition for a saddle point} \qquad (6.13)$$

These derivatives are conveniently arranged in a matrix, called the Hessian matrix [24]:

$$\mathbf{H} = \begin{bmatrix} \dfrac{\partial^2 f}{\partial x^2} & \dfrac{\partial^2 f}{\partial x \partial y} \\[12pt] \dfrac{\partial^2 f}{\partial y \partial x} & \dfrac{\partial^2 f}{\partial y^2} \end{bmatrix} \qquad \text{The Hessian matrix} \qquad (6.14)$$

The conditions for maxima, minima, and saddle points can be rewritten in terms of the Hessian matrix and its determinant[7] $|\mathbf{H}|$, evaluated at each critical point[8]:

---

7 We'll look at computing determinants in Section 7.2.4. For now, note that the determinant of a matrix $\mathbf{A} = \begin{pmatrix} a & b \\ c & d \end{pmatrix}$

is $|\mathbf{A}| = ad - bc$, and the determinant of a matrix A can be computed in Maxima with `determinant(A)`.

8 If the determinant of the Hessian matrix is zero, we have to examine higher-order derivatives to classify the critical point.

- Minima have $H_{1,1} > 0$, $H_{2,2} > 0$, and $|\mathbf{H}| > 0$.
- Maxima have $H_{1,1} < 0$, $H_{2,2} < 0$, and $|\mathbf{H}| > 0$.
- At saddle points, $|\mathbf{H}| < 0$.

Maxima's `hessian` function can be used to compute $\mathbf{H}$:

hessian(f, x)
Computes the Hessian matrix of expression f with respect to the variables in the list x. The i, j element of the matrix is $\partial^2 f / \partial x_i \partial x_j$.

Let's find and classify the critical points of $f(x, y) = 4x^3 + 6xy^2 - y^3 - 12x$.

```
(%i1)   f(x,y) :=4*x^3 + 6*x*y^2 - y^3 - 12*x$
(%i2)   criticalpoints: solve([diff(f(x,y),x)=0, diff(f(x,y),y)
        ➥ = 0], [x,y]);
```

$$(\%o2) \quad [[x = -1, y = 0], [x = 1, y = 0], [x = -\frac{1}{3}, y = -\frac{4}{3}], [x = \frac{1}{3}, y = \frac{4}{3}]]$$

Compute the Hessian matrix, and then evaluate it at each critical point:

```
(%i3)   H: hessian(f(x,y), [x,y]);
```

$$(\%o3) \quad \begin{pmatrix} 24x & 12y \\ 12y & 12x - 6y \end{pmatrix}$$

```
(%i4)   H, criticalpoints[1];
(%i5)   determinant(%);
```

$$(\%o4) \quad \begin{pmatrix} -24 & 0 \\ 0 & -12 \end{pmatrix}$$

$(\%o5)$ 288

$H_{1,1}$ and $H_{2,2}$ are negative and $|H|$ is positive, so the first critical point $(-1, 0)$ is a maximum.

```
(%i6)   H, criticalpoints[2];
(%i7)   determinant(%);
```

$$(\%o6) \quad \begin{pmatrix} 24 & 0 \\ 0 & 12 \end{pmatrix}$$

$(\%o7)$ 288

$H_{1,1}$ and $H_{2,2}$ are positive and $|H|$ is positive, so the second critical point $(1, 0)$ is a minimum.

```
(%i8)   H, criticalpoints[3];
(%i9)   determinant(%);
```

$$(\%o6) \quad \begin{pmatrix} -8 & -16 \\ -16 & 4 \end{pmatrix}$$

$(\%o7)$ -288

The determinant of H is negative, so the third point $(-\frac{1}{3}, -\frac{4}{3})$ is a saddle point.

```
(%i10)   H, criticalpoints[4];
(%i11)   determinant(%);
```

$$(\%o8) \quad \begin{pmatrix} 8 & 16 \\ 16 & -4 \end{pmatrix}$$

$$(\%o9) \quad -288$$

Again, the determinant of H is negative, so the fourth point $(\frac{1}{3}, \frac{4}{3})$ is a saddle point. A plot of the function is consistent with these critical points[9]:

```
(%i12)   plot3d(f(x,y), [x, -1.5, 1.5], [y, -1.5, 1.5],
             [gnuplot_preamble, "set contour base; set cntrparam
         ➥ levels incremental -10,0.5,10; set nokey"]);
```

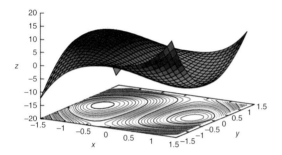

where we have plotted contours in the x, y plane (see Section 3.2.7).

With functions of more than two variables, it's more convenient to examine the eigenvalues of the Hessian matrix to classify critical points, as we'll see in Section 7.2.7.

 Worksheet 6.4: Critical Points

In this worksheet, we'll find and classify critical points for functions of one and two variables.

6.4.2   Numerical Minimization

It isn't always practical to minimize a function analytically. The function may not have a simple closed form. It may be difficult or impossible to compute closed forms for its derivatives.

In cases like these, we'll have to minimize the function numerically. Numerical techniques start with a guess at the coordinates of the minimum, and explore the function around that point. "Steepest descent" methods take a step in the direction of the highest gradient, and take the new coordinates as the guess. The process is repeated until the gradient in the function falls below some preset threshold.

---

9 To actually plot the critical points on the surface in the current version of Maxima, you must use the `draw3d` environment rather than the (much simpler) `plot3d` environment covered in this book.

Maxima includes several functions for numerical minimization. The [Calculus ⟩⟩ Find Minimum...] dialog builds a call to the `lbfgs` function, which uses an efficient steepest descent method by Liu and Nocedal [25]:

> ⚙️ lbfgs(*f*, *x*, *x_0*, *epsilon*, [*iprint*, *verbosity*])
> Minimize *f* for a list of variables *x*, starting from initial estimates *x_0*. The search for a minimum stops when the relative error[10] falls below *epsilon*. A progress message will be printed every *iprint* iterations (with printing suppressed if *iprint* < 0). The *verbosity* flag controls how much information is printed each iteration; it can be 0, 1, 2, or 3, with 3 providing the most information and 0 the least.

For example, let's find the minimum value of $\sin(x)$, using $x = 0$ as an initial guess:

(%i1)   `lbfgs(sin(x), [x], [0], 1e-8, [-1,0]);`

(%o1) $[x = -1.570796326786181]$

This is $-\pi/2$. If we use $x = 6$ as an initial guess, we get a different answer:

(%i2)   `lbfgs(sin(x), [x], [6], 1e-8, [-1,0]);`

(%o2) $[x = 4.712388980385152]$

...which is $3\pi/2$. Figure 6.4 shows why different initial guesses can find different minima. Steepest descent methods find the first minimum that is downhill from the initial guess. They will not find the deepest minimum (the "global" minimum) if they find a shallower local minimum first.

Multivariate functions can be minimized as well. For example, let's use `lbfgs` to minimize the function we examined in the previous section, which had a minimum at [1,0]. Select [Calculus ⟩⟩ Find Minimum...] and fill in the dialog:

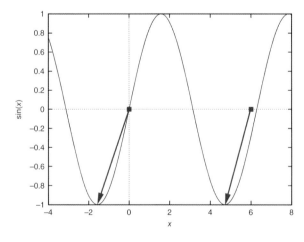

Figure 6.4 Steepest descent methods find the first minimum that is "downhill" from an initial guess. For $\sin(x)$, an initial guess of $x = 0$ finds the minimum at $-\pi/2$; an initial guess of $x = 6$ finds the minimum at $3\pi/2$.

---

10 The relative error is estimated as the length of the gradient vector of f relative to the length of vector x.

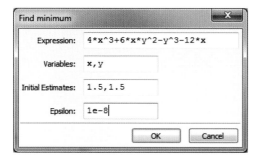

The result is

```
(%i1)   lbfgs(4*x^3 + 6*x*y^2 - y^3 -12*x, [x, y], [1.5,1.5], 1e
    ➡ -8, [-1,0]);
```

(%o1) $[x = 0.99999999999361, y = 1.2703729650836928 \ 10^{-10}]$

The true minimum is at $x = 1, y = 0$.

Study the graph of this function in the previous section to see why starting at $[-1, -1]$ fails to converge to the minimum:

```
(%i2)   lbfgs(4*x^3 + 6*x*y^2 - y^3 -12*x, [x, y], [-1,-1], 1e
    ➡ -8, [-1,0]);
```

```
IFLAG= -1
LINE SEARCH FAILED. SEE DOCUMENTATION OF ROUTINE MCSRCH
ERROR RETURN OF LINE SEARCH: INFO=  3
POSSIBLE CAUSES: FUNCTION OR GRADIENT ARE INCORRECT
OR INCORRECT TOLERANCES
```

There is nothing wrong with the function or gradient; the problem is that the initial guess is bad. There is no minimum downhill from the point $[-1, 1]$.

The `lbfgs` function does an *unconstrained* minimization; it can find any combination of x and y values. Suppose we wanted to minimize this same expression but also require that $x^2 + y^2 = 1$. This minimization is said to be *constrained*.

The augmented Lagrangian method is an efficient way to perform constrained minimization [26]. Each constraint c_i is written as an expression that should be zero at the constrained minimum (for example, the constraint $x^2 + y^2 = 1$ is written as $x^2 + y^2 - 1$). Instead of minimizing only the function $f(x, y)$, we instead minimize an *augmented Lagrangian function* which is $f(x, y)$ plus a quadratic in each constraint c_i. A parameter λ is iteratively adjusted during the calculation. Its final value is a measure of the gradient at the constrained minimum; if the constrained minimum is a true minimum of the function, λ will be close to zero.

Maxima's function for constrained minimization by this method is

`augmented_lagrangian_method(`*f*`,` *x*`,` *constraints*`,` *x_0*`,` *options*`)`

Minimize *f* for a list of variables *x*, starting from initial estimates *x_0*, with the coordinates constrained by the expressions listed in *constraints*. Each *constraint* is held equal to zero. The options can be `lbfgs_tolerance=`*epsilon* and `iprint=[`*iprint*`,` *verbosity*`]`, where *epsilon* and *iprint* arguments have the same meaning as they have in `lbfgs`. You must `load(augmented_lagrangian)` before calling this function.

```
(%i3)    load(augmented_lagrangian)$
(%i4)    augmented_lagrangian_method(4*x^3+6*x*y^2-y^3-12*x, [x,y
    ➥ ], [x^2+y^2-1], [1.5,1.5], lbfgs_tolerance=1e-8,
    ➥ iprint=[-1,0]);
```

(%o4) $[[x = 0.99999999936577, y = 2.906492982835581\ 10^{-13}],$
$\lambda = [7.611271168883604\ 10^{-9}]]$

Once again we obtain a result that is essentially the true minimum at [1,0], which happens to satisfy the constraint $x^2 + y^2 - 1$. The value of λ is essentially zero, indicating that the constrained and unconstrained minima are the same.

If the constraint is $x^2 + y^2 = \frac{1}{2}$, we have a different result:

```
(%i5)    augmented_lagrangian_method(4*x^3+6*x*y^2-y^3-12*x, [x,y
    ➥ ], [x^2+y^2-0.5], [1.5,1.5], lbfgs_tolerance=1e-8,
    ➥ iprint=[-1,0]);
```

(%o5) $[[x = 0.70710690439259, y = 1.8511701940154863\ 10^{-17}],$
$\lambda = [4.242638469566978]]$

The point $[1/\sqrt{2}, 0]$ gives the coordinates of the lowest value of the function that obeys the constraint $x^2 + y^2 = \frac{1}{2}$. Notice that the value of λ is much larger than the previous calculation, indicating that the constrained and unconstrained minima are *not* the same.

It's easier to see how constrained minimization works with a function of only one variable. Suppose we want to find the minimum value of $\sin(x)$ subject to the constraint $(x-1)(x-2)(x-3) = 0$. This is the same as finding the value of x that minimizes $\sin(x)$ where x must be equal to 1, 2, or 3:

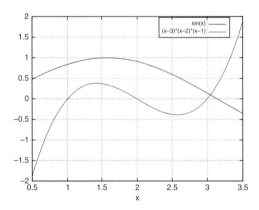

Clearly $\sin(x)$ is lower at $x = 3$ than at $x = 1$ or $x = 2$. Let's see if the augmented Lagrangian method correctly chooses $x = 3$:

```
(%i1)    augmented_lagrangian_method(sin(x), [x], [(x-1)*(x-2)*(x
    ➥ -3)], [0], iprint=[-1,0]);
```

(%o1) $[[x = 3.00000000000019], \lambda = [0.49499624850125]]$

The value of λ is not zero, indicating that this constrained minimum is not the same as the unconstrained minimum (which would be at $-\pi/2$, from our starting guess of $x = 0$).

 Worksheet 6.5: Numerical Minimization

In this worksheet, we'll minimize functions using the steepest descent and augmented Lagrange methods. We'll also use numerical minimization to find optimal forms of trial wavefunctions using the variation theorem.

6.5 Integration

Integration is the inverse of differentiation. If we have $f(x) = \frac{dF(x)}{dx}$, we can find $F(x)$ from $f(x)$ by writing

$$F(x) = \int f(x)dx + c \qquad \text{An indefinite integral} \qquad (6.15)$$

The function $f(x)$ is called the integrand and x is called the *integration variable*. The constant c is called the integration constant. It appears because the derivative of a constant is zero, so $\frac{d}{dx}(F(x) + c) = f(x)$ for any constant c. The integral can be computed symbolically with the integrate function (introduced in Section 1.2.3):

    integrate(*expression*, x)
    Compute the indefinite integral of *expression* with respect to x.

For example,

```
(%i1)    'integrate(diff(f(x),x),x) = integrate(diff(f(x),x),x);
(%i2)    'diff(integrate(f(x),x), x) = diff(integrate(f(x),x), x)
          ➡ ;
```

$$(\%o1) \qquad \int \frac{d}{dx} f(x)\,dx = f(x)$$

$$(\%o2) \qquad \frac{d}{dx} \int f(x)\,dx = f(x)$$

The single-quote operator ' displays the left-hand sides of the equations as nouns (Section 4.1).

Let's compute the derivative and integral of $f(x) = \sin(ax)$ to show why integrals are sometimes called "antiderivatives."

```
(%i3)    'diff(sin(a*x),x) = diff(sin(a*x), x);
(%i4)    'integrate(rhs(%),x) = integrate(rhs(%),x);
```

$$(\%o3) \qquad \frac{d}{dx} \sin(ax) = a\cos(ax)$$

$$(\%o4) \qquad a \int \cos(ax)\,dx = \sin(ax)$$

Note that $\sin(ax)$ is the integral of $a\cos(ax)$ because $a\cos(ax)$ is the derivative of $\sin(ax)$. Integration undoes differentiation.

6.5.1 Integration Constants

Integration constants aren't shown when integrating an expression. If you want it, you'll have to add it yourself. Maxima does insert an integration constant when integrating both sides of an equation. For example,

```
(%i1)  diff(A(t),t) = -k;
(%i2)  integrate(%,t);
```

$$(\%o1) \quad \frac{d}{dt}A(t) = -k$$

$$(\%o2) \quad A(t) = \%c1 - kt$$

Both sides of the equation are integrated with respect to t, and the integration constant appears as `%c1`.

6.5.2 Definite Integration

To find the value of an integration constant, we have to specify integration bounds a and b to compute a *definite integral*:

$$F(x) = \int_a^b f(x)dx \qquad \text{A definite integral} \tag{6.16}$$

The definite integral corresponds to the area under a plot of $f(x)$ from $x = a$ to $x = b$ (see Figure 6.5).

The integration bounds can be added as two additional arguments to `integrate`:

`integrate(expression, x, a, b)`
Compute the definite integral of *expression* with respect to x from $x = a$ to $x = b$.

For example, integrating from 0 to t, we have

```
(%i1)  diff(A(t),t) = -k;
(%i2)  integrate(%,t,0,t);
```

Is it positive, negative, or zero?`positive;`
   $(\%o1) \quad A(t) - A(0) = -kt$

Maxima sometimes needs to ask questions about variables in the integrand. Hit ⇧ + Enter after each answer.

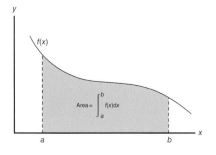

Figure 6.5 The definite integral corresponds to the area under the integrand curve.

```
(%i1)    integrate(a/x+x^c,x);
```

Is c equal to -1? no;

$$(\%o1) \quad a\log(x)+\frac{x^{c+1}}{c+1}$$

The `integrate` function consults the current context (see Section 4.5). If it finds what it needs, it won't ask questions. It does not enter the answers to these questions into the context, so if you want to avoid answering the questions again and again, provide the answers with the `assume` function before computing the integral:

```
(%i2)    assume(c>0);
(%i3)    integrate(a/x+x^c,x);
```

$$(\%o2) \quad [c>0]$$

$$(\%o3) \quad a\log(x)+\frac{x^{c+1}}{c+1}$$

Unfortunately, the `integrate` function does *not* consult the dependencies established by the `depends` function that we used in computing partial derivatives:

```
(%i1)    depends(f,x)$
(%i2)    depends(g,x)$
(%i3)    integrate(f*g, x);
```

$$(\%o3) \quad f\,g\,x$$

The functions f and g were incorrectly treated as constants. You will have to show that f depends on x by typing `f(x)` in the integrand.

```
(%i1)    integrate(f(x)*g(x),x);
```

$$(\%o1) \quad \int f(x)\,g(x)\,dx$$

6.5.3   When Symbolic Integration Fails

Integrals cannot always be computed symbolically. When Maxima doesn't know how to integrate a function, it will simply echo back the integral as a noun:

```
(%i1)    integrate(f(x),x,a,b);
```

$$(\%o1) \quad \int_a^b f(x)\,dx$$

Even for some apparently simple functions, it may not be possible to write the integral as a finite combination of constants, polynomials, exponentials, logarithms, or trigonometric functions. For example, consider the indefinite integral of e^{-x^2}:

```
(%i1)    integrate(exp(-x^2),x);
```

$$(\%o1) \quad \frac{\sqrt{\pi}\,\mathrm{erf}(x)}{2}$$

The result is expressed in terms of the error function, `erf(x)`, which is defined as

$$\text{erf}(x) = \frac{2}{\sqrt{\pi}} \int_0^x e^{-t^2} dt \qquad \text{The error function} \qquad (6.17)$$

This integral has no finite solution (although it can be expressed as an infinite sum). However, `integrate` has no problem with certain definite integrals of e^{-x^2}:

(%i1) **`integrate(exp(-x^2),x,0,inf);`**

(%o1) $\dfrac{\sqrt{\pi}}{2}$

You'll see many other special functions when `integrate` can't write results in a simple form. For example, an integral that is encountered when computing total energy density emitted by a black body is

$$\int_0^\infty \frac{x^3}{e^x - 1} dx = \frac{\pi^4}{15} \qquad (6.18)$$

Symbolic integration with `integrate` gives an ugly result:

(%i1) **`integrate(x^3/(exp(x)-1),x,0,inf);`**

(%o1) $\displaystyle\lim_{x\to\infty-} 6\, li_4\left(e^x\right) - 6\, x\, li_3\left(e^x\right) + 3\, x^2\, li_2\left(e^x\right) + x^3 \log\left(1 - e^x\right) - \frac{x^4}{4} - \frac{\pi^4}{15}$

The `li[s](z)` function represents a *polylogarithm* of order s, defined by the infinite series

$$Li_s(z) = \sum_{k=1}^{\infty} \frac{z^k}{k^s} \qquad \text{A polylogarithm} \qquad (6.19)$$

Definite integrals that can't be computed symbolically can often be computed numerically, as we'll see in the next section. In some cases, though, we can solve a difficult integral with a simple change of integration variables. The `changevar` function we used to change summation indices in Chapter 5 can also change integration variables:

`changevar('integral, y = f(x), y, x)`
Substitute y for x in the noun form of an integral (`'integral`), where y is defined by $y = f(x)$.

Suppose we want to simplify the form of the integral

$$\int x e^{-x^2} dx$$

by making the substitution $y = x^2$.

(%i1) **`'integrate(x*exp(-x^2),x);`**
(%i2) **`changevar(%,y=x^2, y, x);`**

(%o1) $\displaystyle\int x\, e^{-x^2}\, dx$

(%o2) $\dfrac{\int e^{-y} dy}{2}$

The integral in Equation (6.18) could not be computed directly with `integrate`, but it *can* be solved with a simple variable change $y = e^{-x}$:

```
(%i1)   'integrate(x 3/(exp(x)-1), x, 0, inf);
(%i2)   changevar(%, y=exp(-x), y, x);
```

$$(\%o1) \quad \int_0^\infty \frac{x^3}{e^x - 1} dx$$

$$(\%o2) \quad \int_0^1 \frac{\log(y)^3}{y-1} dy$$

This substitution made the range of integration finite – and `integrate` has no trouble solving it now:

```
(%i3)   %, nouns;
```

$$(\%o2) \quad \frac{\pi^4}{15}$$

Here we've evaluated the noun form of the integral using the *nouns* switch.

When will a change in variable solve an "unsolvable" integral? How do we choose the substitution? We'll have to recognize the following pattern in the integrand:

$$\int_a^b f(x)dx = \int g(y)\frac{dy}{dx}dx = \int_{y(a)}^{y(b)} g(y)dy \qquad \text{The substitution rule} \qquad (6.20)$$

In the integral $\int_0^\infty xe^{-x^2}dx$, notice that the factor of x is half the derivative of x^2, which appears in the exponential. That suggests the substitution $y = x^2$.

It's harder to see the pattern in the integral $\int_0^\infty \frac{x^3}{e^x-1}dx$ unless we multiply the integrand by e^{-x}/e^{-x}:

$$\int_0^\infty \frac{x^3 e^{-x}}{1 - e^{-x}}dx$$

The factor of e^{-x} is minus the derivative of e^{-x}, which suggests the substitution $y = e^{-x}$.

The `integrate` function sometimes has trouble with integrands that contain absolute value functions. For example, in Chapter 12 we'll compute the Fourier transform of a function $f(t)$ as follows:

```
(%i1)   f(t) := (cos(5*t)+cos(10*t))*exp(-abs(t));
(%i2)   (1/(2*%pi))*(integrate(f(t)*exp(%i*w*t), t, minf, inf) )
        ➡ ;
```

$$(\%o1) \quad f(t) := (\cos(5\,t) + \cos(10\,t)) \exp(-|t|)$$

$$(\%o2) \quad \frac{\int_{-\infty}^\infty (\cos(10\,t) + \cos(5\,t))\, e^{i\,t\,w - |t|} dt}{2\,\pi}$$

The `integrate` function fails. One way to help it is to break up the integration interval:

```
(%i3)   (1/(2*%pi))*(integrate(f(t)*exp(%i*w*t), t, minf, 0) +
        ➡ integrate(f(t)*exp(%i*w*t), t, 0, inf)), ratsimp;
```

(%o3) $$\frac{2\,w^6 - 119\,w^4 + 881\,w^2 + 333502}{\pi\,w^8 - 246\,\pi\,w^6 + 20381\,\pi\,w^4 - 623496\,\pi\,w^2 + 6895876\,\pi}$$

A better way is to load the `abs_integrate` package, which extends `integrate` for integrands that contain absolute values:

```
(%i4)   load(abs_integrate)$
(%i5)   (1/(2*%pi))*integrate(f(t)*exp(%i*w*t), t, minf, inf),
    ➡ ratsimp;
```

(%o5) $$\frac{2\,w^6 - 119\,w^4 + 881\,w^2 + 333502}{\pi\,w^8 - 246\,\pi\,w^6 + 20381\,\pi\,w^4 - 623496\,\pi\,w^2 + 6895876\,\pi}$$

The Fourier integral is now evaluated correctly.

 Worksheet 6.6: Symbolic Integration

In this worksheet, we'll use `integrate` to compute both indefinite and definite integrals, and see what steps can be taken when symbolic integration fails.

6.5.4 Numerical Integration

A definite integral can be approximated as a sum

$$\int_a^b f(x)\mathrm{d}x \approx \sum_{i=1}^{N} w_i f(x_i) \qquad \text{Approximating a definite integral numerically} \qquad (6.21)$$

where the points x_i and the weights w_i are carefully chosen so that the sum and the integral are as close as possible [27]. This method is called *numerical quadrature*. Maxima's `quadpack` package includes specialized functions for numerical integration that use this approach [28].

Four general purpose functions are listed in Table 6.1. Other routines are available for integrating weighted integrands and Fourier transforms; see the Maxima manual for `quadpack` for more.

Table 6.1 General-purpose `quadpack` functions for numerical integration of a function *f* of *x* over an interval [*a*, *b*].

quadpack function	Choose this function for:
`quad_qags(f, x, a, b)`	Finite *a* and *b*, general-purpose integration, sometimes handles singularities
`quad_qagi(f, x, a, b)`	Integration interval has infinite *a* and/or *b*
`quad_qagp(f, x, a, b, points)`	Finite *a* and *b*, higher accuracy for functions with specific discontinuities or singularities in the integrand are listed in *points*
`quad_qag(f, x, a, b, key)`	Finite *a* and *b*, higher accuracy for smooth functions with strongly oscillating integrands (set the extra "key" argument to 6)

Table 6.2 Error codes returned by `quadpack` functions.

0	No problems.
1	Too many subintervals were done.
2	Excessive round-off error was detected.
3	Extremely bad integrand behavior was detected.
4	The calculation failed to converge.
5	The integral is probably divergent or very slowly convergent.
6	The input was invalid.

Each `quadpack` function returns a list with four elements:

1) The estimated value of the integral,
2) the estimated absolute error in the integral,
3) the number of integrand evaluations, and
4) an error code (see Table 6.2).

6.5.4.1 Numerical Integration over Infinite Intervals

Let's apply `quadpack` to the black body integral we considered earlier:

$$\int_0^\infty \frac{x^3}{e^x - 1} dx$$

We are integrating over an infinite interval, so we must choose `quad_qagi`:

```
(%i1)    quad_qagi(x^3/(exp(x)-1), x, 0, inf);
```

```
(%o1)    [6.493939402266829, 2.628471501138993 10^{-9}, 165, 0]
```

The true value is $\pi^4/15$, so in fact the absolute error is

```
(%i2)    %[1] - %pi^4/15, numer;
```

```
(%o2)    1.7763568394002505 10^{-15}
```

The estimated absolute error ($2.628471501138993 \times 10^{-9}$) is very conservative.

Now consider the calculation of the second virial coefficient for a gas, B, from its pair potential E_p [29]:

$$B = -2\pi N_A \int_0^\infty \left(e^{-E_p/kT} - 1\right) r^2 dr \qquad \text{Second virial coefficient} \qquad (6.22)$$

where N_A is Avogadro's number and k is the Boltzmann constant. The pair potential E_p for noble gas atoms can be approximated by the Lennard-Jones potential

$$E_p \approx 4\epsilon \left[\left(\frac{r_0}{r}\right)^{12} - \left(\frac{r_0}{r}\right)^6\right] \qquad \text{Lennard-Jones potential} \qquad (6.23)$$

For argon, $\epsilon = 111.84k$ and $r_0 = 362.3$ pm.

```
(%i1)    NA : 6.022e23$
(%i2)    k: 1.3806e-23$
```

```
(%i3)   epsilon: 111.84*k$
(%i4)   r0: 362.3$
(%i5)   T: 298$
(%i6)   Ep: 4*epsilon*((r0/r)^12 - (r0/r)^6)$
(%i7)   quad_qagi(-2*%pi*NA*(1e6/1e36)*(exp(-Ep/(k*T))-1)*r^2, r
      ➥ , 0, inf);
```

(%o1) $[-14.29091302667043, 9.761845700885778 * 10^-8, 435, 0]$

The computed value $B = -14.29$ cm^3/mol^{-1} is close to the experimental value of -15.5 cm^3/mol^{-1} [30].

6.5.4.2   Numerical Integration with Strongly Oscillating Integrands
The following integral can't be integrated by `integrate`:

$$\int_{0.0001}^{1} \cos\left(\frac{1}{x}\right) dx$$

Let's see what the integrand looks like over the integration range:

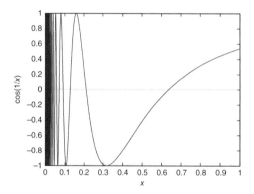

We can try either `quad_qags` or `quad_qag`.

```
(%i1)   quad_qags(cos(1/x),x,1/10000,1);
```

(%o1) $[-0.084481251419635, 6.935920637188286 \ 10^{-5}, 8379, 1]$

The `quad_qags` function gives us an answer that is good to only four significant figures, and it complains that too many subintervals were done. Let's try `quad_qag`:

```
(%i2)   quad_qag(cos(1/x),x,1/10000,1,6);
```

(%o2) $[-0.084410953613813, 6.825603381688606 \ 10^{-10}, 20801, 0]$

The final argument of `quad_qag` is called "key," and it can be set to an integer from 1 (a smooth function) to 6 (a wildly oscillating function). Here we've set it to 6, which gives an answer that is good to about 9 significant figures.

6.5.4.3 Numerical Integration with Discontinuous Integrands

The following integrand is discontinous; it has cusps at $x = \sqrt{2}$ and $x = 3/2$:

$$\int_1^2 \ln\left(|2x - 3|\,|x^2 - 2|\right) dx \qquad \text{Integrand with a discontinuity} \qquad (6.24)$$

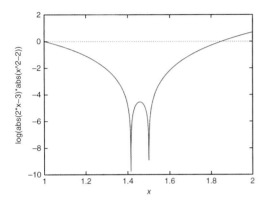

We can try `quad_qags` or `quad_qagp`; both can handle singularities and discontinuities on a finite interval. With `quad_qagp`, we need to supply a list of x values where the integrand is singular or discontinuous; this extra information makes it more likely that we'll find an accurate solution.

```
(%i1)  integrate(-log(abs((x^2-2)*(2*x-3))), x, 1, 2);
(%i2)  quad_qagp(-log(abs((x^2-2)*(2*x-3))), x, 1, 2, [sqrt(2)
    ➥ ,3/2]);
(%i3)  quad_qags(-log(abs((x^2-2)*(2*x-3))), x, 1, 2);
```

$$(\%o1) \quad -\int_1^2 \log\left(|2x - 3|\,|x^2 - 2|\right) dx$$

$(\%o2) \quad [-1.61370563888011, 6.8833827526759706\ 10^{-15}, 861, 0]$

$(\%o3)$

quad_qags: Cannot numerically evaluate $-\log\left(|2x - 3|\,|x^2 - 2|\right)$ at 1.5
- an error. To debug this try: debugmode(true);

Both `integrate` and `quad_qags` fail, but `quad_qagp` gives an extremely accurate result.

Several other `quadpack` functions are available for integrating specialized integrands. We'll see a few of them when we compute Fourier transforms and Fourier series coefficients in Chapter 12.

 Worksheet 6.7: Numerical integration

In this worksheet, we'll numerically integrate functions with infinite intervals, discontinuities, and oscillatory integrands.

6.5.5 Multiple Integration

Just as the integral of $f(x)$ is the area under a curve, the integral of $f(x, y)$ is the volume under a surface. Functions like $f(x, y, z)$ can be seen as analogous to density f at a point $[x, y, z]$; integrating the function over x, y, and z gives the total "mass" for the volume of integration.

Functions of many variables can be integrated with respect to each variable in turn, while holding the others constant. By convention the integrations are nested like

$$\int_{z_0}^{z_1} \int_{y_0}^{y_1} \int_{x_0}^{x_1} f(x, y, z) \mathrm{d}x \, \mathrm{d}y \, \mathrm{d}z =$$
$$\int_{z_0}^{z_1} \left(\int_{y_0}^{y_1} \left(\int_{x_0}^{x_1} f(x, y, z) \mathrm{d}x \right) \mathrm{d}y \right) \mathrm{d}z \qquad \begin{array}{l} \text{Execution order} \\ \text{in a multiple integral} \end{array} \qquad (6.25)$$

Multiple integrals are computed in Maxima by nesting calls to integration functions. For example, suppose we want to integrate the function $f(r, \theta, \phi)$ over all space. The triple integral is

$$\int_0^{2\pi} \mathrm{d}\phi \int_0^{\pi} \sin\theta \int_0^{\infty} f(r, \theta, \phi) r^2 \mathrm{d}r \qquad \begin{array}{l} \text{Integration over} \\ \text{all space} \\ \text{in spherical coordinates} \end{array} \qquad (6.26)$$

and it would be evaluated by

```
(%i1)   integrate(
(%i2)       integrate(
(%i3)           integrate(f(r,theta,phi)*r^2*sin(theta),
(%i4)               r, 0, inf),
(%i5)           theta, 0, %pi),
(%i6)       phi, 0, 2*%pi);
```

$$(\%o1) \qquad \int_0^{2 \cdot \pi} \int_0^{\pi} \int_0^{\infty} r^2 \mathrm{f}(r, \theta, \phi) \, dr \, \sin(\theta) \, d\theta \, d\phi$$

Numerical multiple integration can also be accomplished by nesting calls. For example, to compute $\int_{-\infty}^{\infty} \int_{-\infty}^{\infty} e^{-x^2 - y^2} \mathrm{d}x \mathrm{d}y = \pi$,

```
(%i1)   quad_qagi(integrate(exp(-(x^2+y^2)), x, minf, inf), y,
        ➥ minf, inf);
```

$$(\%o1) \qquad [3.141592653589793, 2.5173518295352398 \ 10^{-8}, 270, 0]$$

Nesting two `quadpack` calls only works if the bounds of the inner integral explicitly depend on the outer integration variable. For example, the integral $\int_{-\infty}^{\infty} \int_{-\infty}^{y} e^{-x^2 - y^2} \mathrm{d}x \mathrm{d}y = \dfrac{\pi}{2}$ is computed by

```
(%i1)   quad_qagi(quad_qagi(exp(-(x^2+y^2)), x, minf, y)[1], y,
        ➥ minf, inf);
```

$$(\%o1) \qquad [1.570796326794898, 1.2586758089540625 \ 10^{-8}, 270, 0]$$

Note that we had to index the value of the inner call (with Ref. [1]). Note also that the reported error (1.26×10^{-8}) is almost certainly an underestimate in nested calls; it is the error in the outer integration alone.

 Worksheet 6.8: Multiple Integration

In this worksheet, we'll use multiple integration to handle chemically significant quantities like partition functions and electron densities.

6.5.6 Discrete Integration

We often have a set of discrete values of the integrand rather than an expression or function. In this case we can fit or interpolate a continuous function between the data points (Section 5.5) and numerically integrate the function.

Suppose we have a list of $[x, y]$ data points:

```
(%i1)   datapoints: [[0,0], [1,1], [2,4], [3,9], [4,16]]$
```

We would like to find the area under the function that connects the points. The simplest such function is just a series of lines connecting adjacent points. We can use the linear interpolation function builder `linearinterpol` from the `interpol` package (see Section 5.5 and Worksheet 5.7):

```
(%i2)   load(interpol)$
(%i3)   f(x) := ''linearinterpol(datapoints)$
(%i4)   wxplot2d([f(x), [discrete, datapoints]], [x, 0, 4], [
        ➥ style, lines, points])$
```

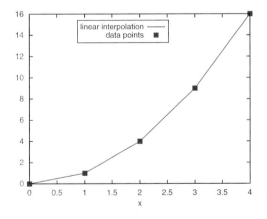

Integrating the function sums the areas of the trapezoids under each line segment. We will use the general-purpose `quad_qags` function for numerical integration (see Section 6.5.4 and Worksheet 6.8):

```
(%i5)   quad_qags(f(x), x, 0, 4);
```

$$(\%o5) \quad [22.0, 2.444711100224595 \cdot 10^{-13}, 147, 0]$$

The data points were exact samples from the function $y = x^2$, so the "true" value of the integral is

```
(%i6)  integrate(x^2, x, 0, 4), numer;
```

$(\%o6)$ 21.33333333333333

Let's use this approach to to estimate the third-law entropy S_T° of solid copper, using a series of heat capacities C_p measured at different temperatures T.

Third-law entropies of a material at temperature T are computed by integrating $(\partial S/\partial T)_P$ from absolute zero to T, taking the molar entropy at absolute zero as zero:

$$S_T^\circ = \int_0^T \frac{C_p(T)}{T} \mathrm{d}T \qquad \text{Third-law entropy of a solid} \qquad (6.27)$$

The C_p vs. T data is read from a spreadsheet stored as a comma-separated value file, with the temperatures ranging from 10 to 300 K, and the heat capacities in $\mathrm{J\,mol^{-1}\,K^{-1}}$ [31]. We have no data for temperatures close to absolute zero, but we can estimate heat capacities at low temperature analytically using the *Debye–Sommerfield equation*:

$$C_p = \gamma T + \alpha T^3 \qquad \text{Debye–Sommerfield equation for } C_p \text{ at low temperature} \qquad (6.28)$$

where $\gamma = 10.81 \times 10^{-6}\ \mathrm{J\,g^{-1}\,K^{-2}}$ and $\alpha = 0.746 \times 10^{-6}\ \mathrm{J\,g^{-1}\,K^{-4}}$ for Cu(s) [31].

First, let's set up a function for C_p that follows Equation (6.28) at temperatures below 10 K, and uses a piecewise linear interpolation of the data otherwise:

```
(%i1)  data: read_matrix("k:/work/SMC-Maxima/Calculus/Cp-copper
        ➥ .csv")$
(%i2)  load(interpol)$
(%i3)  molar_mass : 63.546$
(%i4)  gamma : 10.81e-6 * molar_mass$
(%i5)  alpha : 0.746e-6 * molar_mass$
(%i6)  Cp(x) := if  x < 10 then (gamma*x + alpha*x^3) else ''(
        ➥ linearinterpol(data))$
(%i7)  plot2d(Cp(x), [x,0,300], [xlabel, "T/K"], [ylabel, "Cp /
        ➥ J mol^-1 K^-1"], [title, "Constant pressure heat
        ➥ capacity of Cu(s) at 1 bar"])$
```

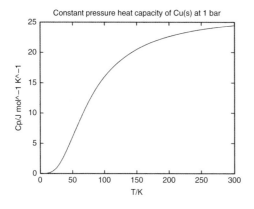

Now apply Equation (6.27):

```
(%i8)   quad_qags(Cp(T)/T, T, 0, 298.15);
```

$$(\%o8) \quad [33.16613955718072, 2.1136825264052818 \, 10^{-4}, 4095, 2]$$

The value of $S^\circ_{298.15 \, K}$ is close to the literature value (33.15 J mol^{-1} K^{-1}), but the `quad_qags` routine is returning an error (the code 2 means "excessive roundoff error detected"), and the estimated absolute error is high. Let's try the integration again with a cubic spline replacing the piecewise linear interpolation:

```
(%i9)   Cp(x) := if  x < 10 then (gamma*x + alpha*x^3) else ''(
        ➥ cspline(data))$
(%i10)  quad_qags(Cp(T)/T, T, 0, 298.15);
```

$$(\%o10) \quad [33.1512257825734, 1.486940135464465 \, 10^{-7}, 1743, 0]$$

This time $S^\circ_{298.15 \, K}$ is essentially identical to the literature value, with a much smaller estimated absolute error and no error message.

Let's plot S°_T as a function of T. Because an integral is being computed for every point on the plot, and because the interpolating function is expensive to evaluate, the graph may take several minutes to build on slower computers:

```
(%i11)  S(x) := quad_qags(Cp(T)/T, T, 0, x)[1]$
(%i12)  datapoints: makelist([x,S(x)], x, 1, 300, 20);

(%i14)  wxplot2d([discrete, datapoints], [xlabel, "T / K"], [
        ➥ ylabel, "S / (J / mol K)"],
(%i15)  [title, "Third-law molar entropy of Cu(s) at 1 bar"]
(%i16)  )$
```

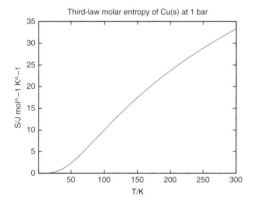

(M) **Worksheet 6.9: Discrete Integration**

In this worksheet, we'll integrate interpolated functions to compute entropies, enthalpies, and fugacity coefficients for materials from from experimental data. We'll also consider a few ways to improve the accuracy of the interpolated integrand.

6.6 Power Series

In Section 5.1, we saw several tools for computing and manipulating series in Maxima. Some of the series we examined were equal to simple expressions. For example,

$$\ln(1+x) = x - \frac{x^2}{2} + \frac{x^3}{3} - \frac{x^4}{4} + \cdots = -\sum_{i=1}^{\infty} \frac{(-1)^i x^i}{i} \qquad \text{Power series expansion of } \ln(1+x)$$

(6.29)

This power series expansion is equal to the function $\ln(1+x)$ when $x^2 < 1$.

The `powerseries` function can expand an expression around a single point. It can be automatically inserted by selecting Calculus ⟩ Get Series... or the Series... button on the *General Math* panel.

> `powerseries(expression, x, c)`
> Computes the general form of a power series expansion for `expression` in the variable x around the point $x = c$.

For example, we can derive Equation (6.29) with

```
(%i1)  powerseries(log(1+x), x, 0);
(%i2)  niceindices(%);
```

$$(\%o1) \quad -\sum_{i1=1}^{\infty} \frac{(-1)^{i1} x^{i1}}{i1}$$

$$(\%o2) \quad -\sum_{i=1}^{\infty} \frac{(-1)^{i} x^{i}}{i}$$

where we have used the `niceindices` function (Section 5.1.2) to use indices like i, j, and k rather than $i1$, $i2$, and $i3$.

6.6.1 Testing Power Series for Convergence

We often want to know if a series is convergent or divergent. One simple test for convergence is the Cauchy ratio test, which computes the ratio $|u_{n+1}/u_n|$ for a series $\sum_i u_i$. For a convergent series, the ratio becomes less than 1 as n approaches infinity. If the limit of the ratio is greater than one, the series is divergent.[11] For example, for the series $1 + x/1! + x^2/2! + x^3/3! \ldots$, we have

```
(%i1)  u[n] := x^n/n!;
(%i2)  limit(abs(u[n+1]/u[n]), n, inf);
```

$$(\%o1) \quad u_n := \frac{x^n}{n!}$$

$$(\%o2) \quad 0$$

where we have defined the terms in the series with a subscripted function. The ratio is less than one, so the series is convergent. On the other hand, for the series $x - x^2/2^2 + x^3/3^2 - x^4/4^2 + \cdots$,

---

11 If the ratio is equal to one, the test fails; we can't tell if the series is convergent or divergent.

```
(%i1)   u[n] := (-1)^(n-1)*x^n/n^2;
(%i2)   limit(abs(u[n+1]/u[n]), n, inf);
```

$$(\%o1) \quad u_n := \frac{(-1)^{n-1} \, x^n}{n^2}$$

$$(\%o2) \quad |x|$$

the limit of the ratio depends on the value of x. The series converges when $|x| < 1$, that is, when $-1 < x < 1$, and diverges when x is outside that interval.

 Worksheet 6.10: Power series

In this worksheet, we'll derive power series expansions for functions, manipulate them, and test them for convergence.

6.7   Taylor Series

The `powerseries` function will sometimes be unable to compute an exact infinite expansion for an expression. An alternative approach is to compute a Taylor series expansion to approximate the expression around some single point c. Using Taylor series, complicated functions are easily written as simple finite closed form expressions. This makes Taylor series indispensable in computationally intensive applications like molecular geometry optimization and in understanding how a function behaves at extreme values of its variables.

The Taylor series is an infinite power series expansion that has the following form:

$$f(x) = \sum_{i=0}^{\infty} \frac{(x-c)^i}{i!} \left(\frac{\mathrm{d}^i f(x)}{\mathrm{d}x^i} \right)_{x=c} \qquad \begin{array}{l} \text{Taylor series expansion of} \\ f(x) \text{ around } x = c \end{array} \qquad (6.30)$$

In Maxima, we can compute a Taylor series truncated after the nth term with the command

 `taylor(expression, x, c, n)`
Computes a Taylor series expansion for *expression* in the variable x around the point $x = c$, out through the term proportional to $(x-c)^n$.

For example, the Taylor series expansion for $\ln(1 + x)$ expanded around $x = 0$ and truncated at four terms is

```
(%i1)   tseries: taylor(log(1+x), x, 0, 4);
```

$$(\%o1) \quad /T/ \quad x - \frac{x^2}{2} + \frac{x^3}{3} - \frac{x^4}{4} + \ldots$$

The trailing dots indicate that the series is truncated. The missing terms will cause the Taylor series to be less accurate when evaluated at points that are far from the expansion point:

```
(%i2)   wxplot2d([log(1+x), tseries], [x, -0.9, +0.9],
(%i3)       [title, "Taylor series expansion around x=0 for y=
                ➡ log(1+x)"],
(%i4)       [ylabel, "y"],
```

```
(%i5)           [legend, "log(x+1)", "4-term Taylor series"],
(%i6)           [gnuplot_preamble, "set key bottom center"]
(%i7)   )$
```

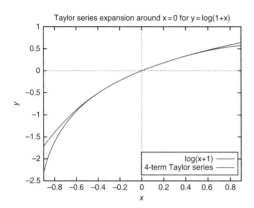

Maxima sometimes marks the output with /T/ to indicate that the truncated series is *not* stored as a general expression.

```
(%i8)   tseries;
```

$$(\%o2) \quad /T/ \quad -\frac{3x^4 - 4x^3 + 6x^2 - 12x}{12}$$

Convert the truncated Taylor series into a rational expression before using it in other expressions. This can be done with `ratexpand` or `ratsimp`:

```
(%i9)   ratexpand(tseries);
```

$$(\%o2) \quad -\frac{x^4}{4} + \frac{x^3}{3} - \frac{x^2}{2} + x$$

6.7.1 Exploring Function Properties with Taylor Series

Taylor series are useful for finding approximate forms of a function and relating them to more exact forms. For example, classical physics predicts the spectrum of a black body at low frequencies v, but completely fails at high frequencies. Quantum mechanics correctly predicts the spectrum at all frequencies. Taking the low-frequency limit of the more general quantum mechanical distribution should yield the classical distribution.

Let's look at a Taylor series expansion of the quantum mechanical distribution around zero frequency:

```
(%i1)   QM_distribution: (8*%pi*nu^3*h)/(c^3*(%e^((nu*h)/(k*T))
        ➥ -1));
(%i2)   classical_distribution: (8*k*T*%pi*nu^2)/c^3;
(%i3)   tseries: taylor(QM_distribution, nu, 0, 8);
```

$$(\%o1) \quad \frac{8\pi h v^3}{c^3 \left(e^{\frac{hv}{kT}} - 1\right)}$$

(%o2) $\dfrac{8\,\pi\,k\,v^2\,T}{c^3}$

(%o3) /T/ $\dfrac{8\,k\,T\,\pi\,v^2}{c^3} - \dfrac{4\,\pi\,h\,v^3}{c^3} + \dfrac{2\,\pi\,h^2\,v^4}{3\,k\,T\,c^3} - \dfrac{\pi\,h^4\,v^6}{90\,k^3\,T^3\,c^3} + \dfrac{\pi\,h^6\,v^8}{3780\,k^5\,T^5\,c^3} + \cdots$

The first term of the Taylor series *is* the classical distribution. The remaining terms will become much smaller than the first at low frequency, so the quantum mechanical distribution approaches the classical results under those conditions.

Notice that Planck's *h* constant only appears in the second and higher terms of the Taylor series. These terms show the size of the error in the classical law due to quantum mechanical effects.[12]

We can use Maxima's `partition` function[13] to split the Taylor series into classical and quantum mechanical terms.

(%i4)   [classical_term,quantum_terms] : **partition**(tseries,h);

(%o4) $[\dfrac{8\,\pi\,k\,v^2\,T}{c^3}, \dfrac{2\,\pi\,h^2\,v^4}{3\,c^3\,k\,T} - \dfrac{\pi\,h^4\,v^6}{90\,c^3\,k^3\,T^3} + \dfrac{\pi\,h^6\,v^8}{3780\,c^3\,k^5\,T^5} - \dfrac{4\,\pi\,h\,v^3}{c^3}]$

Taylor series are also useful in analyzing errors and small perturbations in experimental parameters. For example, suppose we are studying the vapor pressure *P* of a liquid when it is dispersed as small droplets of radius *r*. The relationship is

$$P = P^{\bullet} \exp\left(\dfrac{2\gamma V}{rRT}\right) \qquad \text{The Kelvin equation} \qquad (6.31)$$

where *V* is the molar volume of the liquid, $P^{\bullet}$ is the vapor pressure over a flat surface, γ is the surface tension, *R* is the ideal gas law constant, and *T* is the temperature in kelvins. We can write a Taylor series for Equation (6.31) to study the effect of a small increase d*r* in droplet radius on the vapor pressure.

(%i1)   **taylor**(Po***exp**(2*%gamma*V/((r+dr)*R*T)), dr, 0, 1);

(%o1)   /T/ $\left(e^{\frac{V\gamma}{rRT}}\right)^2 Po - \dfrac{2\left(e^{\frac{V\gamma}{rRT}}\right)^2 V\,\gamma\,Po\,dr}{r^2\,R\,T} + \cdots$

The second term is the shift in vapor pressure caused by `dr`. It is negative, so an increase in droplet size `dr` decreases the vapor pressure. Simplifying the term makes it easier to interpret, so let's substitute the Kelvin equation into the second-term expression:

(%i2)   **part**(tseries, 2);
(%i3)   **ratsubst**(P, Po***exp**(2*%gamma*V/(r*R*T)), %);

(%o2) $-\dfrac{2\gamma\,Po\,V\,dr\,e^{\frac{2\gamma V}{RTr}}}{RT\,r^2}$

(%o3) $-\dfrac{2\gamma\,PV\,dr}{RT\,r^2}$

---

12 This series is an example of the *Bohr correspondence principle*, which states that quantum mechanics predicts classical results when all terms in *h* vanish, or when the values of quantum numbers like *n* become very large.
13 See Section 5.3.3 for more about this function.

The decrease will be proportional to $1/r^2$. When r is smaller, this expression becomes much larger, so the vapor pressure will drop faster when smaller droplets grow, if everything else is held constant. The vapor pressure will drop more slowly with droplet growth as the temperature increases.

6.7.2 The Remainder Term

The error involved in approximating a function $f(x)$ with a truncated Taylor series is called the remainder term:

$$R_n(x) = \frac{(x-c)^{n+1}}{(n+1)!} \frac{d^{n+1}f(\xi)}{dx^{n+1}}, \quad c < \xi < x \qquad \text{The remainder term} \tag{6.32}$$

where the Taylor series is a polynomial in $(x - c)$ of degree n. Let's define the remainder term as a function, and see how the remainder's values at $\xi = c$ and $\xi = x$ can bound the true truncation error:

```
(%i1)   R(f,x,c,n,xi) := (x-c)^(n+1)/(n+1)!*ratsubst(xi,x,diff(f
        ➥ (x),x,n+1))$
```

Specifically, let's explore the error in a four-term Taylor series for $\ln(1 + x)$, expanded around $x = 0$:

```
(%i2)   f(x) := log(1+x)$
(%i3)   tseries : taylor(f(x), x, 0, 4);
(%i4)   remainderx: R(f,x,0,4,x);
(%i5)   remainderc: R(f,x,0,4,0);
(%i6)   truncationerror: f(x)-tseries, ratexpand;
```

$$(\%o3) \quad x - \frac{x^2}{2} + \frac{x^3}{3} - \frac{x^4}{4} + \dots$$

$$(\%o4) \quad \frac{x^5}{5\left(x^5 + 5x^4 + 10x^3 + 10x^2 + 5x + 1\right)}$$

$$(\%o5) \quad \frac{x^5}{5}$$

$$(\%o6) \quad \log(x+1) + \frac{x^4}{4} - \frac{x^3}{3} + \frac{x^2}{2} - x$$

Let's plot the remainder along with the true truncation error, as we move away from the expansion point:

```
(%i7)   wxplot2d([truncationerror, remainderc, remainderx], [x,
        ➥ 0, 0.2],
(%i8)      [legend, "truncation error", "remainder xi=c", "
        ➥ remainder xi=x"]
(%i9)   )$
```

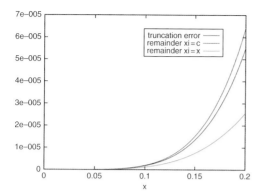

The remainder at extreme values of ξ bounds the true truncation error for this Taylor series.

6.7.3 Taylor Series for Multivariate Functions

Multivariate functions like $f(x, y, z, \ldots)$ can be expanded using lists for the arguments:

```
taylor(expression, [x, y, ], [c, d,], [n, m, ])
```
Computes a Taylor series expansion for `expression` in the variables x, y, ... around the point $x = c$, $y = d$, ... out through terms proportional to $(x-c)^n$, $(y-d)^m$, ...

For example,

(%i1)   **taylor**(f(x,y), [x,y], [a,b], [1,1]);

(%o1)   /T/ $f(a,b) + \left(\left(\frac{d}{dx} f(x,b) \Big|_{x=a} \right) (x-a) + \left(\frac{d}{dy} f(a,y) \Big|_{y=b} \right) (y-b) \right) + \ldots$

(%i2)   **taylor**(**sin**(x*y^2), [x,y], [0,0], [30,30]);

(%o2)   /T/ $y^2 x - \frac{y^6 x^3}{6} + \frac{y^{10} x^5}{120} - \frac{y^{14} x^7}{5040} + \frac{y^{18} x^9}{362880} + \ldots$

6.7.4 Approximating Taylor Series

Taylor series do not always converge quickly – and sometimes they do not converge at all. In these cases, you can try to approximate the function with a *rational function* (a ratio of polynomials).

Taylor series can be efficiently approximated as rational functions using the Padé approximation. We can often use these approximations to efficiently compute integrals or construct plots of complicated functions.

The `pade` function[14] gives a list of rational expressions that match a given Taylor series:

---

14 The `pade` function can also be inserted using ‭ Calculus ⟩⟩ Pade Approximation... ‭.

pade(*Taylor series, numerator degree, denominator degree*)

List rational functions that approximate the given *Taylor series*. The polynomials in the numerator and denominator of the rational functions are limited to *numerator degree* and *denominator degree*. The sum of the numerator and denominator degree limits must be less than or equal to the truncation level of the Taylor series.

For example, let's look at the Taylor series for the Langevin function $L(x)$, which is used in the calculation of average dipole moments [32]:

```
(%i1)   L(x) := (exp(x)+exp(-x))/(exp(x)-exp(-x)) - 1/x;
(%i2)   Lseries: taylor(L(x), x, 0,7);
```

$$(\%o1) \quad L(x) := \frac{\exp(x) + \exp(-x)}{\exp(x) - \exp(-x)} - \frac{1}{x}$$

$$(\%o2) \quad /T/ \quad \frac{x}{3} - \frac{x^3}{45} + \frac{2x^5}{945} - \frac{x^7}{4725} + \dots$$

Find the first rational expression Lapprox that has the same Taylor series as L(x):

```
(%i3)   Lapprox: first(pade(Lseries, 3,4));
```

$$(\%o3) \quad \frac{14x^3 + 315x}{x^4 + 105x^2 + 945}$$

Verify that Lapprox does have the same Taylor series:

```
(%i4)   taylor(Lapprox, x, 0, 7);
```

$$(\%o4) \quad /T/ \quad \frac{x}{3} - \frac{x^3}{45} + \frac{2x^5}{945} - \frac{x^7}{4725} + \dots$$

Ⓜ **Worksheet 6.11: Taylor series**

In this worksheet, we'll use Taylor series expansions to simplify several thermodynamic and kinetic expressions.

7

Matrices and Vectors

[Matrix theory]... is an algebra upon algebra; a calculus which enables us to combine and foretell the results of algebraical operations, in the same way algebra itself enables us to dispense with the performance of special operations of arithmetic. All analysis must ultimately clothe itself under this form.

— J. J. Sylvester, who coined the word *matrix* in 1850 [33].

Many chemical problems are naturally expressed in terms of operations on vectors and matrices. In Chapter 2, we saw how vectors and arrays can be created and edited in Maxima. In this chapter we'll use Maxima to do algebraic and differential vector and matrix calculations. Matrix notation doesn't just make the calculation *look* simpler: it *is* simpler, because symbolic mathematics engines can manipulate matrices as easily as they manipulate ordinary variables.

7.1   Vectors

Measurements like concentration, volume, and temperature can be represented as scalar quantities, which have only a magnitude attached to an appropriate unit. Other quantities like force, velocity, electric field, or displacement have both magnitude and direction; they are vectors. In this book, vectors will be written in bold face; $\mathbf{v}$ is a vector, and v is a scalar.

A vector can be imagined as an arrow that points from the origin of some coordinate system to a point in space (see Figure 7.1). The length of the vector gives its magnitude. The direction of the vector is given by specifying the point's coordinates. In the Cartesian coordinate system, there are three coordinates; each corresponds to a component that is the projection of the vector on an axis. The x component is the x coordinate times a unit vector $\mathbf{i}$ which points in the x direction. Similarly, the y and z components are the y and z coordinates times unit vectors $\mathbf{j}$ and $\mathbf{k}$ that point along the y and z axes. A vector $\mathbf{r}$ that points to (r_x, r_y, r_z) can be written as a sum of its components:

$$\mathbf{r} = r_x\mathbf{i} + r_y\mathbf{j} + r_z\mathbf{k} \qquad \text{Writing a vector as a sum of its components} \qquad (7.1)$$

In Maxima, a vector can be represented as a list.[1] A list $[a, b, c]$ corresponds to a vector $a\mathbf{i} + b\mathbf{j} + c\mathbf{k}$.

---

1 Maxima lists are actually what mathematicians call "n-tuples," ordered sets of numbers. Vectors can also be represented in Maxima as matrices with a single row or column.

Symbolic Mathematics for Chemists: A Guide for Maxima Users, First Edition. Fred Senese.
© 2019 John Wiley & Sons Ltd. Published 2019 by John Wiley & Sons Ltd.
Companion website: http://booksupport.wiley.com

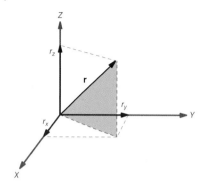

Figure 7.1 Vector components in three dimensions.

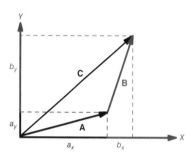

Figure 7.2 Vector addition **C** = **A** + **B**. The components of **C** are the sum of the components for **A** and **B**.

7.1.1   Vector Arithmetic

Lists are a convenient and natural way to represent vectors because vector arithmetic can be understood in terms of operations on a vector's components. In Section 2.3.3, we saw that operations on a list are applied to each item in the list; operations on a vector are applied to its components. For example, the sum of two vectors is obtained by adding the individual components:

```
(%i1)   A : [a[x], a[y]];
(%i2)   B : [b[x], b[y]];
```

$$(\%o1) \quad [a_x, a_y]$$
$$(\%o2) \quad [b_x, b_y]$$

```
(%i3)   C : A+B;
```

$$(\%o3) \quad [b_x + a_x, b_y + a_y]$$

Similarly, vector subtraction is accomplished by subtracting the components:

```
(%i4)   C-A
```

$$(\%o4) \quad [b_x, b_y]$$

```
(%i5)   C-B
```

$$(\%o5) \quad [a_x, a_y]$$

Figure 7.2 shows how the components add in vector addition.

Vector arithmetic can be represented by a set of scalar equations for each component ($c_x = a_x + b_x$ and $c_y = a_y + b_y$, in this case), or by a single vector equation **C** = **A** + **B**. The vector equation is simpler and more compact than the scalar equations.

Multiplying a vector by a scalar constant gives a "scaled" vector. The constant multiplies each component of the vector:

```
(%i6)   k*A
```
$$(\%o6) \quad [k \cdot a_x, k \cdot a_y]$$

Vector scaling and addition can be used to compute the dipole moment μ of a collection of point charges. The dipole moment is a vector that points from the center of positive charge towards the center of negative charge. It can be calculated from the positions $\mathbf{r}_i$ of the point charges q_i:

$$\boldsymbol{\mu} = \sum_i q_i \mathbf{r}_i \qquad \text{Dipole moment for a collection of point charges} \qquad (7.2)$$

Let's use this equation to estimate[2] the dipole moment of a water molecule, using approximate charges and position vectors estimated by a simple quantum mechanical calculation.

```
(%i1)   qO : -0.87616$
(%i2)   qH1: 0.43808$
(%i3)   qH2: 0.43808$
```

The charge unit is the charge of an electron, e ($1\ e = 1.60217662 \times 10^{-19}$ coulombs).

```
(%i4)   rO : [0, 0.110843, 0]$
(%i5)   rH1 : [0.783809, -0.443373, 0]$
(%i6)   rH2 : [-0.783809, -0.443373, 0]$
```

The distance units are Ångstroms, Å($1\ Å = 10^{-10}$ m). The dipole moment (in $e\cdot$ units) is then

```
(%i7)   mu : qO*rO + qH1*rH1 + qH2*rH2;
```
$$(\%o7) \quad [0.0, -0.4855818905600001, 0]$$

Let's convert μ to SI units, which are coulomb meters in this case:

```
(%i8)   mu *1.60217662e-19*1e-10;
```
$$(\%o8) \quad [0.0, -7.779879521506309 \cdot 10^{-30}, 0]$$

Dipole moments are usually expressed in Debye units, D (with 1 D = 3.33564×10^{-30} coulomb meters).

```
(%i9)   %/3.33564e-30;
```
$$(\%o9) \quad [0.0, -2.332349870341617, 0]$$

7.1.2   The Dot Product

The dot product of two vectors[3] is the sum of the products of corresponding components. Dot products can be computed simply with the dot operator (.); for example,

---

2 The calculation is approximate at best because the charge distribution in a molecule can only crudely be represented by a collection of point charges.
3 The dot product is also called the *scalar product* or the *inner product*.

```
(%i10)   A : [a[x], a[y]]$
(%i11)   B : [b[x], b[y]]$
(%i12)   A . B
```
$$(\%o7) \quad a_y\,b_y + a_x\,b_x$$

Leave a space on either side of the dot in a dot product to distinguish it from a simple decimal point.

 The product a*b is a vector with components $a_i b_i$, while the dot product a . b is a scalar $\sum_i a_i b_i$.

The same results are obtained if we choose to represent vectors with one-row or one-column matrices instead of lists (see Section 2.4):

```
(%i13)   Arow : matrix(A);
(%i14)   Brow : matrix(B);
(%i15)   Arow . Brow;
```
$$(\%o8) \quad \begin{pmatrix} a_x & a_y \end{pmatrix}$$
$$(\%o9) \quad \begin{pmatrix} b_x & b_y \end{pmatrix}$$
$$(\%o10) \quad a_y\,b_y + a_x\,b_x$$

```
(%i16)   Acol : transpose(A);
(%i17)   Bcol : transpose(B);
(%i18)   A . B;
```
$$(\%o11) \quad \begin{pmatrix} a_x \\ a_y \end{pmatrix}$$
$$(\%o12) \quad \begin{pmatrix} b_x \\ b_y \end{pmatrix}$$
$$(\%o13) \quad a_y\,b_y + a_x\,b_x$$

7.1.3   Vector Lengths and Angles

We can use the dot product to find the length of a vector. The length of a vector **v** is often written as |**v**|, computed as

$$|\mathbf{v}| = \sqrt{\mathbf{v} \cdot \mathbf{v}} \qquad \text{Length of a vector} \tag{7.3}$$

```
(%i19)   lengthA : sqrt(A.A);
```
$$(\%o14) \quad \sqrt{a_y^2 + a_x^2}$$

Dividing a vector by its length gives us a unit vector that points in the same direction as the original vector. This process is called **normalization**:

```
(%i1)   v : [1, -1/2, 0];
(%i2)   v / sqrt(v . v);
```

$$(\%o1) \quad [1, -\frac{1}{2}, 0]$$

$$(\%o2) \quad [\frac{2}{\sqrt{5}}, -\frac{1}{\sqrt{5}}, 0]$$

The dot product $\mathbf{A} \cdot \mathbf{B}$ can be used to find the angle θ between the vectors:

$$\mathbf{A} \cdot \mathbf{B} = |A||B| \cos(\theta) \qquad \text{The dot product of two vectors} \qquad (7.4)$$

where $|A|$ and $|B|$ are the lengths of vectors $\mathbf{A}$ and $\mathbf{B}$. The dot product is positive for $\theta < \pi/2$, and negative for $\theta > \pi/2$. When the vectors are at right angles to each other ($\theta = \pi/2$), their dot product is zero and they are said to be orthogonal.

For example, the OH bond vectors in the approximate water molecule above are $\mathbf{r}_{H_1} - \mathbf{r}_O$ and $\mathbf{r}_{H_2} - \mathbf{r}_O$:

```
(%i1)   rO : [0, 0.110843, 0]$
(%i2)   rH1 : [0.783809, -0.443373, 0]$
(%i3)   rH2 : [-0.783809, -0.443373, 0]$
(%i4)   bond1 : rH1-rO$
(%i5)   bond2 : rH2 - rO$
```

The angle between the bonds is

```
(%i6)   acos(bond1 . bond2 / (sqrt(bond1 . bond1)*sqrt(bond2 .
        ➡ bond2)));

(%o6)   2.031861682417625
```

or in degrees,

```
(%i7)   %*180/%pi, numer;

(%o2)   116.4170989568807
```

The true bond angle in water is about 104.5°, so this structure is not terribly accurate.

 Worksheet 7.0: Vector Arithmetic

In this worksheet, we'll use vector arithmetic to compute bond lengths and bond angles, estimate dipole moments, and compute force vectors for collections of interacting ions.

7.1.4   The Cross Product

We've seen that vectors can be added by placing them head-to-tail and drawing a new vector from start of the tail of the first vector to the head of the second. The order of the vectors doesn't matter. The sum of the vectors is the same either way.

But there is a noticeable difference that depends on the order of the vectors. Placing a pair of vectors head-to-tail implies rotation, and reversing the order of the vectors reverses the direction of rotation (Figure 7.3).

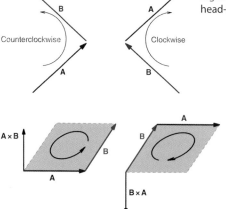

Figure 7.3 Rotation suggested by placing two vectors head-to-tail.

Figure 7.4 The cross product of two vectors points up for counterclockwise rotation and down for clockwise rotation. The length of the cross product vector is equal to the area of the shaded parallelogram.

We can represent the direction of rotation as a new vector **V** that is perpendicular to both **A** and **B**. **V** points out of the screen towards us when **A** and **B** are placed to represent counterclockwise rotation; it points back into the screen for clockwise rotation.

Curl the fingers of your right hand in the direction of the rotation, and your thumb will point in the direction of **V**. This is the famous *right-hand rule* for visualizing vector directions in many problems in physics and engineering.

V can be computed as the cross product[4] of **A** and **B**:

$$
\begin{aligned}
\mathbf{V} &= \mathbf{A} \times \mathbf{B} \\
|\mathbf{V}| &= |\mathbf{A}||\mathbf{B}|\sin(\theta)
\end{aligned}
\qquad \text{The cross product}
\tag{7.5}
$$

where θ is the angle between **A** and **B**. The length of **V** is equal to the area of a parallelogram with sides **A** and **B** (see Figure 7.4).

7.1.5 Angular Momentum

The cross product is most often used in chemistry to represent *angular momentum*, a property that is crucial in understanding the electronic structure and spectroscopy of atoms and molecules.

The angular momentum ℓ of a particle with mass m orbiting a point is defined as $r \times p$, the cross product of the position vector **r** and the momentum vector $p = mv$. Let's compute the components ℓ_x, ℓ_y, and ℓ_z of the angular momentum in terms of the position **r** and momentum **p**.

We must load the vect package to compute cross products (and many other vector functions). The cross product operator is ~, *not* ×. For example, the cross product $r \times p$ is typed as `r~p`:

```
(%i1)   load(vect)$
(%i2)   r: [x, y, z]$
(%i3)   p: [p[x], p[y], p[z]]$
(%i4)   l :  r ~ p;

(%o4)   [x, y, z] [p_x, p_y, p_z]
```

---

4 The cross product is also called the *vector product*.

Maxima will leave the cross product and other vector expressions unexpanded; to evaluate vector expressions use `express` function (also found in the `vect` package).

```
(%i5)   express(%);
```

$$(\%o5) \quad [y\,p_z - p_y\,z, p_x\,z - x\,p_z, x\,p_y - p_x\,y]$$

7.1.6 Vector Algebra

Maxima can simplify vector expressions in various ways. If the variables in the expression haven't been assigned vector or scalar values, we must first tell Maxima which variables are not scalars. This can be done with the `declare` function as follows:

```
(%i1)   declare([u,v,w], nonscalar)$
```

We can now show that the cross product doesn't commute:

```
(%i2)   v~u;
```

$$(\%o2) \quad -u\,v$$

For more complicated expressions, though, we'll need to simplify with the `vectorsimp` function:

 `vectorsimp(expression)`
Simplifies a vector expression using simplification switches. Use the `expandall` switch to apply all simplifications; refer to the Maxima manual for more specialized switches.

For example, let's show that the cross product distributes across sums:

```
(%i3)   u~ (v + w)$
(%i4)   vectorsimp(%), expandall;
```

$$(\%o2) \quad u\,w + u\,v$$

```
(%i5)   (u+v) ~ w$
(%i6)   vectorsimp(%), expandall;
```

$$(\%o2) \quad v\,w + u\,w$$

 Worksheet 7.1: Cross Products

In this worksheet, we'll see how the cross product can be used to compute angular momenta and other quantities such as torsion angles.

7.2 Matrices

A matrix is a two-dimensional array of numbers or expressions. Casting repetitive calculations in matrix form often allows both problems and solutions to be expressed in a simpler and more efficient way.

We have already seen how to build, edit, read, write, and transform matrices in Section 2.4. Let's build three matrices with the `matrix` function, which assembles a matrix from row lists:

```
(%i1)   A : matrix([a11, a12], [a21, a22]);
(%i2)   B : matrix([b11, b12], [b21, b22]);
(%i3)   C : matrix([c11, 0, 0], [0, c22, 0], [0, 0, c33]);
(%i4)   I : matrix([1, 0, 0], [0, 1, 0], [0, 0, 1]);
```

$$(\%o1) \quad \begin{pmatrix} a11 & a12 \\ a21 & a22 \end{pmatrix}$$

$$(\%o2) \quad \begin{pmatrix} b11 & b12 \\ b21 & b22 \end{pmatrix}$$

$$(\%o3) \quad \begin{pmatrix} c11 & 0 & 0 \\ 0 & c22 & 0 \\ 0 & 0 & c33 \end{pmatrix}$$

$$(\%o4) \quad \begin{pmatrix} 1 & 0 & 0 \\ 0 & 1 & 0 \\ 0 & 0 & 1 \end{pmatrix}$$

All of these matrices are square – they have the same number of rows and columns. The matrices **C** and **I** have nonzero elements only along their major diagonals, so they are design matrices. Diagonal matrices have occur quite frequently in chemistry. We saw in Chapter 2 that diagonal matrices could be entered more simply using Algebra ⟩⟩ Enter Matrix ... ; there is also a function `diag_matrix` for building them:

```
(%i5)   diag_matrix(c11,c22,c33);
```

$$(\%o5) \quad \begin{pmatrix} c11 & 0 & 0 \\ 0 & c22 & 0 \\ 0 & 0 & c33 \end{pmatrix}$$

The matrix **I** has ones along its major diagonal and zeros elsewhere. It is called an identity matrix. It can be generated more simply by `ident(N)`, where N is the number of rows and columns.

```
(%i6)   ident(3);
```

$$(\%o6) \quad \begin{pmatrix} 1 & 0 & 0 \\ 0 & 1 & 0 \\ 0 & 1 & 1 \end{pmatrix}$$

7.2.1   Matrix Arithmetic

For matrices, the operators +, -, *, and / act between matching elements. Powers are applied to each matrix element with the ^ operator. For example,

```
(%i7)    A+B;
```

$$(\%o7)\quad \begin{pmatrix} b11 + a11 & b12 + a12 \\ b21 + a21 & b22 + a22 \end{pmatrix}$$

```
(%i8)    A*B;
```

$$(\%o8)\quad \begin{pmatrix} a11\,b11 & a12\,b12 \\ a21\,b21 & a22\,b22 \end{pmatrix}$$

```
(%i9)    A/B;
```

$$(\%o9)\quad \begin{pmatrix} \frac{a11}{b11} & \frac{a12}{b12} \\ \frac{a21}{b21} & \frac{a22}{b22} \end{pmatrix}$$

```
(%i10)   A^2;
```

$$(\%o10)\quad \begin{pmatrix} a11^2 & a12^2 \\ a21^2 & a22^2 \end{pmatrix}$$

 The two matrices must have the same row and column dimensions for these operations to work.

7.2.2   The Transpose

The transpose of a matrix $\mathbf{A}^{\mathrm{T}}$ swaps its row and column indices; the ith column of $\mathbf{A}$ is the ith row of $\mathbf{A}^{\mathrm{T}}$. In Maxima, $\mathbf{A}^{\mathrm{T}}$ is written as transpose(A):

```
(%i11)   transpose(A);
```

$$(\%o11)\quad \begin{pmatrix} a11 & a21 \\ a12 & a22 \end{pmatrix}$$

The transpose function makes a row vector into a column vector, and vice versa.

```
(%i12)   v : [v1, v2];
(%i13)   transpose(v);
(%i14)   transpose(%);
```

$$(\%o12)\quad [v1, v2]$$

$$(\%o13)\quad \begin{pmatrix} v1 \\ v2 \end{pmatrix}$$

$$(\%o14)\quad \begin{pmatrix} v1 & v2 \end{pmatrix}$$

Notice a subtle difference between the output on lines 12 and 14. Line 12 represents the vector as a list, and line 14 represents it as a one-row matrix.

If $\mathbf{A} = \mathbf{A}^{\mathrm{T}}$, $\mathbf{A}$ is a symmetric matrix. Symmetric matrices must be square. For example,

(%i15)   D : **matrix**([1, 2, 2], [2, 4, 3], [2, 3, 5]);

(%o15) $\begin{pmatrix} 1 & 2 & 2 \\ 2 & 4 & 3 \\ 2 & 3 & 5 \end{pmatrix}$

(%i16)   **transpose**(D);

(%o16) $\begin{pmatrix} 1 & 2 & 2 \\ 2 & 4 & 3 \\ 2 & 3 & 5 \end{pmatrix}$

7.2.3   The Matrix Product

We've already seen that the $\boxed{\;.\;}$ represents the dot product when it operates between vectors. When the operands are matrices, the $\boxed{\;.\;}$ operator produces a *matrix product*. The product **A.B** is the dot product of each row of **A** with each column of **B**. If **A** is an $m \times n$ matrix and **B** is an $n \times p$ matrix, the matrix product is an $m \times p$ matrix:

$$\mathbf{A.B} = \mathbf{C}, \text{ with } C_{ij} = \sum_{k=1}^{n} A_{ik} B_{kj} \qquad \text{The matrix product} \qquad (7.6)$$

(%i17)   A . B;

(%o17) $\begin{pmatrix} a12\,b21 + a11\,b11 & a12\,b22 + a11\,b12 \\ a22\,b21 + a21\,b11 & a22\,b22 + a21\,b12 \end{pmatrix}$

(%i18)   B . A;

(%o18) $\begin{pmatrix} a21\,b12 + a11\,b11 & a22\,b12 + a12\,b11 \\ a21\,b22 + a11\,b21 & a22\,b22 + a12\,b21 \end{pmatrix}$

Notice that unlike normal multiplication, matrix products are *not* commutative; $\mathbf{A.B} \neq \mathbf{B.A}$ in general.

 Don't confuse A*B and A.B!

The sizes of the matrices in the product must be compatible; the number of columns in **A** must be equal to the number of rows in **B**. Technically, then, to multiply a vector **v** and a matrix **A**, in the product **v.A**, the vector must be a row vector, and in **A.v**, v must be a column vector. For example,

(%i19)   A . **transpose**(v);

(%o19) $\begin{pmatrix} a12\,v2 + a11\,v1 \\ a22\,v2 + a21\,v1 \end{pmatrix}$

```
(%i20)    v . A;
```
$$(\%o20) \quad \left(a21\,v2 + a11\,v1 \quad a22\,v2 + a12\,v1\right)$$

If the vector is stored as a list, though, Maxima will decide whether to treat the vector as a row or column vector based on context.

```
(%i21)    A . v;
```
$$(\%o21) \quad \begin{pmatrix} a12\,v2 + a11\,v1 \\ a22\,v2 + a21\,v1 \end{pmatrix}$$

To form the matrix product of a matrix with itself, there is a matrix exponentiation operator, `^^`.

```
(%i22)    A^^2;
```
$$(\%o22) \quad \begin{pmatrix} a12\,a21 + a11^2 & a12\,a22 + a11\,a12 \\ a21\,a22 + a11\,a21 & a22^2 + a12\,a21 \end{pmatrix}$$

 `A^^2` is identical to `A . A` (but different from `A^2`!)

A matrix product with an identity matrix leaves the original matrix unchanged:

```
(%i23)    A . ident(2);
```
$$(\%o23) \quad \begin{pmatrix} a11 & a12 \\ a21 & a22 \end{pmatrix}$$

```
(%i24)    ident(2) . A;
```
$$(\%o24) \quad \begin{pmatrix} a11 & a12 \\ a21 & a22 \end{pmatrix}$$

 Worksheet 7.2: Matrix Arithmetic and Symmetry Operations

In this worksheet, we'll represent several important transformations as matrix multiplications, including the rotation of atomic coordinates in a molecule around any specified axis, including the molecule's axis of symmetry.

7.2.4   Determinants

One of the more useful properties of square matrices is the determinant, written as $|\mathbf{A}|$ or $det\mathbf{A}$ for a square matrix $\mathbf{A}$. We've already seen that the determinant of a Hessian matrix can be helpful in classifying the stationary points of a function (Section 6.4.1). In quantum chemistry, determinants are used to represent many-electron wavefunctions, secular equations, and angular momenta in three dimensions. In matrix algebra, the determinant can tell us whether or not the matrix has an inverse (Section 7.2.5) and whether or not a unique solution exists for a system of linear equations (Section 7.2.6).

$|\mathbf{A}|$ is computed in Maxima with the `determinant` function (which can be inserted with Algebra ⟩⟩ Determinant):

```
(%i25)   determinant(A);
```
(%o25) $a11\,a22 - a12\,a21$

For larger matrices, the expression for the determinant is more complex. It can be understood as an expansion in terms of *minors*. The minor M_{ij} of a matrix $\mathbf{C}$ is the matrix with row i and column j omitted; it can be computed with the `minor` function. For example,

```
(%i26)   C: matrix([c11, c12, c13],[c21, c22, c23], [c31, c32,
         ➦ c33]);
```
(%o26) $\begin{pmatrix} c11 & c12 & c13 \\ c21 & c22 & c23 \\ c31 & c32 & c33 \end{pmatrix}$

```
(%i27)   minor(C,2,2);
```
(%o27) $\begin{pmatrix} c11 & c13 \\ c31 & c33 \end{pmatrix}$

The determinant of $\mathbf{C}$ written in terms of minors is

$$|\mathbf{C}| = c_{11}|M_{11}| - c_{12}|M_{12}| + c_{13}|M_{13}| \qquad \text{Expansion of a 3×3 determinant} \qquad (7.7)$$

or, in general,

$$|\mathbf{C}| = \sum_{j=1}^{N}(-1)^{i+j}c_{ij}|M_{ij}| = \sum_{i=1}^{N}(-1)^{i+j}c_{ij}|M_{ij}| \qquad \begin{array}{l}\text{Expansion of a}\\ \text{determinant around}\\ \text{any row or column}\end{array} \qquad (7.8)$$

For example, expanding the determinant around row 1, we have

```
(%i28)   sum((-1)^(1+j)*C[1,j]*determinant(minor(C,1,j)), j, 1,
         ➦ 3);
```
(%o28) $c11\,(c22\,c33 - c23\,c32) - c12\,(c21\,c33 - c23\,c31) + c13\,(c21\,c32 - c22\,c31)$

which is equal to the determinant of $\mathbf{C}$,

```
(%i29)   determinant(C);
```
(%o29) $c11\,(c22\,c33 - c23\,c32) - c12\,(c21\,c33 - c23\,c31) + c13\,(c21\,c32 - c22\,c31)$

The determinant of a matrix $\mathbf{A}$ doesn't change with the matrix is transposed:

```
(%i30)   determinant(transpose(C));
```
(%o30) $c11\,(c22\,c33 - c23\,c32) - c21\,(c12\,c33 - c13\,c32) + (c12\,c23 - c13\,c22)\,c31$

However, swapping any pair of rows or columns in a matrix changes the sign of its determinant. This makes them useful for modeling wavefunctions in quantum chemistry [34]. A wavefunction contains all mechanical information about a system of electrons; it can be used to compute properties like dipole moments, energies, polarizabilities, and more.

An acceptable electronic wavefunction must change sign when the coordinates of any pair of electrons are exchanged. For a molecule that contains two electrons with space and spin coordinates $\mathbf{x}_1$ and $\mathbf{x}_2$ in orbitals $\chi_1(\mathbf{x})$ and $\chi_2(\mathbf{x})$, an antisymmetric wavefunction that represents the behavior of the electrons is

$$\Psi(\mathbf{x}_1, \mathbf{x}_2) = \frac{1}{\sqrt{2}} \left[\chi_1(\mathbf{x}_1)\chi_2(\mathbf{x}_2) - \chi_1(\mathbf{x}_2)\chi_2(\mathbf{x}_1) \right]$$

Since the determinant of a 2×2 matrix $\begin{pmatrix} a & b \\ c & d \end{pmatrix}$ is $ad - bc$, we can represent the wave function as a determinant:

$$\Psi(\mathbf{x}_1, \mathbf{x}_2) = \frac{1}{\sqrt{2}} \begin{vmatrix} \chi_1(\mathbf{x}_1) & \chi_2(\mathbf{x}_2) \\ \chi_1(\mathbf{x}_2) & \chi_2(\mathbf{x}_2) \end{vmatrix}$$

```
(%i1)    Psi(x1, x2) := 1/sqrt(2) * determinant(
(%i2)        matrix([chi[1](x1), chi[2](x1)], [chi[1](x2), chi
         ➡ [2](x2)])
(%i3)    );
```

$$(\%o1) \quad \Psi(x1, x2) := \frac{\mathrm{determinant}\left(\begin{pmatrix} \chi_1(x1) & \chi_2(x1) \\ \chi_1(x2) & \chi_2(x2) \end{pmatrix} \right)}{\sqrt{2}}$$

```
(%i4)    Psi(x1,x2);
```

$$(\%o2) \quad \frac{\chi_1(x1) \cdot \chi_2(x2) - \chi_2(x1) \cdot \chi_1(x2)}{\sqrt{2}}$$

Swapping the labels on any pair of coordinates is equivalent to swapping columns in the matrix, and so changes the sign of the determinant (and the wavefunction):

```
(%i5)    is(equal(Psi(x2,x1), -Psi(x1,x2)));
(%o3)    true
```

The orbitals χ_1 and χ_2 must be different functions or the determinant vanishes, as required by the Pauli principle.

```
(%i6)    ratsubst(chi[1], chi[2], Psi(x1,x2);
(%o4)    0
```

The wavefunction requires that the coordinates of the two electrons cannot be identical, either (two electrons cannot occupy the same space and spin coordinates):

```
(%i7)    Psi(r1,r1);
(%o5)    0
```

In general, if N electrons occupy orbitals $\chi_1, \chi_2, \ldots, \chi_N$, the wavefunction Ψ can be written as a determinant:

$$\Psi(\mathbf{x}_1, \mathbf{x}_2, \ldots, \mathbf{x}_N) = \frac{1}{\sqrt{N!}} \begin{vmatrix} \chi_1(\mathbf{x}_1) & \chi_1(\mathbf{x}_2) & \cdots & \chi_1(\mathbf{x}_N) \\ \chi_2(\mathbf{x}_1) & \chi_2(\mathbf{x}_2) & \cdots & \chi_2(\mathbf{x}_N) \\ \vdots & \vdots & \ddots & \vdots \\ \chi_N(\mathbf{x}_1) & \chi_N(\mathbf{x}_2) & \cdots & \chi_N(\mathbf{x}_N) \end{vmatrix} \qquad \text{The Slater determinant}$$

$$(7.9)$$

where $\mathbf{x}_1, \mathbf{x}_2, \ldots, \mathbf{x}_N$ represent the coordinates of the N electrons. This expression is called a *Slater determinant* after John Slater, a pioneer of quantum chemistry.

The matrix in the determinant shows that every electron is associated with every orbital in the wavefunction – you cannot distinguish one electron from another. A consequence of this is that the wavefunction assumes every electron "feels" only the average positions of all the others, rather than individual electron–electron repulsions.

 Worksheet 7.3: Determinants

In this worksheet we'll look at several useful properties of determinants, and explore the Slater determinant further.

7.2.5 The Inverse of a Matrix

The matrix analog of a reciprocal is called the *inverse* of a matrix. The inverse of $\mathbf{A}$ is written as $\mathbf{A}^{-1}$; in Maxima, it can be computed either as A^^-1 or by using the invert function:

```
(%i8)   A : matrix([a11, a12], [a21, a22]);
(%i9)   invert(A);
```

$$(\%o1) \quad \begin{pmatrix} a11 & a12 \\ a21 & a22 \end{pmatrix}$$

$$(\%o2) \quad \begin{pmatrix} \dfrac{\frac{a12\,a21}{a11\left(a22-\frac{a12\,a21}{a11}\right)}+1}{a11} & -\dfrac{a12}{a11\left(a22-\frac{a12\,a21}{a11}\right)} \\ -\dfrac{a21}{a11\left(a22-\frac{a12\,a21}{a11}\right)} & \dfrac{1}{a22-\frac{a12\,a21}{a11}} \end{pmatrix}$$

Multiplying a matrix by its inverse gives an identity matrix:

```
(%i10)   A . invert(A), ratsimp;
```

$$(\%o3) \quad \begin{pmatrix} 1 & 0 \\ 0 & 1 \end{pmatrix}$$

The inverse of a matrix only exists if the matrix is square, and its determinant is not zero. For example,

```
(%i11)   D : matrix([1, 2, 1], [1, 2, 3], [1, 2, 4]);
```

$$(\%o4) \quad \begin{pmatrix} 1 & 2 & 1 \\ 1 & 2 & 3 \\ 1 & 2 & 4 \end{pmatrix}$$

```
(%i12)   invert(D);
```

Unable to compute the LU factorization
– an error. To debug this try: debugmode(true);

Maxima is unable to compute the inverse because there *is* no inverse:

```
(%i13)   determinant(D);
```
```
(%o5)    0
```

7.2.6   Matrix Algebra

Suppose we have a mixture of H_2O and H_2O_2. If a sample contains 3 mol of H and 2 mol of O, how many moles of H_2O (x) and H_2O_2 (y) are in the sample?

We can set up two equations, one for H and one for O:

$$2x + 2y = 3$$
$$x + 2y = 2$$

This is a system of linear equations because powers of all of the variables are 1. It's easy to find the solutions with simple algebra (or `solve` in Maxima, as we did in Section (5.4). But as the system becomes more complicated, it becomes tedious to write out all of those equations.

There is a better way to represent the problem. Suppose we have a mixture of two substances with unknown concentrations x_1 and x_2. Further, suppose that the absorption spectra of the two substances overlap to the extent that there is no wavelength where both substances do not absorb. If their absorbances are additive and proportional to their concentrations, it is sometimes still possible to determine their concentrations separately by measuring the total absorbance A at two different wavelengths λ and λ'. This gives us two equations which can be solved in two unknowns:

$$
\begin{aligned}
A_\lambda &= A_{\lambda,1} + A_{\lambda,2} &= a_{\lambda,1}x_1 + a_{\lambda,2}x_2 \\
A_{\lambda'} &= A_{\lambda',1} + A_{\lambda',2} &= a_{\lambda',1}x_1 + a_{\lambda',2}x_2
\end{aligned}
\qquad
\begin{array}{l}
\text{Absorbances of a mixture} \\
\text{of two substances}
\end{array}
\qquad (7.10)
$$

where a_{ij} is the molar absorptivity of substance j at wavelength i times the path length. If the molar absorptivities, pathlengths, and total absorbances are known, the equations can be solved simultaneously for c_1 and c_2.

We can generalize this procedure for a set of n substances measured at n wavelengths. Consider the following set of linear equations, where we have replaced the total absorbances A_i with b_i:

$$
\begin{aligned}
a_{11}x_1 + a_{12}x_2 + \cdots + a_{1n}x_N &= b_1 \\
a_{21}x_1 + a_{22}x_2 + \cdots + a_{2n}x_N &= b_2 \\
&\vdots \\
a_{n1}x_1 + a_{n2}x_2 + \cdots + a_{nn}x_N &= b_n
\end{aligned}
\qquad \text{A system of } n \text{ linear equations} \qquad (7.11)
$$

It's a chore to write, manipulate, and solve these equations when n is large. We can hide all of this distracting and tedious complexity by using matrix notation, which compactly represents whole datasets and systems of equations with a single symbol. Equation (7.11) can be written in matrix form as

$$\mathbf{Ax} = \mathbf{b} \qquad \text{A system of linear equation in matrix notation} \qquad (7.12)$$

where

$$
\mathbf{A} =
\begin{bmatrix}
a_{11} & a_{12} & \cdots & a_{1n} \\
a_{21} & a_{22} & \cdots & a_{2n} \\
\vdots & \vdots & \ddots & \vdots \\
a_{n1} & a_{n2} & \cdots & a_{nn}
\end{bmatrix}
\quad
\mathbf{x} =
\begin{bmatrix}
x_1 \\
x_2 \\
\vdots \\
x_n
\end{bmatrix}
\quad
\mathbf{b} =
\begin{bmatrix}
b_1 \\
b_2 \\
\vdots \\
b_n
\end{bmatrix}
$$

and $\mathbf{Ax}$ is the matrix product of $\mathbf{A}$ and $\mathbf{x}$ (which would be typed `A . x` in Maxima).

Equation (7.12) certainly looks simpler – but how do we solve it? We can *not* simply write $\mathbf{x} = \mathbf{b}/\mathbf{A}$, because there is no matrix division operation that works with the matrix product that way. However, we *can* multiply both sides of the equation by the inverse of $\mathbf{A}$:

$$\mathbf{A}^{-1}\mathbf{A}\mathbf{x} = \mathbf{A}^{-1}\mathbf{b} \qquad (7.13)$$

and since $\mathbf{A}^{-1}\mathbf{A} = \mathbf{I}$, and $\mathbf{I}\mathbf{x} = \mathbf{x}$,

$\mathbf{x} = \mathbf{A}^{-1}\mathbf{b}$	Matrix solution of a system of linear equations	(7.14)

Let's try this approach in Maxima. First, let's solve three linear equations in unknowns x, y, and z directly with the `solve` function:

```
(%i1)   eqn1 : 2*x - y + z = 1;
(%i2)   eqn2 : 3*x + 2*z = -1;
(%i3)   eqn3 : 4*x + y + 2*z = 2;
(%i4)   solve([eqn1, eqn2, eqn3], [x,y,z]);
```

$$(\%o1) \quad z - y + 2x = 1$$
$$(\%o2) \quad 2z + 3x = -1$$
$$(\%o3) \quad 2z + y + 4x = 2$$
$$(\%o4) \quad [[x = 3, y = 0, z = -5]]$$

Repeating the solution with Equation (7.14), we have

```
(%i5)   A : matrix([2, -1, 1], [3, 0, 2], [4, 1, 2]);
(%i6)   b : transpose([1, -1, 2]);
(%i7)   A^^-1 . b;
```

$$(\%o5) \quad \begin{pmatrix} 2 & -1 & 1 \\ 3 & 0 & 2 \\ 4 & 1 & 2 \end{pmatrix}$$

$$(\%o6) \quad \begin{pmatrix} 1 \\ -1 \\ 2 \end{pmatrix}$$

$$(\%o7) \quad \begin{pmatrix} 3 \\ 0 \\ -5 \end{pmatrix}$$

The last vector correctly gives us $x = 3$, $y = 0$, and $z = -5$.

Maxima has a function that can extract the coefficient matrix $\mathbf{A}$ from a list of linear equations:

`coefmatrix(`*equation list*`, `*variable list*`)`
Returns the coefficient matrix for the linear equations in *equation list*, with columns ordered by the variables in *variable list*.

```
(%i8)   A : coefmatrix([eqn1,eqn2,eqn3], [x,y,z]);
```

$$(\%o2) \quad \begin{pmatrix} 2 & -1 & 1 \\ 3 & 0 & 2 \\ 4 & 1 & 2 \end{pmatrix}$$

Notice that **A** must be square, or no inverse exists and Equation (7.14) won't work. We often want to solve a system of equations that are *overdetermined*, that is, we have more equations than unknowns. Such systems will have rectangular coefficient matrices.

For example, suppose we'd like to find the slope and intercept of a line $b + mx = y$ given a series of points $(x_1, y_1), (x_2, y_2), \ldots, (x_N, y_N)$. We have a system of equations

$$
\begin{aligned}
b + mx_1 &= y_1 \\
b + mx_2 &= y_2 \\
&\vdots \\
b + mx_N &= y_N
\end{aligned}
\tag{7.15}
$$

In matrix form, the equations are

$$
\begin{pmatrix} 1 & x_1 \\ 1 & x_2 \\ \vdots & \vdots \\ 1 & x_N \end{pmatrix} \begin{pmatrix} b \\ m \end{pmatrix} = \begin{pmatrix} y_1 \\ y_2 \\ \vdots \\ y_N \end{pmatrix}
$$

which we can write in matrix notation as

$$
\mathbf{Ap} = \mathbf{y}
\tag{7.16}
$$

where $\mathbf{A} = \begin{pmatrix} 1 & x_1 \\ 1 & x_2 \\ \vdots & \vdots \\ 1 & x_N \end{pmatrix}$, $\mathbf{p} = \begin{pmatrix} b \\ m \end{pmatrix}$, and $\mathbf{y} = \begin{pmatrix} y_1 \\ y_2 \\ \vdots \\ y_N \end{pmatrix}$.

The coefficient matrix **A** isn't square, so it doesn't have an inverse. However, the matrix $\mathbf{A}^\mathrm{T}\mathbf{A}$ *is* square, and it *may* have an inverse. Multiplying both sides of Equation (7.16) by $\mathbf{A}^\mathrm{T}$,

$$
\mathbf{A}^\mathrm{T}\mathbf{Ap} = \mathbf{A}^\mathrm{T}\mathbf{y}
\tag{7.17}
$$

and then multiplying both sides by $(\mathbf{A}^\mathrm{T}\mathbf{A})^{-1}$, we have

$$
\mathbf{p} = (\mathbf{A}^\mathrm{T}\mathbf{A})^{-1}\mathbf{A}^\mathrm{T}\mathbf{y}
\tag{7.18}
$$

Let's try this in Maxima with a set of (x, y) points.

```
(%i1)   data: matrix([0,5.9], [0.9,5.4], [1.8,4.4], [2.6,4.6],
        ➥ [3.3,3.5], [4.4,3.7], [5.2,2.8], [6.1,2.8],
        ➥ [6.5,2.4], [7.4,1.5])$
(%i2)   A : apply(matrix, makelist([1, data[i,1]], i, 1, length(
        ➥ data)));
(%i3)   ydata : col(data, 2);
```

$$
(\%o2) \quad \begin{pmatrix} 1 & 0 \\ 1 & 0.9 \\ 1 & 1.8 \\ 1 & 2.6 \\ 1 & 3.3 \\ 1 & 4.4 \\ 1 & 5.2 \\ 1 & 6.1 \\ 1 & 6.5 \\ 1 & 7.4 \end{pmatrix}
$$

$$(\%o3) \quad \begin{pmatrix} 5.9 \\ 5.4 \\ 4.4 \\ 4.6 \\ 3.5 \\ 3.7 \\ 2.8 \\ 2.8 \\ 2.4 \\ 1.5 \end{pmatrix}$$

Now compute **p** with Equation (7.18):

```
(%i4)   p = invert(transpose(A) . A) . transpose(A) . ydata;
```

$$(\%o4) \quad p = \begin{pmatrix} 5.761185190439042 \\ \text{-}0.53957727498404 \end{pmatrix}$$

so the intercept is 5.761185190439042 and the slope is -0.53957727498404.

This is a linear least-squares fit of the data. Maxima has several built-in functions for doing this same calculation much more efficiently, as we'll see in Chapter 9. Let's preview one of those functions (plsquares) to check our result:

```
(%i5)   load(plsquares)$
(%i6)   plsquares(data, [x,y], y);
```

Determination Coefficient for y=0.95350386049735
$$(\%o6) \quad y = 5.761185190439042 - 0.53957727498404\,x$$

...which is identical to the solution we got from applying Equation (7.18).

Let's look at another application involving rectangular matrices. Any balanced chemical equation can be written in matrix form as $\mathbf{C}\mathbf{x} = 0$, where $\mathbf{C}$ is a formula composition matrix. Think of this matrix as a table with columns labeled by species in the reaction, and rows labeled by elements that appear in those species. The matrix element C_{ij} is the number of atoms of type i in compound j.

$\mathbf{x}$ is a vector of the coefficients in the equation. For example, consider the equation $N_2 + 3H_2 = 2NH_3$. The formula composition matrix is

		N_2	H_2	NH_3
$\mathbf{C} =$	N	2	0	1
	H	0	2	3

The formula composition matrix is

```
(%i1)   C: matrix([2, 0, 1], [0, 2, 3]);
```

$$(\%o1) \quad \begin{pmatrix} 2 & 0 & 1 \\ 0 & 2 & 3 \end{pmatrix}$$

and the vector of coefficients is

	N_2	-1
$\mathbf{x} =$	H_2	-3 .
	NH_3	2

where coefficients on different sides of the equation have different signs. Reactants have negative coefficients, and products are positive:

```
(%i2)   x : transpose([-1, -3, 2]);
```

$$(\%o2) \quad \begin{pmatrix} -1 \\ -3 \\ 2 \end{pmatrix}$$

The equation is balanced since $\mathbf{Cx = 0}$:

```
(%i3)   C . x;
```

$$(\%o3) \quad \begin{pmatrix} 0 \\ 0 \end{pmatrix}$$

To balance an equation, we have to find $\mathbf{x}$ from $\mathbf{C}$. Since $\mathbf{C}$ isn't square, we could find $\mathbf{x}$ using Equation (7.18), but there is a much easier way [35].

All vectors $\mathbf{x}$ that satisfy $\mathbf{Cx = 0}$ are said to be in the null space of $\mathbf{C}$. Maxima has a function `nullspace` that computes a set of independent vectors that *span* the null space for a matrix. Any solution $\mathbf{x}$ of $\mathbf{Cx = 0}$ can be written as a linear combination of these vectors.

```
(%i4)   nullspace(C);
```

$$(\%o4) \quad \text{span}\left(\begin{pmatrix} -2 \\ -6 \\ 4 \end{pmatrix} \right)$$

This suggests that equation $2N_2 + 6H_2 = 4NH_3$ is indeed balanced, but not with the minimum integer coefficients. To get the minimum integer coefficients we can extract the null space vector and divide it by the absolute value of its smallest element. We'll look at this application further in Worksheet 7.5.

 Worksheet 7.4: Matrix Algebra

In this worksheet we'll use matrices to solve systems of linear equations (both square and overdetermined). We'll also use matrix algebra to balance chemical equations and determine the number of independent equations in reaction mechanisms.

7.2.7   Eigenvalues and Eigenvectors

Calculation of molecular energies and orbitals involves solution of matrix equations of the type [36]

$$\mathbf{Ax} = \lambda\mathbf{x} \qquad \text{A matrix eigenvalue problem} \tag{7.19}$$

where we want to find both λ (a scalar) and $\mathbf{x}$ (a vector). The equation can be rewritten as

$$(\mathbf{A} - \lambda\mathbf{I})\,\mathbf{x} = 0 \tag{7.20}$$

where $\mathbf{I}$ is an identity matrix with the same dimension as $\mathbf{A}$. This represents a system of so-called secular equations:

$$
\begin{aligned}
(A_{11} - \lambda)x_1 + a_{12}x_2 + \cdots + A_{1N}x_N &= 0 \\
A_{21}x_1 + (A_{22} - \lambda)x_2 + \cdots + A_{2N}x_N &= 0 \\
\vdots \\
A_{N1}x_1 + A_{N2}x_2 + \cdots + (A_{NN} - \lambda)x_N &= 0
\end{aligned}
\qquad \text{Secular equations} \qquad (7.21)
$$

One (trivial) solution for the equations is $\mathbf{x} = 0$. Other solutions occur when λ is chosen so that the matrix $\mathbf{A} - \lambda\mathbf{I}$ has a determinant of zero:

$$
|\mathbf{A} - \lambda\mathbf{I}| = 0 \qquad \text{The characteristic polynomial} \qquad (7.22)
$$

The determinant can be written as an Nth degree polynomial called the characteristic polynomial. Each root λ of the characteristic polynomial is called an eigenvalue. Each corresponding vector $\mathbf{x}$ that solves Equations (7.19) and (7.20) with a particular eigenvalue is called an eigenvector.

Eigenvalues provide a convenient way to classify critical points for functions of many variables (Section 6.4.1). At a minimum, the Hessian matrix has all eigenvalues positive; at a maximum, all eigenvalues are negative. If the Hessian matrix has both positive and negative eigenvalues, the point is a a saddle point.

Let's look at another elementary matrix eigenvalue problem in chemistry: the approximate calculation of molecular energies and molecular orbitals in conjugated systems using the Hückel method.

7.2.7.1 Application: Energies and Molecular Orbitals of Ethylene

In the Hückel molecular orbital theory, we want to solve the equation

$$
\mathbf{Hc} = \mathbf{c}E \qquad \text{The Hückel method} \qquad (7.23)
$$

to obtain the energies E as eigenvalues, and the vectors $\mathbf{c}$ as corresponding eigenvectors that represent molecular orbitals, which have the form:

$$
\psi = c_1\phi_1 + c_2\phi_2 + \cdots + c_N\phi_N \qquad
\begin{array}{l}\text{A molecular orbital written} \\ \text{as a linear combination of} \\ \text{atomic orbitals}\end{array} \qquad (7.24)
$$

The matrix $\mathbf{H}$ is the Hamiltonian matrix, which has elements

$$
H_{ij} = \int \phi_i^* \mathcal{H}(\phi_j)\mathrm{d}\tau \qquad \text{Hamiltonian matrix elements} \qquad (7.25)
$$

where $\mathcal{H}$ is the Hamiltonian operator for the system and the integrals are performed over all space.

Computing the integrals for the matrix elements in the Hamiltonian is computationally intensive. The Hückel method makes the approximation

$$
\mathbf{H} \approx \alpha\mathbf{I} + \beta\mathbf{C} \qquad
\begin{array}{l}\text{Hückel approximation} \\ \text{for the Hamiltonian}\end{array} \qquad (7.26)
$$

where $\mathbf{C}$ is a "connection matrix" with each row and column corresponding to an atom index. $\mathbf{C}$ has a 1 wherever two adjacent atoms are bonded, and a zero elsewhere. For example, for butadiene, $CH_2 = CH - CH = CH_2$, the Hamiltonian matrix for the π electrons delocalized across the four carbon atoms is

```
(%i1)   C : matrix([0,1,0,0],[1,0,1,0],[0,1,0,1],[0,0,1,0]);
(%i2)   H : alpha*ident(4) + beta*C;
```

$$(\%o1) \quad \begin{pmatrix} 0 & 1 & 0 & 0 \\ 1 & 0 & 1 & 0 \\ 0 & 1 & 0 & 1 \\ 0 & 0 & 1 & 0 \end{pmatrix}$$

$$(\%o2) \quad \begin{pmatrix} \alpha & \beta & 0 & 0 \\ \beta & \alpha & \beta & 0 \\ 0 & \beta & \alpha & \beta \\ 0 & 0 & \beta & \alpha \end{pmatrix}$$

The parameters α and β are the Coulomb integral $H_{ii} = \int \phi_i^* \mathcal{H} \phi_i d\tau = \alpha$ and the resonance integral $H_{ij} = \int \phi_i^* \mathcal{H} \phi_j d\tau = \beta$ for orbitals ϕ_i and ϕ_j on adjacent atoms, and $H_{ij} = 0$ when the orbitals are on nonadjacent atoms.

Finding the eigenvalues E then involves solution of the equation

$$|\mathbf{H} - E\mathbf{I}| = 0 \tag{7.27}$$

Let's apply this equation to ethylene, $H_2C = CH_2$. The π molecular orbitals can be written as linear combinations of two 2p atomic orbitals ϕ_1 and ϕ_2 centered on each carbon atom and lying perpendicular to the plane of the molecule:

$$\psi = c_1 \phi_1 + c_2 \phi_2 \tag{7.28}$$

The Hamiltonian matrix for ethylene can be written

```
(%i1)   H : alpha*ident(2) + beta*matrix([0,1],[1,0]);
```

$$(\%o1) \quad \begin{pmatrix} \alpha & \beta \\ \beta & \alpha \end{pmatrix}$$

where $\alpha = -11.3$ eV and $\beta = -2.4$ eV.

Setting up the characteristic polynomial with Equation (7.22) and solving for E, we have

```
(%i2)   determinant(H - E * ident(2))=0;
(%i3)   solve(%, E);
(%i4)   %, alpha = -11.3, beta = -2.4;
```

$$(\%o2) \quad (\alpha - E)^2 - \beta^2 = 0$$
$$(\%o3) \quad [E = \alpha - \beta, E = \beta + \alpha]$$
$$(\%o4) \quad [E = -8.9, E = -13.7]$$

In this case there are two eigenvalues E, corresponding to the doubly occupied ground state $E = -13.7$ eV and the unoccupied excited state $E = -8.9$ eV.

Maxima can compute the eigenvalues directly from the matrix:

```
eigenvalues(A)
```
Find the eigenvalues of a matrix A. The function returns two lists: a list of eigenvalues and a list of corresponding algebraic multiplicities.

```
(%i5)    eigenvalues(H);
```
(%o5) $[[\alpha - \beta, \beta + \alpha], [1, 1]]$

There is one eigenvalue $\alpha - \beta$ and one eigenvalue $\beta + \alpha$, with each singly degenerate.

The easiest way to find both the eigenvalues and eigenvectors of a matrix is to use the `eigenvectors` function:

eigenvectors(A)
Find both the eigenvalues and eigenvectors of a matrix A. The function returns two nested lists. The first list has two elements: a list of eigenvalues and a list of corresponding algebraic multiplicities. The second list is a list of corresponding eigenspaces. Each eigenspace is a list of its basis vectors (eigenvectors).

The output of the `eigenvectors` function is a deeply nested list. We can extract the energies, multiplicities, and orbitals as follows:

```
(%i1)    eigenvectors(H);
```
(%o6) $[[[\alpha - \beta, \beta + \alpha], [1, 1]], [[[1, -1]], [[1, 1]]]]$

```
(%i2)    [[energies, multiplicities], [[orbital1], [orbital2]]] :
    ➥ %$
```

```
(%i3)    energies;
```
(%o8) $[\alpha - \beta, \beta + \alpha]$

```
(%i4)    multiplicities;
```
(%o9) $[1, 1]$

```
(%i5)    orbital1;
```
(%o10) $[1, -1]$

```
(%i6)    orbital2;
```
(%o11) $[1, 1]$

We'll explore this calculation further with several different conjugated systems in Worksheet 7.6.

7.2.7.2 Eigenvalues and Eigenvectors for Symmetric Matrices

The eigenvalues of a symmetric matrix are real. Unfortunately, the `eigenvalues` and `eigenvectors` functions will occasionally return real eigenvalues in apparently complex forms. These can be simplified to real forms by taking these steps:

1) Transform the complex output of `eigenvectors` to Cartesian form with `rectform`.
2) Reduce the imaginary part of the result to zero with `trigreduce`.
3) Display the result in numeric form with `float` or `, numer`.

For example, this symmetric matrix must have real eigenvalues:

```
(%i1)    A : matrix([1, 4, 1], [4, 1, 9], [1, 9, 1]);
(%i2)    eigenvalues(A);
```

(%o1) $\begin{pmatrix} 1 & 4 & 1 \\ 4 & 1 & 9 \\ 1 & 9 & 1 \end{pmatrix}$

(%o2) $\left[\left[\left(-\dfrac{\sqrt{3}\,i}{2}-\dfrac{1}{2}\right)\left(\dfrac{10\,\sqrt{9062}\,i}{3^{\frac{3}{2}}}+36\right)^{\frac{1}{3}}+\dfrac{98\left(\dfrac{\sqrt{3}\,i}{2}-\dfrac{1}{2}\right)}{3\left(\dfrac{10\,\sqrt{9062}\,i}{3^{\frac{3}{2}}}+36\right)^{\frac{1}{3}}}+1,\right.$

$\left(\dfrac{\sqrt{3}\,i}{2}-\dfrac{1}{2}\right)\left(\dfrac{10\,\sqrt{9062}\,i}{3^{\frac{3}{2}}}+36\right)^{\frac{1}{3}}+\dfrac{98\left(-\dfrac{\sqrt{3}\,i}{2}-\dfrac{1}{2}\right)}{3\left(\dfrac{10\,\sqrt{9062}\,i}{3^{\frac{3}{2}}}+36\right)^{\frac{1}{3}}}+1,$

$\left.\left(\dfrac{10\,\sqrt{9062}\,i}{3^{\frac{3}{2}}}+36\right)^{\frac{1}{3}}+\dfrac{98}{3\left(\dfrac{10\,\sqrt{9062}\,i}{3^{\frac{3}{2}}}+36\right)^{\frac{1}{3}}}+1\right],[1,1,1]\right]$

The eigenvalues can be simplified by placing them in rectangular form, reducing the resulting trigonometric expressions, and converting the final result to floating point numbers[5]:

```
(%i3)   rectform(%)$
(%i4)   trigreduce(%)$
(%i5)   float(%);
```

(%o5) $[[0.26119111858828, -8.50939209923565, 11.24820098064737], [1.0, 1.0, 1.0]]$

The same treatment can be used to obtain real eigenvectors. Imaginary parts that are the same order of magnitude as the unit round-off error can often be safely removed using the `realpart` function (Section 2.1.3).

If **x** is an eigenvector of **A** corresponding to the eigenvalue λ, then $c\mathbf{x}$ is also an eigenvector of **A** with corresponding eigenvalue $c\lambda$, where c is a nonzero constant. This makes it possible for us to obtain normalized eigenvectors. Use `uniteigenvectors` in place of `eigenvectors` for normalized eigenvectors[6]:

```
(%i1)   A : matrix([1, 2, 0], [2, 4, 0], [0, 0, 9]);
(%i2)   eigenvectors(A);
(%i3)   uniteigenvectors(A);
(%i4)   [[lambda, multiplicity], [[vector1],[vector2],[vector3
        ➥ ]]] : %$
```

(%o1) $\begin{pmatrix} 1 & 2 & 0 \\ 2 & 4 & 0 \\ 0 & 0 & 9 \end{pmatrix}$

(%o2) $[[[0,5,9],[1,1,1]],[[[1,-\dfrac{1}{2},0]],[[1,2,0]],[[0,0,1]]]]$

(%o3) $[[[0,5,9],[1,1,1]],[[[\dfrac{2}{\sqrt{5}},-\dfrac{1}{\sqrt{5}},0]],[[\dfrac{1}{\sqrt{5}},\dfrac{2}{\sqrt{5}},0]],[[0,0,1]]]]$

---

5 In Worksheet 7.6, we'll use big floats with 100 digits to perform this calculation to minimize round-off error (see Section 2.1.1).

6 You may have to `load(eigen)` if you use `uniteigenvectors` before the `eigenvalues` or `eigenvectors` functions, which load the `eigen` package automatically.

If **A** is a real symmetric matrix, its eigenvectors are **orthogonal** (Section 7.1). For the matrix above,

```
(%i5)  vector1 . vector2;
(%i6)  vector2 . vector3;
(%i7)  vector1 . vector3;
```

(%o5)   0
(%o6)   0
(%o7)   0

7.2.7.3   Matrix Diagonalization

Let's form a matrix **X** where each column is an unit eigenvector of **A**:

```
(%i8)  X : addcol(transpose(vector1), vector2, vector3);
```

$$(\%o8) \quad \begin{pmatrix} \frac{2}{\sqrt{5}} & \frac{1}{\sqrt{5}} & 0 \\ -\frac{1}{\sqrt{5}} & \frac{2}{\sqrt{5}} & 0 \\ 0 & 0 & 1 \end{pmatrix}$$

The operation $\mathbf{X}^{-1}\mathbf{AX}$ transforms **A** into a diagonal form,[7] with the eigenvalues along the diagonal:

```
(%i9)  invert(X) . A . X;
```

$$(\%o9) \quad \begin{pmatrix} 0 & 0 & 0 \\ 0 & 5 & 0 \\ 0 & 0 & 9 \end{pmatrix}$$

Maxima includes a function called `similaritytransform` that can automate this process. The function is identical to `uniteigenvectors`, but the left-and right-hand matrices are stored in global system variables `leftmatrix` and `rightmatrix`, respectively.

```
(%i10)  similaritytransform(A);
```

$$(\%o10) \quad [[[0, 5, 9], [1, 1, 1]], [[[-\frac{2}{\sqrt{5}}, -\frac{1}{\sqrt{5}}, 0]], [[-\frac{1}{\sqrt{5}}, \frac{2}{\sqrt{5}}, 0]], [[0, 0, 1]]]]$$

```
(%i11)  leftmatrix;
```

$$(\%o11) \quad \begin{pmatrix} \frac{2}{\sqrt{5}} & -\frac{1}{\sqrt{5}} & 0 \\ \frac{1}{\sqrt{5}} & \frac{2}{\sqrt{5}} & 0 \\ 0 & 0 & 1 \end{pmatrix}$$

---

7 The matrix **X** must have an inverse or diagonalization of **A** will not be possible. **X** will not be invertable if its determinant is zero, so this can happen if two or more of the eigenvalues of **A** are the same.

```
(%i12)   rightmatrix;
```

$$(\%o12) \quad \begin{pmatrix} \frac{2}{\sqrt{5}} & \frac{1}{\sqrt{5}} & 0 \\ -\frac{1}{\sqrt{5}} & \frac{2}{\sqrt{5}} & 0 \\ 0 & 0 & 1 \end{pmatrix}$$

```
(%i13)   leftmatrix . A . rightmatrix;
```

$$(\%o2) \quad \begin{pmatrix} 0 & 0 & 0 \\ 0 & 5 & 0 \\ 0 & 0 & 9 \end{pmatrix}$$

 Worksheet 7.5: Eigenvalues and Eigenvectors

In this worksheet, we'll look at the properties and behavior of eigenvalues and eigenvectors. We'll use the `eigenvectors` function to compute the energies and wavefunctions for simple conjugated and aromatic systems using the Hückel molecular orbital theory.

7.3   Vector Calculus

A field is a function of position. A scalar function $f(x, y, z)$ defines a scalar field, which associates a *value* with every point in space. A vector field is a function that associates a vector with every point.

In this section, we'll look at derivatives that can tell us how a field changes from point to point.

7.3.1   Derivative of a Vector with Respect to a Scalar

Consider a position vector **r** with components that are functions of time, t:

$$\mathbf{r} = t\mathbf{i} + 2\sin(t)\mathbf{j} + \cos(t)\mathbf{k}$$

In Maxima, we would write the vector as a list (see Section 7.1):

```
(%i1)   r: [t, 2*sin(t), cos(t)]$
```

If **r** describes the trajectory of an object, we can compute the object's velocity $\mathbf{v} = d\mathbf{r}/dt$ and acceleration $\mathbf{a} = d^2\mathbf{r}/dt^2$:

```
(%i2)   v: diff(r,t);
(%i3)   a: diff(r,t,2);
```

$$(\%o2) \quad [1, 2\cos(t), -\sin(t)]$$
$$(\%o3) \quad [0, -2\sin(t), -\cos(t)]$$

The derivative of a vector with respect to a scalar is calculated like the derivative of any other quantity in Maxima. Each component of v is the derivative of the corresponding component in **r**.

7.3.2 The Jacobian

Suppose we want to find how an electric field vector $\mathbf{E} = [E_x, E_y, E_z]$ changes with position $\mathbf{r} = [x, y, z]$. This corresponds to a derivative of a vector with respect to another vector. Because each component of $\mathbf{E}$ is a function of x, y, and z, the derivative $d\mathbf{E}/d\mathbf{r}$ is a matrix with elements dE_i/dr_j, where i and j are x, y, or z. The matrix is called the Jacobian matrix of $\mathbf{E}$ with respect to x:

```
(%i1)   E(x,y,z) := [E_ x(x,y,z), E_ y(x,y,z), E_ z(x,y,z)]$
(%i2)   r : [x, y, z]$
(%i3)   jacobian(E(x,y,z), r);
```

$$(\%o3) \quad \begin{pmatrix} \frac{d}{dx}E\_x(x,y,z) & \frac{d}{dy}E\_x(x,y,z) & \frac{d}{dz}E\_x(x,y,z) \\ \frac{d}{dx}E\_y(x,y,z) & \frac{d}{dy}E\_y(x,y,z) & \frac{d}{dz}E\_y(x,y,z) \\ \frac{d}{dx}E\_z(x,y,z) & \frac{d}{dy}E\_z(x,y,z) & \frac{d}{dz}E\_z(x,y,z) \end{pmatrix}$$

where we have defined a vector function $\mathbf{E}(x, y, z)$ and used the `jacobian` function to compute the Jacobian matrix:

`jacobian(f, x)`
Compute the derivative $d\mathbf{f}/d\mathbf{x}$, where f is a vector-valued function or expression and x is a vector.

More generally, the Jacobian has the form

$$\nabla \mathbf{v} = \begin{pmatrix} \frac{\partial v_x}{\partial x} & \frac{\partial v_x}{\partial y} & \frac{\partial v_x}{\partial z} \\ \frac{\partial v_y}{\partial x} & \frac{\partial v_y}{\partial y} & \frac{\partial v_y}{\partial z} \\ \frac{\partial v_z}{\partial x} & \frac{\partial v_z}{\partial y} & \frac{\partial v_z}{\partial z} \end{pmatrix} \qquad \text{The Jacobian matrix} \qquad (7.29)$$

The symbol ∇ (called "nabla" or "del" or "grad") is the three-dimensional generalization of the ordinary derivative operator d/dx (Equation 7.39). Applying ∇ to a vector field gives the Jacobian matrix; applying it to a scalar field gives the gradient vector, as we'll see in the next section.

The Jacobian is useful in transforming integrals from one coordinate system to another. Suppose we have a new coordinate system $[u, v, w]$, which can be written in terms of $[x, y, z]$ coordinates as functions $x(u, v, w)$, $y(u, v, w)$, and $z(u, v, w)$. Then

$$\iiint f(x, y, z)\, dx\, dy\, dz =$$
$$\iiint f(x(u, v, w), y(u, v, w), z(u, v, w)) \, |\mathbf{J}|\, du\, dv\, dw$$

General integral coordinate transformation from $[x, y, z]$ to $[u, v, w]$

$$(7.30)$$

where $|\mathbf{J}|$ is the determinant of the Jacobian. For example, let's compute $|\mathbf{J}|$ for the transformation of Cartesian to spherical coordinates:

```
(%i1)   jacobian([r*sin(theta)*cos(phi), r*sin(theta)*sin(phi),
        ➡ r*cos(theta)], [r, theta, phi]);
(%i2)   determinant(%);
(%i3)   trigrat(%);
```

$$(\%o1) \quad \begin{pmatrix} \cos(\phi)\sin(\theta) & \cos(\phi)\,r\cos(\theta) & -\sin(\phi)\,r\sin(\theta) \\ \sin(\phi)\sin(\theta) & \sin(\phi)\,r\cos(\theta) & \cos(\phi)\,r\sin(\theta) \\ \cos(\theta) & -r\sin(\theta) & 0 \end{pmatrix}$$

$$(\%o2) \quad \cos(\phi)^2\,r^2\sin(\theta)^3 - \sin(\phi)\,r\sin(\theta)\left(-\sin(\phi)\,r\sin(\theta)^2 - \sin(\phi)\,r\cos(\theta)^2\right) + \cos(\phi)^2\,r^2\cos(\theta)^2\sin(\theta)$$

$$(\%o3) \quad r^2\sin(\theta)$$

where the expression is simplified using the `trigrat` function.

The volume element $dx\,dy\,dz$ is replaced with $r^2\sin(\theta)\,dr\,d\theta\,d\phi$ in the transformed integral over spherical coordinates.

We can also use the Jacobian to write experimentally measurable expressions for any thermodynamic partial derivative [37], using Equation (7.31):

$$\left(\frac{\partial X}{\partial Y}\right)_{A,B,\dots} = \text{determinant}(\text{jacobian}([X,A,B,\dots],[Y,A,B,\dots])) \qquad \begin{matrix}\text{Jacobian form} \\ \text{for partial derivatives}\end{matrix} \qquad (7.31)$$

For example, let's write the constant volume heat capacity $C_V = (\partial U/\partial T)_V$ in terms of other experimentally measurable quantities like C_P (the constant pressure heat capacity), the coefficient of thermal expansion ($\alpha = (1/V)(\partial V/\partial T)_P$), and the coefficient of isothermal compressibility ($\kappa = -(1/V)(\partial V/\partial P)_T$).

Begin by writing the derivative in terms of S, P, V, and T. Since $dU = T\,dS - P\,dV$, and V is held constant in the derivative, we can write

$$C_V = T\left(\frac{\partial S}{\partial T}\right)_V \qquad (7.32)$$

Applying Equation (7.31), we have

```
(%i1)   eqn: Cv = T*'(determinant(jacobian([S,V],[T,V])))$
```

Notice that we've suppressed execution of the right-hand side with the single quote operator `'`. Before evaluating the equation, we must define the derivatives that appear in the Jacobian in terms of experimentally measurable quantities. In this case, the definitions are $(\partial V/\partial T)_P = V\alpha$, $(\partial S/\partial T)_P = C_P/T$, and $(\partial S/\partial V)_T = \alpha/\kappa$. Use `gradef` to define these derivatives:

```
(%i2)   gradef(V,T,alpha*V)$
(%i3)   gradef(S,T,Cp/T)$
(%i4)   gradef(S,V,alpha/kappa)$
```

Now we can evaluate the equation, using the *nouns* and *expand* switches:

```
(%i5)   eqn, nouns, expand;
```

$$(\%o2) \quad Cv = Cp - \frac{TV\,\alpha^2}{\kappa}$$

We'll explore this powerful technique for systematically deriving thermodynamic derivatives in Worksheet 7.7.

7.3.3   The Gradient

Think of the scalar field $f(x, y, z)$ as terrain, possibly with hills and valleys; the altitude at a point corresponds to the magnitude of the field. At any given point in the field, the gradient is the collection of slopes when we move in the x, y, and z directions.

A scalar field can describe the temperature distribution within an object, or the concentration of a solute in an unmixed solution, or the electron density in a molecule. In these cases the gradient describes how quickly the temperature, or concentration, or electron density changes when we step away from a point in different directions.

In problems involving the flow of energy or molecules through a surface, we often want to calculate a *flux*, which measures how much flow we have through a unit area per second. For example, to compute the flux of a chemical substance A diffusing through another substance B, we can use Fick's first law of diffusion:

$$\mathbf{J}_A = -D_{AB}\nabla c_A \qquad \text{Fick's first law of diffusion} \qquad (7.33)$$

where $\mathbf{J}_A$ is the molar flux of A (in moles per square meter per second), D_{AB} is a diffusion constant that depends on the nature of A and B (with units of square meters per second), and c_A is the concentration of A at a given point in space (with units of moles per cubic meter). The symbol ∇c_A is the gradient of c_A. A gradient is the derivative of a scalar field (c_A, in this case) with respect to a vector ($[x, y, z]$). Fick's law tells us that the flux is proportional to the gradient in the concentration.

The gradient of a scalar field $f(x, y, z)$ is defined as

$$\operatorname{grad} f = \nabla f = \mathbf{i}\frac{\partial f}{\partial x} + \mathbf{j}\frac{\partial f}{\partial y} + \mathbf{k}\frac{\partial f}{\partial z} \qquad \text{The gradient of } f(x, y, z) \qquad (7.34)$$

The gradient is a vector, so the molar flux $\mathbf{J}_A$ is also a vector.

There are two ways to compute gradients in Maxima.

1) If we have a Cartesian scalar function $f(x, y, z)$, we can use the `grad` noun and expand it with `express`. For example,

```
(%i1)   load(vect)$
(%i2)   grad(x^2 + y^3 + z^4);
(%i3)   express(%), diff;
```

$$(\%o2) \quad \operatorname{grad}\left(z^4 + y^3 + x^2\right)$$
$$(\%o3) \quad [2x, 3y^2, 4z^3]$$

We must load the `vect` package to do differential Cartesian vector operations. Because `grad` is a noun, it won't be evaluated directly. In the third line, we pass it to `express`, which expands it into an expression containing symbolic partial derivatives (which are also nouns). We use the `diff` switch to evaluate the derivatives.

2) If we have a more general scalar function $f(x_1, x_2, \ldots, x_n)$, we can use the `jacobian` function to compute the gradient. For example,

```
(%i1)   jacobian([x^2 + y^3 + z^4], [x,y,z]);
```

$$(\%o1) \quad \left(2x \ \ 3y^2 \ \ 4, z^3\right)$$

This approach gives the gradient as a row matrix. Notice that we have to put the scalar function f in square brackets, because `jacobian` expects its first argument to be vector-valued.

Let's use Equation (7.33) to compute the flux of a protein through water. If we suspend a tiny pellet of protein in a large water bath, the concentration of the diffused protein varies with time t and position x, y, z as [38]

```
(%i1)   c(x,y,z,t) := n[0]/(8*(%pi*D*t)^(3/2))*exp(-(x^2+y^2+z
        ➥ ^2)/(4*D*t));
```

$$(\%o1) \quad c\left(x, y, z, t\right) := \frac{n_0}{8\left(\pi D t\right)^{\frac{3}{2}}} \exp\left(\frac{-\left(x^2 + y^2 + z^2\right)}{4 D t}\right)$$

The molar flux of the protein is then

```
(%i2)   load(vect)$
(%i3)   J: express(-D*grad(c(x,y,z,t))), diff;
```

$$(\%o3) \quad \left[\frac{n_0 \, x \, e^{\frac{-z^2-y^2-x^2}{4tD}}}{16\,\pi^{\frac{3}{2}}\,t\,(t\,D)^{\frac{3}{2}}}, \; \frac{n_0 \, y \, e^{\frac{-z^2-y^2-x^2}{4tD}}}{16\,\pi^{\frac{3}{2}}\,t\,(t\,D)^{\frac{3}{2}}}, \; \frac{n_0 \, z \, e^{\frac{-z^2-y^2-x^2}{4tD}}}{16\,\pi^{\frac{3}{2}}\,t\,(t\,D)^{\frac{3}{2}}}\right]$$

Here, we've used the `diff` switch to compute the derivatives in the gradient expression. Using the Jacobian method gives equivalent results:

```
(%i4)   J: -D*jacobian([c(x,y,z,t)], [x, y, z]);
```

$$(\%o2) \quad \left(\frac{n_0 x \, \%e^{\frac{-z^2-y^2-x^2}{4Dt}}}{16\pi^{\frac{3}{2}}t\,(Dt)^{\frac{3}{2}}} \quad \frac{n_0 y \, \%e^{\frac{-z^2-y^2-x^2}{4Dt}}}{16\pi^{\frac{3}{2}}t\,(Dt)^{\frac{3}{2}}} \quad \frac{n_0 z \, \%e^{\frac{-z^2-y^2-x^2}{4Dt}}}{16\pi^{\frac{3}{2}}t\,(Dt)^{\frac{3}{2}}}\right)$$

Explicitly write the gradient in the coordinate system of interest. For example, in spherical coordinates, the gradient operator is

$$\nabla f(r, \theta, \phi) = \mathbf{r}\frac{\partial f}{\partial r} + \boldsymbol{\theta}\frac{1}{r}\frac{\partial f}{\partial \theta} + \boldsymbol{\phi}\frac{1}{r\sin\theta}\frac{\partial f}{\partial \phi} \qquad \begin{array}{l}\text{Gradient of a scalar field} \\ \text{in spherical coordinates}\end{array} \qquad (7.35)$$

where the unit vectors $\mathbf{r}$, θ, and ϕ can be written in terms of the Cartesian unit vectors $\mathbf{i}$, $\mathbf{j}$, and $\mathbf{k}$:

$$\begin{aligned} \mathbf{r} &= \mathbf{i}\sin\theta\cos\phi + \mathbf{j}\sin\theta\sin\phi + \mathbf{k}\cos\theta \\ \boldsymbol{\theta} &= \mathbf{i}\cos\theta\cos\phi + \mathbf{j}\cos\theta\sin\phi + \mathbf{k}\sin\theta \qquad \text{Unit vectors in spherical coordinates} \\ \boldsymbol{\phi} &= -\mathbf{i}\sin\phi + \mathbf{j}\cos\phi \end{aligned}$$

$$(7.36)$$

For example, the spherical gradient of a function $f(r, \theta, \phi)$ would be computed in Maxima as

```
(%i1)   jacobian([f(r,theta,phi)], [r,theta,phi]) * matrix([1,
        ➥ 1/r, 1/(r*sin(theta))]);
```

$$(\%o1) \quad \left(\frac{d}{dr} f\left(r, \theta, \phi\right) \quad \frac{\frac{d}{d\theta} f(r,\theta,\phi)}{r} \quad \frac{\frac{d}{d\phi} f(r,\theta,\phi)}{r\sin(\theta)}\right)$$

In cylindrical coordinates, we have

$$\nabla f(r, \theta, z) = \mathbf{r}\frac{\partial f}{\partial r} + \boldsymbol{\theta}\frac{1}{r}\frac{\partial f}{\partial \theta} + \mathbf{k}\frac{\partial f}{\partial z} \qquad \begin{array}{l}\text{Gradient of a scalar field} \\ \text{in cylindrical coordinates}\end{array} \qquad (7.37)$$

with unit vectors

$$\mathbf{r} = \mathbf{i}\cos\theta + \mathbf{j}\sin\theta$$
$$\theta = -\mathbf{i}\sin\theta + \mathbf{j}\cos\theta$$

Unit vectors in cylindrical coordinates   (7.38)

 Worksheet 7.6: Vector Derivatives

In this worksheet, we'll compute the derivatives of a vector with respect to both scalars and vectors. We'll also use the vector derivatives to transform integrals from one coordinate system to another, link thermodynamic derivatives with experimental quantities, and use gradients to solve problems involving diffusion and thermal conductivity.

7.3.4 The Laplacian

We can think of ∇ as an operator that differentiates the scalar field in the x, y, and z directions:

$$\nabla = \left[\frac{\partial}{\partial x}, \frac{\partial}{\partial y}, \frac{\partial}{\partial z}\right] \qquad \nabla \text{ as an operator} \qquad (7.39)$$

By convention, ∇ operates on whatever quantity is placed to its right. $\nabla\phi$ means "apply ∇ to ϕ," but $\phi\nabla$ means "apply ∇ to whatever is to the right of it, and then multiply by ϕ."

The ∇ operator has x, y, and z components, so it can be thought of as a sort of vector. We've already seen what happens when ∇ operates directly on a scalar f: it yields the gradient of f, ∇f, which is a vector. We've also seen how it operates on a vector $\mathbf{v} = [v_x, v_y, v_z]$: we get the Jacobian matrix (Equation 7.29). Applying ∇ to the gradient of a scalar field $\nabla(\nabla f)$ yields the Hessian matrix (Equation 6.14), which we've computed with the `hessian` function (Section 6.4.1 and Worksheet 7.6).

But what happens when this "vector" ∇ is dotted or crossed with $\mathbf{v}$ or its gradient[8]? The answer turns out to be profoundly important – it leads us to mathematical constructs that describe the behavior of light, sound, fluid flow, diffusion, and electrons in atoms and molecules.

The dot product of ∇ with the gradient of a scalar field ∇f is

$$\nabla \cdot \nabla f = \left[\frac{\partial}{\partial x}, \frac{\partial}{\partial y}, \frac{\partial}{\partial z}\right] \cdot \left[\frac{\partial f}{\partial x}, \frac{\partial f}{\partial y}, \frac{\partial f}{\partial z}\right]$$
$$= \frac{\partial^2 f}{\partial x^2} + \frac{\partial^2 f}{\partial y^2} + \frac{\partial^2 f}{\partial z^2}$$

The Laplacian of $f(x, y, z)$ (7.40)

The result is a scalar, as we'd expect from the dot product of two vectors. The operator $\nabla \cdot \nabla$ is usually written ∇^2 (usually read as "del squared"). It is called the Laplacian, and it appears in equations that describe the behavior of electric potentials, diffusing solutes, heat flow, and quantum mechanical particles.

In Maxima, we write ∇^2 as `laplacian` and again use the `express` function to expand it in terms of partial derivatives.

```
(%i1)   load(vect)$
(%i2)   f(x,y,z) := z*exp(-(x^2+y^2+z^2)/2);
(%i3)   express(laplacian(f(x,y,z))), diff, ratsimp;
```

---

8 Keep in mind that ∇ is not really a vector; it's a vector operator. $\nabla \times \mathbf{v}$ is not a cross product; it is merely a symbol for curl $\mathbf{v}$. Nor is the symbol $\nabla \cdot \mathbf{v}$ actually a dot product; it is a symbol for the divergence of $\mathbf{v}$ [39].

$(\%o2)\quad \mathrm{f}\left(x,y,z\right) := z \exp\left(\dfrac{-\left(x^2+y^2+z^2\right)}{2}\right)$

$(\%o3)\quad \left(z^3+\left(y^2+x^2-5\right)z\right)\%e^{-\frac{z^2}{2}-\frac{y^2}{2}-\frac{x^2}{2}}$

The `diff` switch forces evaluation of the derivatives (which are otherwise just written out).

The Laplacian appears in calculations of the kinetic energy of quantum mechanical particles. For example, we can find the energies of a quantum mechanical particle trapped in a cubic box with side length L by applying the Hamiltonian operator $-\left(h^2/8\pi^2 m\right)\nabla^2$ to the wavefunction ψ for a particle. For a particle of mass m, the wavefunction is

```
(%i1)   load(vect)$
(%i2)   declare(L, constant)$
(%i3)   psi(x,y,z):=(2/L)^(3/2)*sin(nx*%pi*x/L)*sin(ny*%pi*y/L)*
    ➡ sin(nz*%pi*z/L);
```

$(\%o3)\quad \psi\left(x,y,z\right) := \left(\dfrac{2}{L}\right)^{\frac{3}{2}} \sin\left(\dfrac{nx\,\pi\,x}{L}\right) \sin\left(\dfrac{ny\,\pi\,y}{L}\right) \sin\left(\dfrac{nz\,\pi\,z}{L}\right)$

Computing $\mathcal{H}\psi$, we have

```
(%i4)   -h^2/(8*%pi^2*m)*laplacian(psi(x,y,z));
(%i5)   express(%), diff, ratsimp;
(%i6)   %/psi(x,y,z), factor;
```

$(\%o4)\quad -\dfrac{h^2\,\text{laplacian}\left(\sin\left(\frac{\pi\,nx\,x}{L}\right)\sin\left(\frac{\pi\,ny\,y}{L}\right)\sin\left(\frac{\pi\,nz\,z}{L}\right)\right)}{2^{\frac{3}{2}}\,\pi^2\,L^{\frac{3}{2}}\,m}$

$(\%o5)\quad \dfrac{\left(h^2\,nz^2+h^2\,ny^2+h^2\,nx^2\right)\sin\left(\frac{\pi\,nx\,x}{L}\right)\sin\left(\frac{\pi\,ny\,y}{L}\right)\sin\left(\frac{\pi\,nz\,z}{L}\right)}{2^{\frac{3}{2}}\,L^{\frac{7}{2}}\,m}$

$(\%o6)\quad \dfrac{h^2\left(nz^2+ny^2+nx^2\right)}{8\,L^2\,m}$

The last line gives the energy of the particle in the box in terms of the quantum numbers nx, ny, and nz.

The Laplacian can operate on vector fields as well as scalar fields. Vector field Laplacians are encountered in electromagnetic wave equations and problems dealing with fluid flow. For a Cartesian vector $\mathbf{v}=(v_x,v_y,v_z)$, the vector Laplacian is

$$\nabla^2\mathbf{v} = \left(\nabla^2 v_x,\nabla^2 v_y,\nabla^2 v_z\right) \qquad \text{The vector Laplacian in Cartesian coordinates} \qquad (7.41)$$

```
(%i1)   express(laplacian([vx(x,y,z), vy(x,y,z), vz(x,y,z)])),
    ➡ diff;
```

(%o1) $\left[\dfrac{d^2}{dz^2}\,\text{vx}\,(x,y,z)+\dfrac{d^2}{dy^2}\,\text{vx}\,(x,y,z)+\dfrac{d^2}{dx^2}\,\text{vx}\,(x,y,z)\,,\dfrac{d^2}{dz^2}\,\text{vy}\,(x,y,z)+\dfrac{d^2}{dy^2}\,\text{vy}\,(x,y,z)\right.$

$\left.+\dfrac{d^2}{dx^2}\,\text{vy}\,(x,y,z)\,,\dfrac{d^2}{dz^2}\,\text{vz}\,(x,y,z)+\dfrac{d^2}{dy^2}\,\text{vz}\,(x,y,z)+\dfrac{d^2}{dx^2}\,\text{vz}\,(x,y,z)\right]$

```
(%i2)   v : [x^2*y, y^2*z, z^2*x]$
(%i3)   express( laplacian(v)), diff;
```

(%o8) $[2\,y, 2\,z, 2\,x]$

Maxima does not include Laplacian operators for cylindrical or spherical coordinates. We'll build them ourselves in Worksheet 7.7, using Equations (7.42) and (7.43).

$$\nabla^2 f(r,\theta,z) = \frac{1}{r}\frac{\partial}{\partial r}\left(r\frac{\partial f}{\partial r}\right)+\frac{1}{r^2}\frac{\partial^2 f}{\partial\phi^2}+\frac{\partial^2 f}{\partial z^2}\qquad \text{The Laplacian in cylindrical coordinates}\qquad (7.42)$$

$$\nabla^2 f(r,\theta,\phi) = \frac{1}{r}\frac{\partial^2(rf)}{\partial r^2}+\frac{1}{r^2\sin\theta}\frac{\partial}{\partial\theta}\left(\sin\theta\frac{\partial f}{\partial\theta}\right)+\frac{1}{r^2\sin^2\theta}\frac{\partial^2 f}{\partial\phi^2}\qquad \text{The Laplacian in spherical coordinates}$$

$$(7.43)$$

In either coordinate system, the Laplacian is undefined at $r = 0$. In spherical coordinates, we cannot have $\sin\theta$ equal to zero, so we also have to avoid points with $\theta = 0$ or $\theta = \pi$ in our calculations.

7.3.5   The Divergence

The divergence of a vector field $\mathbf{v}(x,y,z)$ is the dot product of the operator ∇ with a vector,

$$\text{div}\,\mathbf{v} = \nabla\cdot\mathbf{v} = \frac{\partial v_x}{\partial x}+\frac{\partial v_y}{\partial y}+\frac{\partial v_z}{\partial z}\qquad \text{The divergence of }\mathbf{v}(x,y,z)\qquad (7.44)$$

which is a scalar.

In Maxima, the divergence of $\mathbf{v}$ is written as `div(v)`. You must use the `express` function to expand it in terms of partial derivatives. For example,

```
(%i1)   load(vect)$
(%i2)   v : [-y, 2*x, z]$
(%i3)   express(div(v)), diff;
```

(%o3)   1

which we can check using Equation (7.44):

```
(%i4)   diff(v[1],x) + diff(v[2],y) + diff(v[3],z);
```

(%o4)

The Laplacian is the divergence of the gradient of a scalar field, that is, $\nabla\cdot\nabla f = \nabla^2 f$. Typing `div(grad(f))` is equivalent to `laplacian(f)`.

```
(%i5)   express( div(grad(f)) );
(%i6)   express( laplacian(f));
```

$$(\%o5) \quad \frac{d^2}{dz^2}f + \frac{d^2}{dy^2}f + \frac{d^2}{dx^2}f$$

$$(\%o6) \quad \frac{d^2}{dz^2}f + \frac{d^2}{dy^2}f + \frac{d^2}{dx^2}f$$

The divergence is particularly important in problems involving fluid flow. For example, suppose we have a vector field $\mathbf{v}(x, y, z, t)$ that shows wind velocity. The net mass of air flowing into a volume element $dx\,dy\,dz$ in unit time is $-(\nabla \cdot \mathbf{v})\,\rho\,dx\,dy\,dz$, where ρ is the air density (also a function of position and time). Conservation of mass requires that this mass must equal the time rate of change of the density times the volume, so

$$-(\nabla \cdot \mathbf{v})\,\rho\,dx\,dy\,dz = \left(\frac{\partial\rho}{\partial t}\right) dx\,dy\,dz \tag{7.45}$$

or

$$\nabla \cdot \mathbf{v} = -\frac{1}{\rho}\left(\frac{\partial\rho}{\partial t}\right) \qquad \begin{array}{l}\text{Hydrodynamic}\\\text{continuity}\end{array} \tag{7.46}$$

In other words, $\nabla \cdot \mathbf{v}$ describes how the mass of the fluid "diverges" out of the volume element. If the density is constant, there will be no divergence.

Maxima's `div` operator applies to vector fields in Cartesian coordinates. In cylindrical coordinates, the divergence is

$$\mathbf{div}\,\mathbf{v}(r, \theta, z) = \frac{1}{r}\frac{\partial}{\partial r}\left(r\,v_r\right) + \frac{1}{r}\frac{\partial v_\theta}{\partial\theta} + \frac{\partial v_z}{\partial z}\,(\mathbf{k}\cdot\mathbf{v}) \qquad \begin{array}{l}\text{The divergence in}\\\text{cylindrical coordinates}\end{array} \tag{7.47}$$

where v_r, v_θ, and v_z are the cylindrical coordinates of $\mathbf{v}$ and $\mathbf{k}$ is the unit vector in the z direction.

In spherical coordinates,

$$\mathbf{div}\,\mathbf{v}(r, \theta, \phi) = \frac{1}{r^2}\frac{\partial}{\partial r}\left(r^2 v_r\right) + \frac{1}{r\sin\theta}\frac{\partial}{\partial\theta}\left(\sin\theta v_\theta\right) + \frac{1}{r\sin\theta}\frac{\partial v_\phi}{\partial\phi} \qquad \begin{array}{l}\text{The divergence in}\\\text{spherical coordinates}\end{array}$$
$$\tag{7.48}$$

where v_r, v_θ, and v_ϕ are the spherical coordinates of $\mathbf{v}$.

7.3.6 The Curl

The curl or rotation of a vector field $\mathbf{v}(x, y, z)$ is defined as

$$\mathbf{curl}\,\mathbf{v} = \nabla \times \mathbf{v} = \left(\frac{\partial v_z}{\partial y} - \frac{\partial v_y}{\partial z}\right)\mathbf{i} + \left(\frac{\partial v_x}{\partial z} - \frac{\partial v_z}{\partial x}\right)\mathbf{j} + \left(\frac{\partial v_y}{\partial x} - \frac{\partial v_x}{\partial y}\right)\mathbf{k} \tag{7.49}$$

The curl is more compactly written as a determinant:

$$\mathbf{curl}\,\mathbf{v} = \nabla \times \mathbf{v} = \begin{vmatrix} \mathbf{i} & \mathbf{j} & \mathbf{k} \\ \frac{\partial}{\partial x} & \frac{\partial}{\partial y} & \frac{\partial}{\partial z} \\ v_x & v_y & v_z \end{vmatrix} \qquad \text{The curl of } \mathbf{v} \tag{7.50}$$

If the vector field showed wind velocity, the curl of the field would indicate the circulation of air around a particular point.

Again, the `express` function can be used to write out the curl of a vector field in terms of partial derivatives, using the noun `curl`. For example,

```
(%i7)    v : [x*y, y*z, z*x] $
(%i8)    express( curl(v) ), diff;
(%o7)    [−y, −z, −x]
```

If the vector field is the gradient of a scalar field, it will have no curl. For example, electrostatic field E is the gradient of electrostatic potential ϕ, so

```
(%i9)    declare(phi, scalar)$
(%i10)   E: express(grad(phi(x,y,z)));
(%i11)   express(curl(E));
```

$$(\%o8) \quad [\frac{d}{dx}\phi\left(x,y,z\right), \frac{d}{dy}\phi\left(x,y,z\right), \frac{d}{dz}\phi\left(x,y,z\right)]$$

$$(\%o9) \quad [0,0,0]$$

If the vector field is the curl of another vector field, it will have zero divergence.

```
(%i12)   declare(vec,nonscalar)$
(%i13)   express(div(curl(vec)));
```

$$(\%o10) \quad 0$$

Again, Maxima's `curl` operator is in Cartesian coordinates. In cylindrical coordinates, we have

$$\mathbf{curl}\,\mathbf{v}(r,\theta,z) = \mathbf{r}\left(\frac{1}{r}\frac{\partial v_z}{\partial \theta} - \frac{\partial v_\theta}{\partial z}\right)$$
$$+\boldsymbol{\theta}\left(\frac{\partial v_r}{\partial z} - \frac{\partial v_z}{\partial r}\right) \qquad \text{The curl of } \mathbf{v}(r,\theta,z) \qquad (7.51)$$
$$+\mathbf{k}\left(\frac{1}{r}\frac{\partial}{\partial r}\left(rv_\theta\right) - \frac{1}{r}\frac{\partial v_r}{\partial \theta}\right)$$

where the unit vectors $\mathbf{r}$ and $\boldsymbol{\theta}$ are given in terms of Cartesian unit vectors $\mathbf{i}, \mathbf{j}$, and $\mathbf{k}$ in Equation (7.38).

In spherical coordinates,

$$\mathbf{curl}\,\mathbf{v}(r,\theta,\phi) = \mathbf{r}\frac{1}{r\sin\theta}\left[\frac{\partial}{\partial \theta}\left(v_\phi \sin\theta\right) - \frac{\partial v_\theta}{\partial \phi}\right]$$
$$+\boldsymbol{\theta}\frac{1}{r}\left[\frac{1}{\sin\theta}\frac{\partial v_r}{\partial \phi} - \frac{\partial}{\partial r}\left(rv_\phi\right)\right] \qquad \text{The curl of } \mathbf{v}(r,\theta,\phi) \qquad (7.52)$$
$$+\boldsymbol{\phi}\frac{1}{r}\left[\frac{\partial}{\partial r}\left(rv_\theta\right) - \frac{\partial v_r}{\partial \theta}\right]$$

where the unit vectors are given in Equation (7.36).

 Worksheet 7.7: Vector Operators

In this worksheet, we'll solve several quantum mechanical problems by applying Laplacian operators to functions in Cartesian, spherical, and cylindrical coordinates. We'll also explore problems involving electromagnetic fields using the div and curl operators.

8

Error Analysis

Every careful measurement in science is always given with the probable error... every observer admits that he is likely wrong, and knows about how much wrong he is likely to be.

– Bertrand Russell [40]

Error analysis is an essential part of the process of science. To draw valid conclusions from experimental data, we must have some estimate of the precision (reproducibility) and accuracy (correctness) of the measurements. To have confidence in computed results, we must propagate errors in the data through our calculations. To compare our results with accepted values, we have to objectively compare the size of the experimental error with the size of any observed discrepancies. To design efficient experiments, we have to know the relative sizes of contributing errors so that we can identify and minimize the errors that limit the accuracy and precision of the final results.

In this chapter, we'll use symbolic mathematics to estimate errors in datasets and propagate them through calculations. We'll also see how statistics and assumptions about the distribution of errors can be used to objectively test hypotheses about the data.

 Throughout this chapter we will use functions from the `stats` package, which must be loaded before some of the functions will work.

```
(%i1)   load(stats)$
```

8.1   Classifying Experimental Errors

Some of the uncertainty in a measurement can be estimated by repeating the measurement under identical conditions. Consider a dataset consisting of 20 replicate readings x_i from a digital thermometer.

```
(%i1)   data: [23.9, 24.0, 24.0, 24.0, 24.0, 24.0, 24.0, 24.0,
        ➡ 24.0, 24.0, 24.1,
(%i2)   24.1, 24.1, 24.1, 24.1, 24.1, 24.1, 24.1, 24.2, 24.2]$
```

How can we estimate the "true" temperature from this data? How uncertain may that estimate be, based on the scatter in the data?

Symbolic Mathematics for Chemists: A Guide for Maxima Users, First Edition. Fred Senese.
© 2019 John Wiley & Sons Ltd. Published 2019 by John Wiley & Sons Ltd.
Companion website: http://booksupport.wiley.com

The answers to both questions lie in the *distribution* of the measurements. Note that the values occur more frequently than others; these measurements are more reproducible and *perhaps* more reliable than the others. The frequencies for each unique value can be found with the `discrete_freq` function, found in the `stats` package:

```
(%i3)   [uniqueData, frequencies] : discrete_freq(data);
(%i4)   uniqueData;
(%i5)   frequencies;
```

```
(%o2)   [[23.9, 24.0, 24.1, 24.2], [1, 9, 8, 2]]
(%o3)   [23.9, 24.0, 24.1, 24.2]
(%o4)   [1, 9, 8, 2]
```

The first list gives the unique sample values, and the second gives their corresponding frequency (the number of times that value occurs in the data). For example, the reading 24.0 occurs nine times in the dataset.

The sum of the frequencies is the number of points in the dataset:

```
(%i6)   N: apply("+", frequencies);
(%o7)   20
```

The `histogram` and `wxhistogram` functions automatically compute and plot these frequencies[1]:

```
(%i7)   wxhistogram(data,
(%i8)       ylabel="frequency",
(%i9)       xlabel="reading",
(%i10)      fill_density=0.6
(%i11)   )$
```

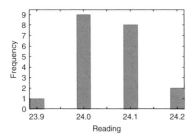

For this data the frequencies rise as the values approach the average or mean value. For the observed values $x_1, \ldots, x_N$, the mean value $\bar{x}$ is

$$\bar{x} = \frac{1}{N} \sum_{i=1}^{N} x_i \qquad \text{The sample mean} \tag{8.1}$$

which can be computed directly in Maxima with the `mean` function:

```
(%i12)   mean(data);
(%o3)    24.05500000000001
```

---

1 See Section 3.1.10 and Worksheet 3.6 for more about Maxima's commands for plotting histograms.

Figure 8.1 A series of measurements that contain both random and systematic errors. Systematic error shifts the average measurement away from the true value, affecting accuracy; random error scatters the measurements around their mean, affecting precision.

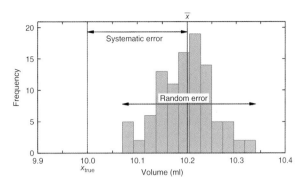

If the true temperature lies between the lowest and highest readings, the uncertainty can be written as the range

```
(%i13)   [apply(min,data),  apply(max,data)];
(%o2)    [23.9, 24.2]
```

We can say that the temperature may be somewhere between 23.9 and 24.2 °C, but is probably closest to 24.055 °C.

Suppose a more trustworthy thermometer gives a temperature of 23.952 °C for the same sample. If we call this the "true" value, μ, we can partition the error δx_i for each observed value into two parts:

$$\delta x_i = (x_i - \overline{x}) + (\overline{x} - \mu) \tag{8.2}$$

The first term arises from random error that scatters the observed values around their mean. Random error affects the reproducibility or precision of the data.

The second term arises from systematic error or bias the data. It is *not* random; it has the same sign and size for all of the data. It tells us something about the correctness or accuracy of the mean as an estimate of the true value (see Figure 8.1).

Separating these two types of errors and estimating their sizes is an important goal in data analysis. Statistical methods can help describe and draw conclusions about random error in large datasets, but systematic error can only be detected and corrected through *calibration* (comparison with a known value). We also have to handle the two types of errors differently when we propagate errors through a calculation.

8.1.1   Systematic Error

Systematic errors are also called *determinate errors* because they have definite causes, such as a procedural flaws, contaminants, or miscalibrated instruments. Finding the cause of a systematic error requires a detailed understanding of the measurement process, and it can be one of the most difficult challenges in interpreting experimental results.

A notorious example of systematic error occurred in September 2011, when researchers at Italy's Gran Sasso National Laboratory reported measurements that showed neutrinos traveling at speeds about 0.002% greater than the speed of light. This apparent violation of Einstein's theory of special relativity immediately made international news. The researchers (and the scientific community) spent the following months searching for some systematic error in their measurements. Within a few months, the probable source of the error was located: a loose fiber-optic cable had introduced a small delay in timing [41].

The simplest systematic errors are *offsets* that have consistent sign and magnitude. Such errors are estimated by repeating the measurement on a series of standards, (samples where the true value is known). The discrepancy between the observed and true values can then be used to correct measurements on further samples. For example, when no one's standing on a bathroom scale, the mass should read zero. If instead it consistently reads 1.2 kg, you can correct masses read from that scale by subtracting 1.2 kg from them.

Another type of systematic error can occur when an instrument has *scaling errors* that cause the slopes of calibration curves to change. For example, the true reading R might be related to the instrument reading r by $r = aR + b$, where a is scale parameter and b is an offset. Calibration of the instrument with several standards (which have known R) can help us find a and b, using a linear fit (Chapter 9).

Instrument *drift* is another type of systematic error. Drift often occurs when an important variable isn't being properly controlled in an experiment. For example, the absorbance read from a spectrophotometer may drift slowly over time due to lamp and detector instabilities. You can recalibrate the instrument periodically by forcing it to read zero absorbance for a "blank," or you can correct the readings for drift over time.

8.1.2   Random Error

Random error arises from uncontrolled changes from run to run.

For different datasets, the means differ slightly. They also differ from the true or correct value. As the sample size becomes larger and larger, random errors on the either side of the mean tend to balance each other, and the mean would be expected to converge towards the true value (unless systematic error is present).

Each sample can be imagined as a subset of all possible measurements of x, called the population. The distribution of measurements in the entire population would give the true mean, called the population mean μ. The means for larger and larger datasets should approach μ.

8.2   Probability Density

In practice, we cannot know the population mean because we can only make a finite number of measurements. But if we can make some reasonable assumptions about what the population distribution looks like, we can find simple mathematical forms for the population distribution that tell us how probable a particular error is. Such probabilities can be quite valuable in finding intervals around the mean where the population mean is likely to be. We can also use these probabilities to test hypotheses about the data.

8.2.1   Discrete Probability Distributions

When a dataset contains discrete values x_k that occur with frequency f_k, Equation (8.1) can be written as

$$\bar{x} = \frac{1}{N} \sum_{j}^{M} x_j f_j \qquad \text{A weighted average} \qquad (8.3)$$

This sum can be written as the dot product of the unique values with their frequencies, divided by the total number of measurements:

```
(%i14)   uniqueData . frequencies / N;
(%o8)    24.055
```

The fraction f_i/N estimates the probability $P(x_i)$ of obtaining a measurement x_i. For example, $P(24.0\,°C)$ is 9/20, or 0.45.

```
(%i15)   P: frequencies/N;
```
$$(\%o9) \quad [\frac{1}{20}, \frac{9}{20}, \frac{2}{5}, \frac{1}{10}]$$

Probabilities must be real numbers between 0 and 1. A probability of zero means that an event is impossible; a probability of one means that the event will always occur. The sum of probabilities for all readings must be one.

```
(%i16)   apply("+", P);
(%o10)   1
```

This requirement is called normalization. We can write

$$\sum_i^M P(x_i) = \frac{1}{N} \sum_i^M f(x_i) = 1 \qquad \text{Normalized discrete probability density} \qquad (8.4)$$

The function $P(x)$ is a normalized probability density function. It predicts the probability that a measurement will take on a value of x. For the temperature readings, $P(23.9) = 1/20, P(24.0) = 9/20, P(24.1) = 8/20$, and $P(24.2) = 2/20$.

The histogram function can plot normalized probability densities automatically using the option *frequency=density*. For example, let's plot the normalized probability density for 1000 random numbers between 1 and 10, generated with the random_discrete_uniform function.

```
(%i1)   data: random_discrete_uniform(10,1000)$
(%i2)   histogram( data,
            frequency=density,
            xlabel="uniform random number",
            ylabel="probability",
            title="Normalized probability density of 1000 random
               ➥ numbers"
        )$
```

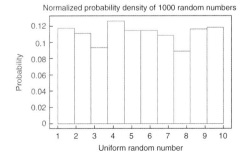

The data consists of integers between 1 and 10. By default, `histogram` sorts the numbers into 10 bins dividing the interval [1,10]. The bars show the probability that a number will fall into a particular bin, not the probability of each integer occurring,[2] not the probabilities of the individual integers. In this case, we can use the output of the `discrete_freq` function with `plot2d` and the *impulses* style to get a better plot:

```
(%i3)   [uniqueData, frequencies] : discrete_freq(data);
```
$$(\%o3)\quad [[1, 2, 3, 4, 5, 6, 7, 8, 9, 10], [103, 96, 98, 93, 110, 99, 106, 109, 89, 97]]$$

```
(%i4)   N : apply("+", frequencies);
(%o4)   1000
```

```
(%i5)   P : frequencies/N;
```
$$(\%o5)\quad [\frac{103}{1000}, \frac{12}{125}, \frac{49}{500}, \frac{93}{1000}, \frac{11}{100}, \frac{99}{1000}, \frac{53}{500}, \frac{109}{1000}, \frac{89}{1000}, \frac{97}{1000}]$$

```
(%i6)   plot2d([discrete, uniqueData, P], [x, 0, 11], [style,
        ➥ impulses],
        [xlabel, "uniform random number"],
        [ylabel, "probability"],
        [title, "Normalized probability density of 1000
        ➥ random numbers"]
        )$
```

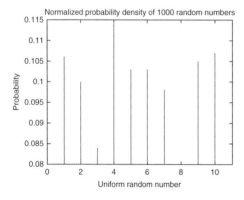

8.2.2 The Poisson Distribution

When the sample size is large, the probability density often converges to a smoothly varying limiting distribution. Limiting distributions can be used to compute probabilities directly.

Suppose we are interested in measuring a count: the number of photons emitted by a chemiluminescent reaction or the number of particles emitted by a radioactive substance. Each of these events occurs randomly, but they have a fixed average rate r. If we measure the number of

---

2 Recall from Section 3.1.10 that the number of bins in `histogram` can be set with the *nclasses=b* option, where *b* is the number of bins you want.

events that occur over time t, and we repeat the counting experiment many times, the average count μ will be

$$\mu = rt \qquad \text{Average number of counts for many counting experiments} \qquad (8.5)$$

The probability of counting exactly n events is

$$P(n) = \frac{\mu^n}{n!} \exp(-\mu) \qquad \text{The Poisson distribution} \qquad (8.6)$$

This probability density function is called the Poisson distribution. For example, suppose we want to know how likely it will be for 10 molecules to arrive over a 1 μs interval at the surface of a sensor. If an average of 15 molecules arrive each microsecond, the probability of having 10 molecules arrive is $\frac{15^{10}}{10!} \exp(-15) = 0.0486$.

Let's plot the Poisson distribution in this case. After building a list of 30 $(n, P(n))$ points with $\mu = 15$, we plot them using the *impulses* style:

```
(%i1)   P(n,mu) := mu^n / n! * exp(-mu);
(%i2)   points: makelist([n, P(n,mu)], n, 0, 30), mu=15$
(%i3)   plot2d([discrete,points],
            [style, impulses],
            [title, "The Poisson Distribution with μ=15 and 30
              ➥ points"],
            [ylabel, "P(n)"],
            [xlabel, "n"])$
```

$$(\%o1) \quad P(n, \mu) := \frac{\mu^n}{n!} \exp(-\mu)$$

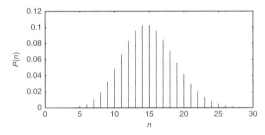

The Poisson distribution is discrete (the count n is an integer). It isn't quite symmetric; it tails off faster on the left than it does on the right. The mean of the distribution is μ. We can describe the width of the distribution in terms of the root-mean-square deviation from the mean (which is called the standard deviation when there are a large number of trials):

$$\sigma = \sqrt{\mu} \qquad \text{Standard deviation } \sigma \text{ for the Poisson distribution} \qquad (8.7)$$

The Poisson distribution as $P(n, \mu)$ is a built-in Maxima function `pdf_poisson(n, μ)`. A list of built-in functions for performing common tasks with Poisson-distributed variables is given in Table 8.1.

The Poisson-distributed random numbers generated by `random_poisson` can be used to simulate counting experiments. For example, photon counts arriving at a detector [42], particle counts in radioactive decay [43], the number of mutations along a DNA strand induced by

Table 8.1 Functions for performing common tasks with Poisson-distributed variables.

Task	Maxima function
Probability of counting exactly n events	pdf_poisson(n, μ)
Probability of counting 0 to n events	cdf_poisson(n, μ)
Given the probability P of counting 0 to n events, calculate n	quantile_poisson(P, μ)
Generate an array of N Poisson-distributed random numbers	random_poisson(μ, N)
Standard deviation of a Poisson-distributed variable	std_poisson(μ) or sqrt(μ)

To access the Maxima functions you must first type load(stats).

radiation [44], and molecule counts in molecular clusters and micelles [45] all follow the Poisson distribution, and accurate simulations of these processes will require Poisson-distributed random numbers.

For example, suppose a radioactive isotope sample emits an average of 15 alpha particles every minute. We can simulate 100 measurements of the alpha particle count with

```
(%i1)   data: random_poisson(15, 100)$
```

Here is an impulse plot of the simulated data, superimposed with the Poisson distribution:

```
(%i2)   [uniqueData, frequencies] : discrete_freq(data)$
(%i3)   mu: 15$
(%i4)   pdist : makelist([n, pdf_poisson(n,mu)], n, 0, 30)$
(%i5)   wxplot2d([[discrete, uniqueData, frequencies/100], [
          ➥ discrete, pdist]],
        [style, impulses, lines],
        [ylabel, "P(n)"],
        [xlabel, "n"],
          [legend, "simulated data", "Poisson Distribution"],
          [title, concat("Simulated radioactive decay with μ=",
            ➥ mu)]
        )$
```

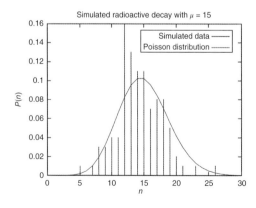

The width of the distribution is characterized by the standard deviation σ. The interval $[\mu - \sigma, \mu + \sigma]$ encompasses roughly two-thirds of all the data. For the Poisson distribution,

$$\sigma = \sqrt{\mu} \qquad \text{Standard deviation of the Poisson distribution} \qquad (8.8)$$

```
(%i6)   sigma: std_poisson(mu);
(%i7)   %, numer;
```

```
(%o5)   √15
(%o6)   3.872983346207417
```

The cumulative distribution function `cdf_poisson` gives the sum of all probabilities for 0 to n events. Let's use it to compute the probability of getting a count between $\mu - \sigma$ and $\mu + \sigma$:

```
(%i8)   cdf_poisson(mu+sigma, mu) - cdf_poisson(mu-sigma, mu),
        ➥ numer;
```

```
(%o7)   0.6347199126087909
```

Ⓜ **Worksheet 8.0: Discrete Probability Distributions**

In this worksheet, we'll plot the distributions of discrete datasets and use the Poisson distribution to analyze the results of counting experiments.

8.2.3 Continuous Probability Distributions

When x is a continuous variable, the normalization requirement becomes

$$\int P(x)\mathrm{d}x = 1 \qquad \text{Normalized continuous probability density} \qquad (8.9)$$

where the integration is over all possible values of x. The mean becomes

$$\bar{x} = \int xP(x)\mathrm{d}x \qquad \text{Mean of a continuous distribution} \qquad (8.10)$$

To compute the probability that x will fall in a particular range (say, between a and b), we can write

$$\text{probability that } a \leq x \leq b = \int_a^b P(x)\mathrm{d}x \qquad \begin{array}{l}\text{Computing probabilities}\\\text{from probability density functions}\end{array}$$

$$(8.11)$$

We'll need the form of the probability density function $P(x)$ to compute the integrals in Equations (8.9)–(8.11).

8.2.4 The Normal Distribution

When continuous experimental data contains only small and purely random errors, the distribution for the measurements can often be approximated by the normal distribution, which takes the form:

$$P(x) = \frac{1}{\sigma\sqrt{2\pi}}\exp\left(-\frac{1}{2}\left[\frac{x-\mu}{\sigma}\right]^2\right) \qquad \text{The normal distribution} \qquad (8.12)$$

The parameter σ is once again the standard deviation, and is a measure of how the measurements scatter around μ.

Let's plot the normal distribution with $\mu = 0$ and $\sigma = 1$:

```
(%i1)  P(x) := 1/(sigma*sqrt(2*%pi))*exp(-(1/2)*((x - mu)/sigma
       ➡ )^2)$
(%i2)  plot2d(P(x), [x, -4,+4], [ylabel, "P(x)"]), mu=0, sigma
       ➡ =1$
```

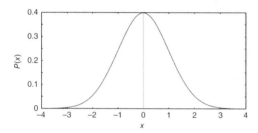

The built-in function `pdf_normal(x, `μ`, `σ`)` computes the probability distribution function for the normal distribution; an identical plot can be made with

```
(%i3)  load(stats)$
(%i4)  plot2d(pdf_normal(x,0,1), [x, -4, +4], [ylabel, "P(x)"])
       ➡ $
```

The curve has a bell shape, with points of inflection at one standard deviation from the mean.

Let's use Equations (8.11) and (8.12) to estimate the probability that a measurement will fall within one standard deviation of the population mean. The probability that $\mu - \sigma \leq x \leq \mu + \sigma$ is obtained by integrating $P(x)$ over this interval:

```
(%i5)  integrate(P(x),x,mu-sigma,mu+sigma);
(%i6)  %, numer;
```

$$(\%o5) \quad \mathrm{erf}\left(\frac{1}{\sqrt{2}}\right)$$

$$(\%o6) \quad 0.68268949213709$$

The integration results in the *error function*, `erf(x)`, which is defined as

$$\mathrm{erf}(x) = \frac{2}{\sqrt{\pi}}\int_0^x \exp(-t^2)\mathrm{d}t \qquad \text{The error function} \qquad (8.13)$$

Table 8.2 Functions for performing common tasks with a normally distributed variable X with mean μ and standard deviation σ.

Task	Maxima function[3]
Probability that $X = c$	`pdf_normal(c,`μ`,`σ`)`
Probability that $X \leq c$	`cdf_normal(c,`μ`,`σ`)`
Find the value c for which $X \leq c$ with probability P	`quantile_normal(P,`μ`)`
Generate an array of N normally distributed random numbers	`random_normal(`μ`,N)`
Estimate the population standard deviation σ for data in an array X	`std(X)`
Estimate the sample standard deviation s for data in an array X	`std1(X)`
Estimate the population variance σ^2 for data in an array X	`var(X)`
Estimate the sample variance s^2 for data in an array X	`var1(X)`

To access these functions you must first type `load(stats)`.

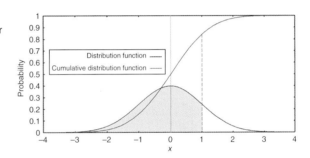

Figure 8.2 A cumulative distribution function (in red) gives the running area under a distribution function. The probability that the independent variable X is less than or equal to 1 is given by the shaded area under the distribution function, or the value of the cumulative distribution function at 1.

The error function has no simple closed form but it can be computed numerically. The total area under $P(x)$ is one, so this result tells us that there is about a 68.3% chance that x will be found within one standard deviation of the mean.

A list of built-in functions for performing common tasks with normally distributed variables is given in Table 8.2.

The cumulative distribution function `cdf_normal` gives the total area under the normal distribution curve (see Figure 8.2).

```
(%i1)   cdf_normal(c,mu,sigma) = integrate(pdf_normal(x,mu,sigma
        ➥ ), x, minf, c);
```

$$(\%\text{o1}) \quad \text{cdf\_normal}(c, \mu, \sigma) = \int_{-\infty}^{c} \text{pdf\_normal}(x, \mu, \sigma)\, dx$$

For example, we could have computed the probability that x is within one standard deviation of the mean more simply as follows:

```
(%i2)   cdf_normal(1,0,1) - cdf_normal(-1,0,1);
(%i3)   %, numer;
```

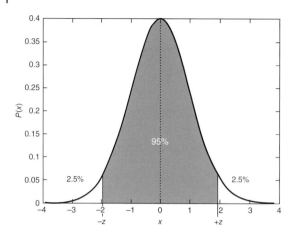

Figure 8.3 The 95% confidence interval for the standard normal distribution with mean zero and standard deviation 1 is [−1.96, 1.96].

(%o7) $\mathrm{erf}\left(\dfrac{1}{\sqrt{2}}\right)$

(%o8)   0.68268949213709

The standard deviation gives us a 68% confidence interval, that is, we can be 68% certain that a measurement x will fall within one standard deviation of the mean.

We can build intervals with different confidence levels with the `quantile_normal` function, which finds the value of c for which $x \le c$ with probability P. For example, we can be 2.5% confident that x is less than

(%i1)   `quantile_normal(0.025,mu,sigma), mu=0, sigma=1, `**`numer;`**

(%o1)   −1.959963984540054

We can therefore be 95% confident that x falls within ±1.96 of the mean when $\mu = 0$ and $\sigma = 1$ (see Figure 8.3).

  Worksheet 8.1: Continuous Probability Distributions

In this worksheet, we'll explore the normal distribution, which is the limiting distribution for many types of experimental data.

8.3   Estimating Precision

The standard deviation can be used as an estimate of the precision of the data. For a series of measurements $x_1, x_2, ..., x_N$ it is given by the root-mean-square deviation from the mean μ:

$$\sigma = \sqrt{\frac{\sum_i^N (x_i - \mu)^2}{N}} \qquad \text{The population standard deviation} \qquad (8.14)$$

The standard deviation can be computed only if μ is known. In practice, we have only a finite sample of experimental measurements, so we would like to use the sample mean as an estimate of μ. The best estimate of the standard deviation from the sample data is

$$s = \sqrt{\frac{\sum_i^N (x_i - \bar{x})^2}{N - 1}} \qquad \text{Sample standard deviation} \qquad (8.15)$$

where $N - 1$ in the denominator is the *degrees of freedom*,[4] that is, the number of independent deviations considered.

 Take care not to confuse the standard deviation σ (only to be used with a population mean, or a mean that is known *a priori*) with the best estimate of the standard deviation s (only to be used with a sample mean, computed from experimental data).

When $s = 0$ all of the data have exactly the same value as their mean. When s is very large some of the data is far from the mean, and the mean is not necessarily a good estimate of any particular datum.

The sample standard deviation is a valid statistic even when the data isn't normally distributed – about 95% of the data will fall within two standard deviations of the mean even for non-normal distributions, although other statistics may be more appropriate for skewed distributions [46].

The square of the standard deviation is called the variance. The population variance is σ^2; the best estimate of the variance made from a finite sample is s^2.

Let's estimate the sample standard deviation and variance of a random dataset using Maxima's built-in functions std1 and var1.

First, make a list of 1000 uniformly distributed random numbers between 0.0 and 1.0 using the built-in random function, and compute the mean.

```
(%i1)   data: makelist(random(1.0), i, 1, 1000)$
(%i2)   mean(data);


(%o2)   0.50428450393149
```

The std function estimates the standard deviation with a factor of N for data stored as a list[5]; std1 uses a factor of $N - 1$:

```
(%i3)   std(data);
(%i4)   std1(data);


(%o3)   0.27919074482082
(%o4)   0.28059725621531
```

We can compute the variances with var and var1, which uses factors of N and $N - 1$, respectively.

```
(%i5)   var(data);
(%i6)   var1(data);
```

---

4 The degrees of freedom is $N - 1$ and not N because we have computed a parameter (the mean) from the data. This gives us only $N - 1$ independent deviations.

5 When applied to matrices, std and std1 return a list of standard deviations for each column. This is convenient when each column in the matrix represents a different dataset, but if you want a single standard deviation for all entries in the matrix, convert it into a list with list_matrix_entries.

```
(%o5)    0.077947471993603
(%o6)    0.078734820195558
```

The difference between using N and $N − 1$ in the calculation is small. The standard deviation and variance using a factor of $N − 1$ is slightly higher. Which factor should you use, N or $N − 1$? For large sample sizes, the difference between the two is negligible. It's always safer to err on the side of caution and use the higher standard deviation, computed with a factor of $N − 1$. In any case, make it clear whether you've used N or $N − 1$ in your calculations.

8.3.1   Standard Error of the Mean

How can we judge whether a sample mean (or other calculated parameter) is representative of the entire population? We could take several different samples and compare their means. As we saw in Section 8.1.2, the sample means will be different for different samples. If we calculate the standard deviation of the sample means around the average sample mean, we get a **standard error of the mean**.

If the data is normally distributed, it's possible to estimate the standard error from the data for a single sample[6]

$$s_E = \frac{s}{\sqrt{N}} \qquad \text{Standard error of the mean} \qquad (8.16)$$

Equation (8.16) makes it obvious that a standard deviation and a standard error aren't quite the same thing. If you sample the entire population, s_E will be zero, because the sample mean will *be* the population mean. This isn't so for the sample standard deviation s. It converges to the population standard deviation σ as N increases – and the latter is *not* zero.

The essential difference between standard deviation and standard error is that a standard deviation describes the scatter in the data around the mean, while a standard error infers the uncertainty in an estimated parameter (such as a mean, or the slope and intercept of a linear fit) when many different samples are taken from the population [46–48].

Let's look again at the random dataset used in the previous section. We had 100 random numbers between 0 and 1, with $\bar{x} = 0.504$ and $s = 0.281$. We can make two statements:

- There is a 68% chance that any individual random number will be on the interval 0.50 ± 0.28;
- There is a 68% chance that the true (population) mean μ will be on the interval 0.504 ± 0.028.

The true mean of an infinite number of random numbers on the interval from 0 to 1 should be 0.5, so in this case the population mean *is* on the interval $\bar{x} \pm s_E$.

Confidence intervals for the mean are computed from standard errors, as we'll see in the following section.

8.3.2   Confidence Interval of the Mean

If the number of measurements were large enough, roughly 68% of the sample means fall within one standard error of the population mean μ. We could say we are 68% confident that sample mean falls within one standard error of the population mean.

We'd like a higher confidence level (say 95%) that the population mean lies somewhere on the interval. In other words, if the experiment is repeated 100 times, 95 times out of 100 the interval will embrace the population mean.

---

6 The standard error is the standard deviation for the Student t-distribution, which takes into account that we are estimating σ with s.

The confidence interval of the mean can be written as

$$\bar{x} \pm ts_{\text{E}} \qquad \text{Confidence interval of the mean} \qquad (8.17)$$

where t is the number of standard errors that widens the interval to the desired level of confidence. We can obtain t values for a given level of confidence and a given number of degrees of freedom using the Student's t-distribution, which we'll explore further in the next section.

Maxima has built-in functions for finding t. If we want a confidence level of 0.95 (95%) with N measurements, we can read the t value we need from the Student's t distribution using `quantile_student_t(1-(1-0.95)/2, N-1)`. In the following examples, we define a function `CI(confidence_level, data array)` to compute ts_{E}.

```
(%i1)   load(stats)$
(%i2)   CI(confidence_level,data) := quantile_student_t(1-(1-
        ➡ confidence_level)/2, length(data)-1)*std1(data)/
        ➡ sqrt(length(data)))$
(%i3)   data: [0.4999, 0.4990, 0.5009, 0.4991, 0.4998, 0.4999 ]$
(%i4)   mean(data);
(%i5)   CI(0.95,data), numer;

(%o4)   0.49976666666667
(%o5)   7.1996689949853766 10^{-4}
```

When reporting confidence intervals, round the error estimate to one or two figures, and round the mean accordingly. If the data above were molarities (M), the confidence interval would be reported as $0.49977 \pm 0.00072\,M$ ($N = 6$, 95% confidence) or $0.4998 \pm 0.0007\,M$ ($N = 6$, 95% confidence). It's important to provide the number of data points used to compute the mean and the confidence interval, so that your mean can be used by others in statistical hypothesis testing.

 Worksheet 8.2: Estimating Precision

In this worksheet, we'll compute standard deviations, standard errors, and confidence intervals for several large datasets.

8.4 Hypothesis Testing

Statistics can objectively determine when experimental data does not support a hypothesis. However, statistics never *prove* a hypothesis. They can only *reject* a hypothesis at some chosen level of confidence.

Hypotheses often involve comparisons. For example, a new experimental procedure determines the concentration of mercury in fish. We have computed a sample mean and an estimated standard deviation for several runs of the procedure. Our hypothesis might be, "Any difference we see between the sample mean and the true value is purely due to random error in the data." This is an example of a null hypothesis, that is, a proposition that we'll try to reject. We can expect the difference between the sample mean and true value to be one standard error or less about 68% of the time. The difference will be less than two standard errors about 95% of the time if the error is purely random.

We begin by computing a *statistic*, which is a measure of how large the difference is. In this case, the statistic is

$$t = \frac{|\bar{x} - \mu|}{s_E} \qquad \text{\textit{t}-Statistic for comparing a mean and "true" value} \qquad (8.18)$$

where t is the difference between the sample mean $\bar{x}$ and the true value μ in units of standard deviation. If $t > 1$, the difference is more than one standard deviation, and we should compute the probability that t could have a value this high purely by chance. If that probability is low enough, we can reject the null hypothesis.[7]

To compute probabilities for a statistic, we must know its probability distribution. The t-statistic follows the well-known Student's t distribution [49], which is applied when we are testing hypotheses about the mean of a normally distributed variable when the sample size is small and the population standard deviation is unknown. In Maxima, the t distribution is given by the function pdf_student_t(t, ν). Here ν is the degrees of freedom, which is $N - 1$ for computing a mean from N measurements.

```
(%i1)  load(stats)$
(%i2)  plot2d([pdf_normal(x,0,1), pdf_student_t(x,9),
        ➥ pdf_student_t(x,2)],
          [x, -4, +4],
            [legend, "normal distribution", "t distribution,
              ➥ N=10", "t distribution, N=3"],
            [ylabel, "probability"],
            [xlabel, "t"]
        )$
```

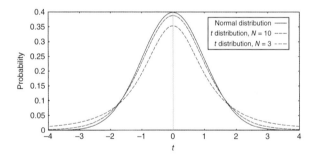

The t distribution approaches the normal distribution for larger datasets. For small datasets, the t distribution does not tail off as quickly as the normal distribution does.

We want to know the probability p that we'll be wrong if we reject the null hypothesis. This is equal to the chance that we might get t-value at least as large as the one we've computed using Equation (8.18). Looking at the probability distribution, P is equal to the area under the tails outside the interval $[-t, t]$ (see Figure 8.4).

Integrating the probability distribution between $-t$ and $+t$ and subtracting it from one yields p:

$$p = 1 - \int_{-t}^{+t} \text{pdf\_student\_t}(x, \nu)\,dx \qquad (8.19)$$

---

7 The probability isn't zero, so there's a chance that we might reject the null hypothesis and be wrong. This is called a type I error. A type II error is failure to reject a false null hypothesis.

Figure 8.4 The probability of finding *t* with an absolute value greater than or equal to two is equal to the area under both tails on the *t*-distribution.

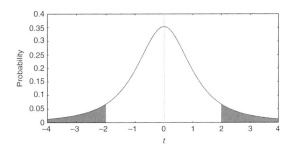

where `pdf_student_t(x,v)` is the probability density function for the *t* distribution at *x*, with *v* degrees of freedom.

There is an easier way. Maxima provides a cumulative distribution function for the *t*-distribution that gives the total area under the distribution function from 0 to *t*: `cdf_student_t(t,v)`. *p* can then be calculated as

$$p = 2(1 - \texttt{cdf\_student\_t}(t, v))$$

Calculating the *p*-value from the cumulative distribution function

(8.20)

where the factor of 2 gives us the area under *both* tails.

This probability is called the *p*-value. If it is less than 0.05, it means that the null hypothesis is true less than 5% of the time, and we can reject the null hypothesis with at least 95% confidence. Such a result that is said to be *significant*. If it is less than 0.01, the null hypothesis is rejected with 99% confidence, a *highly significant* result.

On the other hand, if the *p*-value is *greater* than 0.05, we cannot reject the hypothesis at that level. But we cannot say that we've "proven" the null hypothesis.

 Failure to reject a hypothesis is not the same thing as accepting the hypothesis! You can never prove anything with statistics; you can only disprove.

8.4.1 Comparing a Mean with a True Value

Consider the determination of the specific heat of water experimentally. A worker obtains five measurements: 4.23, 4.18, 4.27, 4.31, and 4.43 J(g °C)$^{-1}$. The literature value is 4.184 J(g °C)$^{-1}$. Does the literature value agree with the experimental values?

```
(%i1)   numer: true$
(%i2)   data: [4.23, 4.18, 4.27, 4.31, 4.43]$
(%i3)   N: length(data)$
(%i4)   avg: mean(data);
(%i5)   stderr: std1(data)/sqrt(N);

(%o4)   4.284
(%o5)   0.042379240200834
```

The average experimental specific heat is 4.28(4), so the worker might be tempted to report the presence of a systematic error equal to 4.28(4) − 4.184 = +0.10 J(g °C)$^{-1}$.

But the standard error is about 0.042. Random error causes the mean in a small sample to differ from the true mean, so the discrepancy between the literature value and the sample mean might be the result of random and not systematic error. How can we tell if there is a significant systematic error present?

If we can reject the hypothesis that the sample mean $\bar{x}$ equals the true value μ, we can estimate the systematic error as the difference $\bar{x} - \mu$. Our null hypothesis is that the average is equal to 4.184 J(g °C)$^{-1}$. First, compute the t-statistic using Equation (8.18):

```
(%i6)   t: abs(avg - 4.184)/stderr;
(%o6)   2.359645890915038
```

Now compute the p-value:

```
(%i7)   pvalue: 2*(1-cdf_student_t(t,N-1));
(%o7)   0.077692923987415
```

Since the p-value is greater than 0.05, we can *not* reject the null hypothesis with 95% confidence. We *could* reject it with 90% confidence, because it is not greater than 0.1.

The data doesn't give strong evidence of a systematic error in this case. That does *not* mean that the data is free of systematic error!

The `stats` package contains `test_mean` which automates testing the mean against a known value[8]:

```
(%i8)   load(stats)$
(%i9)   test_mean(data,  mean=4.184);
```

$$
(\%o9) \quad
\begin{array}{|c}
\text{MEAN TEST} \\
mean\_estimate = 4.284 \\
conf\_level = 0.95 \\
conf\_interval = [4.166336366127895, 4.401663633872105] \\
method = \text{Exact t-test. Unknown variance.} \\
\text{hypotheses=H0: mean = 4.184 , H1: mean} \# 4.184 \\
statistic = 2.359645890915038 \\
distribution = [student\_t, 4] \\
p\_value = 0.077692923987415
\end{array}
$$

The `test_mean` function computes the 95% confidence interval. Any level of confidence can be selected with the `'conflevel=c` option, where c is between 0 and 1 (it is 0.95 by default). The true mean in this case is within the confidence interval.

When the population standard deviation σ is known, it can be included in the test with `'dev = σ`. One-sided tests (with alternative hypotheses like $\bar{x} > \mu$ or $\bar{x} < \mu$) are possible with the options `'alternative=greater` and `'alternative=less`, respectively.

8.4.2   Comparing Variances

In designing experiments and in interpreting data, we often want to compare the precisions of two different methods. Suppose we have replicate measurements for two methods A and B:

---

8 The mean test can also be accessed from the Statistics pane. Choose View ⟩ Statistics and then Mean Test... .

```
(%i1)   A : [0.1053, 0.1089, 0.1047, 0.1017, 0.1042, 0.1002]$
(%i2)   B : [0.1095, 0.1055, 0.1084, 0.1168, 0.1042, 0.1109,
           ➡ 0.1145, 0.1156, 0.1180]$
(%i3)   load(stats)$
(%i4)   varA : var1(A);
(%i5)   varB : var1(B);
```

```
(%o4)   9.1586666666666692 10⁻⁶
(%o5)   2.4926111111111107 10⁻⁵
```

$$(\%o4) \quad 9.1586666666666692 \times 10^{-6}$$
$$(\%o5) \quad 2.4926111111111107 \times 10^{-5}$$

We might be tempted to say that A has better precision than B because the variance for A is smaller. But, the variances for both methods are only estimates, and they will fluctuate if the experiment is repeated. We don't know how large those fluctuations will be. It may be that the next time we try the experiment, Method B will have the smaller variance.

How can we be confident that A is significantly more precise than B? And how do we account for different numbers of measurements in the two datasets?

Statisticians have developed a test called the *F-test* that tests the null hypothesis that two normal populations have the same variance. The test is made by computing a statistic called F that is the ratio of variances (squared standard deviations) for the two datasets:

$$F = \frac{s_{\text{big}}^2}{s_{\text{small}}^2} \qquad \text{The } F \text{ statistic} \qquad (8.21)$$

where s_{big} is the bigger standard deviation, and s_{small} is the smaller. The distribution of the F-statistic is well known [50]. The `stats` package contains a cumulative distribution function for F called `cdf_f`.

The following code uses the F-statistic to test the hypothesis that data collected by two different methods have equal variances. The p-value is computed as it was in Equation (8.20), using the cumulative F-distribution in place of the t-distribution.

```
(%i6)   F : varB/varA;
(%i7)   df_numerator: length(B)-1;
(%i8)   df_denominator : length(A)-1;
(%i9)   pvalue : 2*(1-cdf_f(F, df_numerator, df_denominator));
```

```
(%o6)   2.721587324695491
(%o7)   8
(%o8)   5
(%o9)   0.28506345235727
```

Since the p-value is greater than 0.05, we can *not* reject the null hypothesis with 95% confidence.

The `stats` package provides a function `test_variance_ratio(x1, x2)` for performing the F test on two datasets `x1` and `x2` which can be either lists or column matrices.

```
(%i10)  test_variance_ratio(A, B);
```

$$
\begin{array}{c}
\text{VARIANCE RATIO TEST}\\
ratio\_estimate = 0.36743263423006\\
conf\_level = 0.95\\
conf\_interval = [0.076273950246951, 2.482805510622588]\\
method = Variance\ ratio\ F - test.\ Unknown\ means.\\
hypotheses = H0\ :\ var1 = var2\ , H1\ :\ var1 \neq var2\\
statistic = 0.36743263423006\\
distribution = [f, 5, 8]\\
p\_value = 0.28506345235727
\end{array}
$$

(%o10)

The p-value is identical to the one we calculated by using the cumulative distribution function for F in the previous example.

The options for changing the confidence level and the alternative hypothesis are the same as those for `test_mean`.

The F-test assumes that both variances come from a normal distribution. It can give invalid results if the distributions are non-normal [51]. You can check the normality of the distribution graphically or by performing a Shapiro–Wilk test for normality. The Shapiro–Wilk test has a null hypothesis that a sample comes from a normally distributed population. If the p-value given by the test is less than some chosen α level (e.g. 0.05 for 95% confidence), then the null hypothesis can be rejected.

The `stats` package includes the Shapiro–Wilk test with the function `test_normality`[9]:

```
(%i11)   test_normality(A);
(%i12)   test_normality(B);
```

$$
\begin{array}{c}
SHAPIRO - WILK\ TEST\\
statistic = 0.9652736031850186\\
p_{value} = 0.8593131992848643
\end{array}
$$

(%o11)

$$
\begin{array}{c}
SHAPIRO - WILK\ TEST\\
statistic = 0.9398998052174259\\
p_{value} = 0.5808441963595483
\end{array}
$$

(%o12)

The tests give p-values for both sets that are above 0.05, so we cannot reject the hypothesis that they are normally distributed with 95% confidence.

8.4.3   Comparing Two Sample Means

Suppose we want to know whether the means $\bar{x}_A$ and $\bar{x}_B$ are different for two different datasets A and B. We could compute t comparing two experimental means using

$$
t = \frac{\bar{x}_A - \bar{x}_B}{s_{AB}} \qquad \begin{array}{l} t\text{-Statistic for testing} \\ \text{the hypothesis } \bar{x}_A = \bar{x}_B \end{array} \tag{8.22}
$$

---

9 The normality test can also be accessed from the Statistics pane. Choose View ⟩ Statistics and then Normality Test… .

where s_{AB} is the pooled standard error for datasets A and B, calculated as

$$s_{AB} = \begin{cases} \sqrt{\dfrac{(N_A - 1)s_A^2 + (N_B - 1)s_B^2}{N_A + N_B - 2} \left(\dfrac{1}{N_A} + \dfrac{1}{N_B}\right)} & \text{if } s_A \text{ and } s_B \text{ are not} \\ & \text{significantly different,} \\[2ex] \sqrt{s_A^2/N_A + s_B^2/N_B} & \text{otherwise.} \end{cases} \tag{8.23}$$

where s_A and s_B are the standard deviations from datasets A and B, respectively. Be careful when the variances are significantly different for the two datasets. Different variances sometimes are a symptom of completely different distributions. The test implicitly assumes that both distributions are normal.

To decide which expression to use for s_{AB}, you must first apply the F-test (described in the previous section). The degrees of freedom for the t-statistic in Equation (8.22) is $N_A + N_B - 2$ if s_A and s_B are not significantly different. If they *are* significantly different, the number of degrees of freedom can be estimated as

$$\frac{\left(s_A^2/N_A + s_B^2/N_B\right)^2}{\dfrac{\left[s_A^2/N_A\right]^2}{N_A - 1} + \dfrac{\left[s_B^2/N_B\right]^2}{N_B - 1}} \quad \begin{array}{l} \text{Approximate degrees} \\ \text{of freedom for } t \\ \text{when } s_A \neq s_B \end{array} \tag{8.24}$$

The number of degrees of freedom estimated by Equation (8.24) is often not an integer, but it can be used anyway to read values from the cumulative distribution function for t. The following code compares the means for two datasets. First, we input the datasets and compute their means:

```
(%i1)   load(stats)$
(%i2)   A : [0.1053, 0.1089, 0.1047, 0.1017, 0.1042, 0.1002]$
(%i3)   B : [0.1095, 0.1055, 0.1084, 0.1168, 0.1042, 0.1109,
          ➥ 0.1145, 0.1156, 0.1180]$
(%i4)   NA: length(A)$
(%i5)   NB: length(B)$
(%i6)   meanA : mean(A);
(%i7)   meanB: mean(B);

(%o6)   0.10416666666667
(%o7)   0.11148888888889
```

The variances for each set are

```
(%i8)   varA : var1(A);
(%i9)   varB : var1(B);

(%o8)   9.1586666666666692 10^{-6}
(%o9)   2.4926111111111107 10^{-5}
```

We applied the F-test to this same data in the previous section. We couldn't conclude that the variances were significantly different or that the two sets are sampled from a population that doesn't have a normal distribution, so we can compute the pooled standard error for the two datasets as

```
(%i10)    sAB : sqrt(((NA-1)*varA +  (NB-1)*varB)/(NA+NB-2)*(1/NA
         ➥ +1/NB));
```

```
(%o10)    0.0022889656447175
```

Now perform a *t*-test to see if the means are significantly different.

```
(%i11)    t: abs(meanA - meanB)/sAB;
(%i12)    pvalue : 2*(1-cdf_student_t(t,NA+NB-2));
```

```
(%o11)    3.198921853248648
(%o12)    0.0069820667770486
```

The *p*-value is less than 0.05, so we can reject the null hypothesis with better than 95% confidence. The means for the two datasets are significantly different.

The `stats` package has a built-in function `test_means_difference(x1, x2)` for performing the *t* test on two datasets `x1` and `x2` which can be either lists or column matrices.[10] The test has an option `'varequal` that can be set to `true` if the variances for the two sets are known to be equal; by default, the test assumes unequal variances. Since the *F*-test showed no evidence in the data that the variances were significantly different, we'll use that option:

```
(%i13)    test_means_difference(A, B, 'varequal=true);
```

$$
\begin{array}{c}
\text{DIFFERENCE OF MEANS TEST} \\
\text{diff\_estimate} = -0.0073222222222222 \\
\text{conf\_level} = 0.95 \\
\text{conf\_interval} = [-0.012267231648683, -0.002377212795761] \\
\text{method=Exact t-test. Unknown equal variances} \\
\text{hypotheses=H0: mean1 = mean2 , H1: mean1} \neq \text{mean2} \\
\text{statistic} = 3.198921853248648 \\
\text{distribution=[student\_t,13]} \\
\text{p\_value} = 0.0069820667770486
\end{array}
$$

(%o13)

If the population standard deviations σ_A and σ_B are known, they can be set with the options `'dev1=`σ_A and `'dev2=`σ_B. The confidence level and alternative hypothesis can be changed with the same options used for `test_mean`.

 Worksheet 8.3: Hypothesis Testing

In this worksheet, we'll use Maxima to test hypotheses that compare a sample mean to a "true" value, compare variances for two normally distributed datasets, and compare means for two normally distributed datasets. The Shapiro–Wilk test and rankit normal probability plots will be used to test the normality of datasets.

---

10 The mean difference test can also be accessed from the Statistics pane. Choose ⎡View⎤⟩⟩⎡Statistics⎤ and then ⎡Mean Difference Test...⎤.

8.5 Propagation of Error

To estimate the error in results calculated from measurements, we must *propagate* the error in the measurements through the calculation. For example, suppose we have a function $f(x, y)$ with two independent variables x and y with means $\bar{x}$ and $\bar{y}$.

We can approximate the function $f(x, y)$ with a Taylor series expansion around the mean [52]:

$$
\begin{aligned}
f_i &= f(x_i, y_i) \\
&\approx f(\bar{x}, \bar{y}) + \left(\frac{\partial f}{\partial x}\right)(x_i - \bar{x}) + \left(\frac{\partial f}{\partial y}\right)(y_i - \bar{y}) + \cdots
\end{aligned}
\qquad
\begin{aligned}
&\text{Taylor series} \\
&\text{expansion of } f(x_i, y_i)
\end{aligned}
\qquad (8.25)
$$

where any variables held constant in partial derivatives are replaced by their average values. The variance in f is then

$$
\begin{aligned}
s_f^2 &= \frac{1}{N-1} \sum_i^N \left(f_i - \bar{f}\right)^2 \\
&= \frac{1}{N-1} \sum_i^N \left(\frac{\partial f}{\partial x}\right)^2 (x_i - \bar{x})^2 + \left(\frac{\partial f}{\partial y}\right)^2 (y_i - \bar{y})^2 \\
&\quad + 2 \left(\frac{\partial f}{\partial x}\right)^2 \left(\frac{\partial f}{\partial y}\right)^2 (x_i - \bar{x})(y_i - \bar{y})
\end{aligned}
\qquad (8.26)
$$

The first two terms contain the variances s_x^2 and s_y^2, respectively. The third term contains the covariance s_{xy}, defined as

$$
s_{xy} \equiv \frac{1}{N-1} \sum_i^N (x_i - \bar{x})(y_i - \bar{y})
\qquad \text{Definition of covariance}
\qquad (8.27)
$$

Rewriting Equation (8.26) in terms of variances and covariances, we have

$$
s_f^2 = \left(\frac{\partial f}{\partial x}\right)^2 s_x^2 + \left(\frac{\partial f}{\partial y}\right)^2 s_y^2 + 2 \left(\frac{\partial f}{\partial x}\right)^2 \left(\frac{\partial f}{\partial y}\right)^2 s_{xy}
\qquad (8.28)
$$

If the errors are independent and random, the covariance is zero and an expansion over variances can be used (Section 8.5.2). For correlated errors, the covariance in Equation (8.28) cannot be neglected (Section 8.5.3).

8.5.1 Propagation of Independent Systematic Errors

Suppose an experimental result $f(x_1, x_2, \ldots, x_p)$ is computed from a set of independent measurements $x_1, x_2, \ldots x_p$. Each measurement x_i contains a systematic error δx_i, which may be positive or negative. Equation (8.28) cannot correctly propagate systematic errors because the squared terms strip off the signs, which are important in determining how the errors contribute to the total. How, then, do we estimate the systematic error δf in f?

If the systematic errors are small and independent of each other, we can approximate the error in f using a total differential expansion (Equation 6.6):

$$
\delta f \approx \frac{\partial f}{\partial x_1} \delta x_1 + \frac{\partial f}{\partial x_2} \delta x_2 + \cdots
\qquad
\begin{aligned}
&\text{Propagation of systematic} \\
&\text{error (independent } x_i\text{'s,} \\
&\text{small } \delta x_i\text{'s)}
\end{aligned}
\qquad (8.29)
$$

where the signs of the systematic errors must be included in the calculation.

The partial derivatives are called sensitivity coefficients because they show how sensitive the calculation of f is to errors in each contributing measurement. They are conveniently listed in Maxima using the `jacobian` (Section 7.3.2). For example, if f is a function of three variables x_1, x_2, and x_3,

```
(%i1)   jacobian([f(x_1,x_2,x_3)], [x_1,x_2,x_3]);
```
$$(\%o1) \quad \left(\frac{d}{dx_1} f\left(x_1, x_2, x_3\right) \quad \frac{d}{dx_2} f\left(x_1, x_2, x_3\right) \quad \frac{d}{dx_3} f\left(x_1, x_2, x_3\right) \right)$$

Equation (8.29) provides an estimate of the systematic error in f, but more importantly, it clearly shows how each measurement contributes to the total error. The error associated with the largest term (called the limiting error) is a "weak link" in an experimental design. Redesigns that minimize the limiting error will lead to the biggest improvements in the experiment's accuracy (and precision).

For example, suppose we want to find the systematic error in the standard potential E^0 from systematic errors in temperature T and an equilibrium constant K in the equation

$$E^0 = \frac{RT}{nF} \ln K \tag{8.30}$$

```
(%i1)   declare([R,n,F], constant);
(%i2)   E0(T,K) := R*T/(n*F)*log(K);
```
$$(\%o2) \quad E0(T,K) := \frac{RT}{nF} \log(K)$$

The sensitivity coefficients are

```
(%i3)   sensitivities: jacobian([E0(T,K)], [T,K]);
```
$$(\%o3) \quad \left(\frac{R \log(K)}{Fn} \quad \frac{RT}{FnK} \right)$$

If the systematic errors in T and K are

```
(%i4)   systematicErrors: [del(T), del(K)]$
```

the propagated error is

```
(%i5)   delta_E0 : sensitivities . systematicErrors;
```
$$(\%o4) \quad \frac{R \log(K)\, del(T)}{Fn} + \frac{RT\, del(K)}{FnK}$$

or equivalently,

```
(%i6)   delta_E0 : diff(E0(T,K));
```
$$(\%o5) \quad \frac{R \log(K)\, del(T)}{nF} + \frac{RT\, del(K)}{nFK}$$

We'd also like to know whether the error in T or the error in K is most important in determining the propagated error in E^0. Suppose we have a systematic error of -0.01 kelvins in a temperature measured at 298.15 K, and a systematic error of $+1 \times 10^{-6}$ in the equilibrium constant measured at $K = 1.8 \times 10^{-5}$, the propagated error in E^0 will be

```
(%i7)   delta_E0, R=8.314, T=298.15, n=1, F=96487, del(T)=-0.01,
    ➥   K=1.8e-5, del(K)=1e-6;
```

(%o4)   0.0014366752853966

...so E^0 will be about 1.4 mV too high.

Which error is limiting? We can evaluate the two terms in the propagated error in E^0 separately, and see which contributes the most to the total error. One approach is to extract the terms from the propagation equation using `part`:

```
(%i8)   part(delta_E0, 1);
(%i9)   %, R=8.314, T=298.15, n=1, F=96487, del(T)=-0.01, K=1.8e
    ➥   -5, del(K)=1e-6;
```

(%o5) $\dfrac{R \log(K) \, \text{del}(T)}{Fn}$

(%o6) $9.413869638787222 \times 10^{-6}$

```
(%i10)  part(delta_E0, 2);
(%i11)  %, R=8.314, T=298.15, n=1, F=96487, del(T)=-0.01, K=1.8e
    ➥   -5, del(K)=1e-6;
```

(%o7) $\dfrac{RT \, \text{del}(K)}{FnK}$

(%o8)   0.001427261415757793

A more direct approach uses matrix notation:

```
(%i12)  systematicErrors * list_matrix_entries(sensitivities), R
    ➥   =8.314, T=298.15, n=1, F=96487, del(T)=-0.01, K
    ➥   =1.8e-5, del(K)=1e-6;
```

(%o2) $[9.413869638787222 \times 10^{-6}, 0.001427261415757793]$

Most of the error in E^0 comes from the error in K; improving the temperature measurement will do little to improve the overall experimental error.

8.5.2   Propagation of Independent Random Errors

For p measurements with random errors, Equation (8.28) can be written as

$$s_f^2 = \sum_i^p \sum_j^p \left(\frac{\partial f}{\partial x_i} \right) \left(\frac{\partial f}{\partial x_j} \right) s_{ij} \qquad \begin{array}{l}\text{Propagation of error in a function} \\ f(x_1, \ldots, x_p)\end{array} \qquad (8.31)$$

where s_{ij} is again the covariance, which is the expected product of the deviations of x_i and x_j from their means:

$$s_{ij} = \frac{1}{N-1} \sum_{k=1}^{N} (x_{ki} - \bar{x}_i)(x_{kj} - \bar{x}_j) \qquad \begin{array}{l}\text{Covariance in the} \\ \text{variables } x_i \text{ and } x_j\end{array} \qquad (8.32)$$

The covariance is a measure of how strongly errors in x_i affect errors in x_j; we'll see how to compute it from experimental data in Section 8.5.3.

When the errors in x_i and x_j are independent, s_{ij} is zero. The covariance s_{ii} is just the variance of the ith variable (s_i^2) and Equation (8.31) simplifies to

$$s_f^2 = \sum_i^p \left(\frac{\partial f}{\partial x_i} \right)^2 s_i^2 \qquad \text{Propagation of independent variances in a function } f(x_1, \ldots, x_p) \qquad (8.33)$$

For example, suppose we want to find the standard deviation in the volume V of a sphere from the standard deviation in its radius r.

```
(%i1)   V : (4/3)*%pi*r^3;
```
$$(\%o1) \quad \frac{4\pi r^3}{3}$$

Applying Equation (8.33),

```
(%i2)   varianceV : diff(V,r)^2*s_r^2;
```
$$(\%o2) \quad 16\pi^2 r^4 s_r^2$$

stddevV : sqrt(varianceV);

$$(\%o3) \quad 4\pi r^2 \left| s_r \right|$$

We can also apply Equations (8.31) and (8.33) to propagate confidence intervals instead of variances. For example, if the confidence intervals for the volume and radius of a sphere are $V \pm \lambda_V$ and $r \pm \lambda_r$, respectively, we can estimate the confidence interval $\pm \lambda_V$ in V as

$$\lambda_V = \sqrt{\left(\frac{\partial V}{\partial r} \right)^2 \lambda_r^2} = \left| (4\pi r^2) \lambda_r \right| \qquad \begin{array}{l} \text{Size of the confidence} \\ \text{interval in the volume} \\ \text{of a sphere} \end{array} \qquad (8.34)$$

Let's estimate the standard deviation in the density of an ideal gas from replicate pressure and temperature measurements. The calculation has two parts: the symbolic derivation of the error propagation equation, and the numerical calculation of the final result. We'll avoid assigning values to any variables until after the symbolic work is done.

In the symbolic derivation, the errors in non-measured quantities like the molar mass and the gas law constant are negligible. We declare them as constants *before* deriving the error propagation equation:

```
(%i1)   declare([M,R],constant)$
```

Calculate the density ρ of an ideal gas from its pressure P and temperature T. M is the molar mass, and R is the ideal gas law constant.

```
(%i2)   rho: P*M/(R*T);
```
$$(\%o2) \quad \frac{M P}{R T}$$

Compute the propagated variance `var_rho` from the variances in P and T (`var_P` and `var_T`). We apply Equation (8.33) because the pressure and temperature measurements are independent. Each term in the error propagation equation is computed separately, so we can easily identify the limiting error in the calculation later.

```
(%i3)   P_contribution : diff(rho,P)^2 * var_P$
(%i4)   T_contribution : diff(rho,T)^2 * var_T$
(%i5)   stdev_rho : sqrt(P_contribution + T_contribution);
```

$$(\%o5) \quad \sqrt{\dfrac{M^2\,P^2\,var\_T}{R^2\,T^4} + \dfrac{M^2\,var\_P}{R^2\,T^2}}$$

Fill in the pressure and temperature data (which were replicate readings for O_2 around STP).

```
(%i6)    P_data : [760.0,760.1,760.2,759.9,759.8,760.0]/760$
(%i7)    T_data : [273.0, 273.1, 272.9, 273.2, 273.4, 273.2]$
(%i8)    [rho, stdev_rho], R = 0.08205746, M = 2*15.9994,
(%i9)          P=mean(P_data), T=mean(T_data),
(%i10)         var_P=var1(P_data), var_T=var1(T_data);
```

$$(\%o8) \quad [1.427713011042433, 9.5314931649092757\ 10^{-4}]$$

Both the average density and its standard deviation were computed from the average pressure and temperature measurements. In general, when the variances are relatively small, the average value of any function $f(x_1, x_2, \ldots)$ is equal to the value of the function when all of the variables are at their average values:

$$\bar{f} = f(\bar{x}_1, \bar{x}_2, \ldots) \qquad \text{The average value of a function} \tag{8.35}$$

If the pressure and temperature readings are the only source of error, we expect a precision of about ± 0.001 g l^{-1} in the calculated density (a relative error of about 0.07%). Evaluating `T_contribution` shows that it contributes about 92% of the total variance in the density, so the temperature readings are the limiting error. Improving the precision of the pressure readings will have little effect on the precision of the density.

8.5.3   Covariance and Correlation

Variables that depend on each other often have correlated errors. For example, the parameters in least-squares fits (such as slopes and intercepts) are *never* independent of each other, so Equation (8.33) won't work for propagating errors through least-squares fits. We must use Equation (8.31) instead.

Covariances can be computed from raw data by applying Equation (8.31). For example, consider the following measurements of X and Y. Place the data in a matrix with each variable in a column:

```
(%i1)    X: [0.60,0.48,0.36,0.24,0.12,0.50,0.55]$
(%i2)    Y: [0.920,0.687,0.553,0.322,0.200,0.750,0.919]$
(%i3)    data: transpose(matrix(X,Y));
```

$$(\%o3) \quad \begin{pmatrix} 0.6 & 0.92 \\ 0.48 & 0.687 \\ 0.36 & 0.553 \\ 0.24 & 0.322 \\ 0.12 & 0.2 \\ 0.5 & 0.75 \\ 0.55 & 0.919 \end{pmatrix}$$

Maxima's built-in function `cov1` from the `stats` package can compute the covariance matrix C:

$$C = \begin{pmatrix} s_1^2 & s_{12} \\ s_{21} & s_2^2 \end{pmatrix} \qquad \text{The covariance matrix} \tag{8.36}$$

```
(%i4)   load(stats)$
(%i5)   C: cov1(data);
```

$$(\%o5) \quad \begin{pmatrix} 0.030690476190476 & 0.048515238095238 \\ 0.048515238095238 & 0.078410952380952 \end{pmatrix}$$

The diagonal elements are the variances of X and Y. The off-diagonal element is the covariance of X and Y (s_{XY}).

Suppose we wanted to compute the error in the product $F = XY$ for the data in the example above.

```
(%i6)   F(x,y) := x*y$
```

The sensitivity coefficients are computed using the Jacobian matrix **J** and evaluated at the mean values of X and Y:

```
(%i7)   jacobian([F(x,y)], [x,y]);
(%i8)   J: %, x=mean(X), y=mean(Y);
```

$$(\%o7) \quad \begin{pmatrix} y & x \end{pmatrix}$$
$$(\%o8) \quad \begin{pmatrix} 0.6215714285714287 & 0.4071428571428571 \end{pmatrix}$$

Applying Equation (8.31), we have

$$
\begin{aligned}
s_F^2 &= \left(\frac{\partial F}{\partial X}\right)^2 s_X^2 + \left(\frac{\partial F}{\partial Y}\right)^2 s_Y^2 + 2\left(\frac{\partial F}{\partial X}\right)\left(\frac{\partial F}{\partial Y}\right) s_{XY}^2 \\
&= Y^2 s_X^2 + X^2 s_Y^2 + 2XY s_{XY}^2
\end{aligned}
\tag{8.37}
$$

```
(%i9)   varianceF : J[1,1]^2*C[1,1] + J[1,2]^2*C[2,2] + 2*J
        ➥ [1,1]*J[1,2]*C[1,2];
```

```
(%o9)   0.04941045713654028
```

The calculation is much simpler in matrix form:

$$s_F^2 = \mathbf{JCJ}^{\mathrm{T}} \qquad \begin{array}{l}\text{Computing variances from} \\ \text{the Jacobian and covariance matrices}\end{array} \tag{8.38}$$

```
(%i10)  varianceF: J . C . transpose(J);
(%o10)  0.04941045713654028
```

When the covariance for a pair of variables isn't zero, the variables are said to be correlated. One measure of the strength of the correlation is Pearson's r, also called the *correlation coefficient*:

$$r_{ij} = \frac{s_{ij}^2}{s_i s_j} \qquad \text{The linear correlation coefficient} \tag{8.39}$$

The correlation coefficient is always between -1 and $+1$. It is positive if the signs of the errors in x_i and x_j are the same, and negative if they are opposite. When the absolute value of r_{ij} approaches 1, it can indicate a strong linear relationship between x_i and x_j. When r is close to zero, the variables are not linearly correlated.

The function `cor(data)` computes a correlation matrix for the variables stored as columns in the matrix `data`.

How close does r have to be to one before we can say that we have a *significant* correlation? It is possible to obtain high values of r with uncorrelated variables, purely by chance. Suppose we obtain an r value of r_0. If the probability that we can obtain an r value that is at least as large as r_0 purely by chance is less than 5%, the correlation is considered *significant*. If that probability is less than 1%, we can say that the correlation is *highly significant*.

The probability P can be calculated by

$$P(N, r_0) = \frac{2\Gamma\left(\frac{N-1}{2}\right)}{\sqrt{\pi}\,\Gamma\left(\frac{N-2}{2}\right)} \int_{|r_0|}^{1} \left(1 - r^2\right)^{\frac{N-4}{2}} dr \qquad \begin{array}{l}\text{Probability that } |r| \geq |r_0| \\ \text{purely by chance}\end{array} \tag{8.40}$$

where N is the number of data points and r_0 is the correlation coefficient we obtain for those points. The gamma function $\Gamma(n)$ is an extension of the factorial function $n!$ to complex and real numbers; in Maxima, it is implemented as `gamma(n)`.

The probability would be rather tedious to compute by hand, but it easily done in Maxima, as we'll see in in the example below. The probability decreases rapidly as the number of data points increases. For example, with just three points, there is a 29% chance that you can get $r = 0.9$ even with uncorrelated variables! If you have six points, that probability falls to about 1.4%.

```
(%i1)   load(stats)$
(%i2)   data : [[0.60, 0.92], [0.48, 0.687], [0.36, 0.553],
          ➡ [0.24, 0.322], [0.12, 0.200], [0.50, 0.750],
          ➡ [0.55, 0.919]]$
(%i3)   plot2d([discrete, data], [style, points])$
```

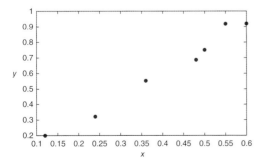

We can see from the plot of the data that some relationship exists between x and y; otherwise, the points would be scattered randomly.

Pearson's r between x and y can be computed directly with the `cor` function; r_{xy} is the off-diagonal element of the correlation matrix:

```
(%i4)   cor(apply(matrix, data));
(%i5)   r : %[1,2];
```

$$(\%o6) \quad \begin{pmatrix} 1.0 & 0.98898109345887 \\ 0.98898109345887 & 1.0 \end{pmatrix}$$

(%o7)   0.98898109345887

Finally, compute the probability P that r can be larger than this value, purely by chance:

```
(%i6)   N : length(data)$
(%i7)   P : 2*gamma((N-1)/2)/(sqrt(%pi)*gamma((N-2)/2)) *
         ➥ integrate((1-x^2)^((N-4)/2), x, abs(r), 1);
```

(%o8) $2.4335015270298375 \, 10^{-5}$

... so the probability of getting this value of r for seven data points purely by chance is very small!

When reporting Pearson's r as evidence of linear correlation, keep in mind that changes in one variable don't necessarily *cause* changes in second variable, even when the r value is quite high. For example, US spending on science, space, and technology correlates strongly with the number of suicides by hanging, strangulation, and suffocation ($r = 0.992$); the correlation coefficient between the divorce rate in Maine and per capita consumption of margarine in the United States is 0.993 [53].

 Worksheet 8.4: Propagation of Error

In this worksheet, we'll use Maxima to estimate the systematic error in a result calculated from measurements with systematic errors, estimate the random error in a result calculated from measurements with independent random errors, and identify the limiting error in the calculated result. We'll also identify correlated variables in datasets and use covariances to propagate error in calculations.

9

Fitting Data to a Straight Line

When the world throws you too much information, the only way you can stay sane or survive is to look for patterns. Amidst all the blurs, is there a constellation that emerges, is there a straight line that's emerging?

– Douglas Coupland

Chemistry often focuses on relationships between variables. A kinetics experiment may study the effect of a catalyst concentration on a reaction rate, or the temperature dependence of a rate constant, or the decay of reactant concentrations with time. A fluorescence experiment might relate the lifetime of an excited state to the concentration of a substance that quenches it. Chemoinformatic studies relate the biological activity of compounds with structural descriptors and molecular properties. Spectrophotometry quantitatively relates absorbance and concentration to find the concentrations in unknown samples by simply measuring their absorbances.

In each case model equations relate the variables. The simplest and most widely used model equations are linear.

For example, suppose we have measured the absorbance A of several solutions of $KMnO_4$, each with a precisely known concentration c. We propose the following model equation to relate the concentration and absorbance:

$$A = abc \qquad \text{The Beer–Lambert law} \tag{9.1}$$

where a is the absorptivity and b is the length of the light path through the solution. We usually know b beforehand, but a is unknown. Fitting the data to the model equation involves finding a value for the unknown parameter that makes both sides of model equation equal, or as close to equal as possible.

We choose the variable that can be measured more precisely as the independent variable. In this case, that is the concentration, and the dependent variable is the absorbance.

We have N pairs of concentration and absorbance observations (c_i, A_i). We cannot simply write $A_i = abc_i$ for each point, because experimental error and possible deviations from the Beer–Lambert law will make one side of the equation different from the other. If experimental errors in the concentrations are negligible compared with the error in the absorbances, we can write a series of equations like

$$
\begin{aligned}
A_1 &= abc_1 + e_1 \\
A_2 &= abc_2 + e_2 \\
&\ \vdots \\
A_N &= abc_N + e_N
\end{aligned}
\tag{9.2}
$$

where e_i is the difference between the observed absorbance A_i and the predicted absorbance abc_i. Solving these equations simultaneously yields an estimate of a.

Symbolic Mathematics for Chemists: A Guide for Maxima Users, First Edition. Fred Senese.
© 2019 John Wiley & Sons Ltd. Published 2019 by John Wiley & Sons Ltd.
Companion website: http://booksupport.wiley.com

More generally, we can write

$$y_i = \alpha x_i + e_i \qquad i = 1, 2, \ldots, N \tag{9.3}$$

where y is the dependent variable and x is the independent variable. The model equation is $y = \alpha x$, where α is the parameter we want to find. e_i is the difference between the observed and predicted values; it is called a residual. We want to find the value of α that makes the residuals as small as possible.

The residuals can be partitioned into two parts:

$$e_i = (y_i - y_{i,\text{true}}) - (y_{i,\text{model}} - y_{i,\text{true}}) \qquad \text{Experimental and model error in residuals}$$

$$\tag{9.4}$$

The first term is the experimental error, the difference between the observed y_i and its true (and usually unknown) value. The second contribution comes from flaws in the model. The model parameters may have incorrect values, or the model equation might not have the correct form, or it might omit other independent variables that affect y.

Optimizing α or the form of the model equation makes only the second contribution smaller. Experimental error is inherent in the experimental data and cannot be removed by altering the model. Even if the model is correct, then, the residuals will not be zero. However, the form of the model equation and the model parameters (like α) can be optimized so that all of the residuals are as small as possible.

We require a collective function of the residuals to minimize. The most common choice is the sum of the *squared* residuals (SSE):

$$\text{SSE} = \sum_{i=1}^{N} e_i^2 = \sum_{i=1}^{N} (y_i - \alpha x_i)^2 \qquad \text{Sum of squared residuals for the fit } y = \alpha x \tag{9.5}$$

The SSE is called a figure of merit. When it is small, the residuals will be small (and the fit will be "good"). To minimize the SSE, find the derivative dSSE/dα, set it to zero, and solve for α (see Section 6.4):

$$\frac{\text{dSSE}}{\text{d}\alpha} = 2\alpha \sum_{i=1}^{N} x_i^2 - 2 \sum_{i=1}^{N} x_i y_i \tag{9.6}$$

$$\alpha = \frac{\sum_{i=1}^{N} x_i y_i}{\sum_{i=1}^{N} x_i^2} \qquad \text{Least-squares fitted } \alpha \text{ for the model } y = \alpha x \tag{9.7}$$

Optimization of model parameters to minimize the SSE is the basis for the ordinary least-squares method.

 Worksheet 9.0: Fitting data to a proportionality

In this worksheet, we'll use least-squares fitting to find proportionality constants, along with their uncertainties.

9.1   The Ordinary Least-Squares Method

Suppose we have a set of N data points (x_i, y_i), with $i = 1, 2, \ldots, N$. In a linear model with only one independent variable x, we have two parameters to fit: the slope (m) and the intercept (b). The y values can be predicted from the x values by

$$\hat{y}_i = b + mx_i \qquad \text{The predicted } y \text{ value} \atop \text{for a linear model} \tag{9.8}$$

where the hat marks the predicted value. Each of the data points will satisfy

$$y_i = b + mx_i + e_i \qquad \text{Residuals } e_i \atop \text{for a linear model} \tag{9.9}$$

Figure 9.1 compares y_i, $\hat{y}_i$, and the residual e_i with the mean $\bar{y}$. It shows how the variation of y_i around its mean has two parts: variation explained by the model ($\hat{y}_i - \bar{y}$) and unexplained variation found in the residuals ($e_i = y_i - \hat{y}_i$).

As before, the SSE is the sum of squared residuals. The ordinary least-squares method becomes

$$\text{Choose } m, b \text{ that minimize } \sum_{i=1}^{N}(y_i - mx_i - b)^2 \qquad \text{The ordinary linear} \atop \text{least-squares method} \tag{9.10}$$

The SSE will be at its minimum value when its derivatives with respect to m and b are zero. This gives us two normal equations, which can be solved simultaneously to give expressions for m and b:

$$\begin{aligned} \frac{\partial}{\partial m} \sum_{i=1}^{N}(y_i - mx_i - b)^2 &= 0 \\ \frac{\partial}{\partial b} \sum_{i=1}^{N}(y_i - mx_i - b)^2 &= 0 \end{aligned} \qquad \text{Normal equations for} \atop \text{ordinary linear least-squares} \tag{9.11}$$

Let's use Maxima to solve the normal equations.

First, declare `sum` as a linear operation (Section 4.5), that is, $\sum_i^N a_i + b_i$ should be written as $\sum_i^N a_i + \sum_i^N b_i$.

```
(%i1)   declare(sum,linear)$
```

Define the figure of merit:

```
(%i2)   SSE: sum((y[i]-(m*x[i]+b))^2,i,1,N);
```

Figure 9.1   Residuals e_i reflect the variation in the y_i data around $\bar{y}$ that the model does not explain.

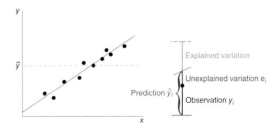

$$(\%o2) \quad \sum_{i=1}^{N} \left(-x_i\, m + y_i - b\right)^2$$

Setting the derivatives of the figure of merit with respect to m and b to zero gives us two equations in two unknowns:

```
(%i3)   eqn1: diff(SSE,m)=0,expand;
(%i4)   eqn2: diff(SSE,b)=0,expand;
```

$$(\%o3) \quad 2 \left(\sum_{i=1}^{N} x_i^2\right) m - 2 \left(\sum_{i=1}^{N} x_i y_i\right) + 2b \sum_{i=1}^{N} x_i = 0$$

$$(\%o4) \quad 2bN + 2 \left(\sum_{i=1}^{N} x_i\right) m - 2 \sum_{i=1}^{N} y_i = 0$$

where we've used the `diff` operator to calculate the derivatives. Solving both equations simultaneously to find m and b,

```
(%i5)   solve([eqn1,eqn2],[m,b]);
```

$$(\%o5) \quad [[m = \frac{\left(\sum_{i=1}^{N} x_i y_i\right) N - \left(\sum_{i=1}^{N} x_i\right) \sum_{i=1}^{N} y_i}{\left(\sum_{i=1}^{N} x_i^2\right) N - \left(\sum_{i=1}^{N} x_i\right)^2},$$

$$b = -\frac{\left(\sum_{i=1}^{N} x_i\right)\left(\sum_{i=1}^{N} x_i y_i\right) - \left(\sum_{i=1}^{N} x_i^2\right) \sum_{i=1}^{N} y_i}{\left(\sum_{i=1}^{N} x_i^2\right) N - \left(\sum_{i=1}^{N} x_i\right)^2}]]$$

9.1.1 Using Built-In Functions

Now that we see where these formulas come from, we can just use Maxima's built-in functions to compute them.

All of Maxima's least-squares functions accept data placed in a matrix, with the x and y data listed in separate columns. For example,

```
(%i1)   data: matrix([0,5.9],  [0.9,5.4],  [1.8,4.4],  [2.6,4.6],
    ➥   [3.3,3.5],  [4.4,3.7],  [5.2,2.8],  [6.1,2.8],
    ➥   [6.5,2.4],  [7.4,1.5])$
```

The `plsquares` function computes a fit equation from the points:

plsquares(*data matrix*, *variable list*, *dependent variable*)

Computes the equation of a line from the points in *data matrix*, with columns labeled by items in the *variable list*, and given a data matrix, a list of variable names for the columns in the matrix, and the name of the *dependent variable*. `load(plsquares)` before using this function.

For example,

```
(%i2)    load(plsquares)$
(%i3)    plsquares(data,[x,y],y);
```

```
Determination Coefficient for y = 0.95350386049735
```
$$(\%o3) \quad y = 5.761185190439042 - 0.53957727498404\,x$$

The determination coefficient is a measure of how well the fit explains the variation of the data around its mean. If it is zero, the fit explains *none* of the variation. If it is one, the fit explains *all* of the variation. In this case, it's 0.95, so the fit explains roughly 95% of the variation in the data. We'll look at the determination coefficient in more detail in Section 9.1.3.

The fit parameters alone can be computed with lsquares_estimates, found in the lsquares package.

lsquares_estimates(*data matrix, variable list, model equation, parameter list*)
Find the best fit of *model equation* with variables in *variable list* to the data in *data matrix*. The fit parameters appearing in the model equation are listed in *parameter list*. The *variable list* gives a name for each column in the data matrix; not all of the variables need to appear in the model equation. Each row in the data matrix corresponds to a data point. *lsquares_estimates* returns a list of solutions, each of which is a list of parameter equations.

For example,

```
(%i4)    load(lsquares)$
(%i5)    lsquares_estimates(data, [x, y], y=m*x+b, [b,m]), numer;
```

$$(\%o5) \quad [[b = 5.761185190439038, m = -0.53957727498404]]$$

where [x,y] is a list of variable names, y=m*x+b is the equation we want to fit, and [b,m] is a list of fit parameters. The result is a list of solution lists (in case more than one solution is found).

The lsquares_estimates function does not directly provide statistics for evaluating the goodness of a fit. It also doesn't provide error estimates for fit parameters. It is best used to obtain nonlinear curve fits, as we'll see in Chapter 10.

A more convenient function for fitting lines is called linear_regression found in the stats package.

linear_regression(*data matrix*)
Performs linear regression on the data in a *data matrix* (which must list the dependent variable data in its last column). The function prints an inference_result object that contains the items in Table 9.1.

For example,

```
(%i6)    load(stats)$
(%i7)    result: linear_regression(data);
```

Table 9.1 Items that can be extracted from the `linear_regression` output.

Label	Meaning
`'b_estimation`	A list of estimated fit parameters, with the y-intercept first.
`'b_covariances`	The covariance matrix of the fit parameters.
`'b_conf_int`	95% confidence intervals of the fit parameters. Change the confidence level by calling `linear_regression` with the option `'conflevel=95/100`, where any desired level can replace the 95.
`'b_statistics`	Statistics for testing the fit parameters, with p-values given by `'b_p_values` and probability distribution and degrees of freedom given by `'b_distribution`
`'v_estimation`	The estimated variance of the fit.
`'v_conf_int`	The 95% confidence interval for the variance estimate.
`'v_distribution`	The probability distribution for the variance test, with degrees of freedom.
`'residuals`	A list of the residuals.
`'adc`	The adjusted determination coefficient.
`'aic`	Akaike's information criterion [54].
`'bic`	Bayesian information criterion [55].

$$(\%o7) \quad \begin{vmatrix} \text{LINEAR REGRESSION MODEL} \\ b\_estimation = [5.761185190439042, -0.53957727498404] \\ b\_statistics = [30.40440791394536, -12.80848528119326] \\ b\_p\_values = [1.4868004605261831 \ 10^{-9}, 1.3024677540940388 \ 10^{-6}] \\ b\_distribution = [student\_t, 8] \\ v\_estimation = 0.10008294027945 \\ v\_conf\_int = [0.045662061388214, 0.36732221318839] \\ v\_distribution = [chi2, 8] \\ adc = 0.94769184305952 \end{vmatrix}$$

The output is an `inference_result` object like those returned by the hypothesis testing functions we used in Chapter 8. It provides a trove of useful but cryptically labeled statistical information about the fit. We can extract items from the inference result using the `take_inference` function, which has the form:

 `take_inference('label, results)`
Extract an item labeled *label* from *results*, which is an `inference_result` returned by some statistical functions in Maxima.

For example, the intercept and slope are listed as `b_estimation`. To extract them as b and m, respectively, we could type

```
(%i8)   [b,m] : take_inference('b_estimation, result);
(%o8)   [5.761185190439042, -0.53957727498404]
```

and use the values to plot the fitted line together with the data:

```
(%i9)   datapoints : makelist([data[i,1], data[i,2]], i, 1,
        ➡ length(data))$
```

```
(%i10)  plot2d([[discrete, datapoints], m*x+b], [x, 0,8], [style
        ➡ , points, lines], [legend, false])$
```

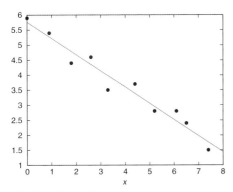

The labels for items in the `linear_regression` output are given in Table 9.1. We'll use some of them in the following sections.

9.1.2   Error Estimates for the Slope and the Intercept

In a good fit, the error in the slope and the intercept comes from the experimental error in the data. If we can assume that all of the error is concentrated in the y values, and that the errors are independent of each other, we can apply Equation (8.33) to propagate the error in the points into the error in the slope and the intercept:

$$\sigma_m^2 = \sum_{j}^{N} \left(\frac{\partial m}{\partial y_j} \right)^2 \sigma_j^2, \qquad \sigma_b^2 = \sum_{j}^{N} \left(\frac{\partial b}{\partial y_j} \right)^2 \sigma_j^2 \tag{9.12}$$

where σ_j^2 is the variance in y_j; it is also the *error variance* in y for that point. In ordinary least-squares, we assume that $\sigma_j^2 = \sigma^2$ for all points; all points have identical error variances.

Let's use Maxima to simplify Equation (9.12) using the solutions of the normal equations, which we computed previously in Section 9.1:

```
(%i1)  declare('sum, linear)$
(%i2)  SSE: sum((y[i]-(m*x[i]+b))^2, i, 1, N)$
(%i3)  eqn1: diff(SSE,m)=0, expand$
(%i4)  eqn2: diff(SSE,b)=0, expand$
(%i5)  solve([eqn1,eqn2],[m,b])$
(%i6)  [M, B] : [m, b], first(%)$
```

where the expressions for the slope and intercept are now in M and B, respectively.

The derivatives in Equation (9.12) can be computed by noting that

$$\frac{\partial}{\partial y_j} \sum_{i}^{N} x_i y_i = x_j \quad \text{and} \quad \frac{\partial}{\partial y_j} \sum_{i}^{N} y_i = 1$$

For the derivative $\partial m / \partial y_j$, we have

```
(%i7)  dmdyj : M, sum(x[i]*y[i], i, 1, N) = x[j], sum(y[i],
        ➡ i, 1, N) = 1;
```

(%o7)
$$\frac{x_j N - \sum_{i=1}^{N} x_i}{\left(\sum_{i=1}^{N} x_i^2\right) N - \left(\sum_{i=1}^{N} x_i\right)^2}$$

Applying Equation (9.12) gives the variance in the slope σ_m^2 as

(%i8)   `sigma^2*sum(dmdyj^2,j,1,N);`

(%o8)
$$\frac{\sigma^2 \sum_{j=1}^{N} \left(x_j N - \sum_{i=1}^{N} x_i\right)^2}{\left(\left(\sum_{i=1}^{N} x_i^2\right) N - \left(\sum_{i=1}^{N} x_i\right)^2\right)^2}$$

which we can simplify by making the sums all run over the same index:

(%i9)   **changevar(%, j=i, i, j), ratsimp;**

(%o9)
$$\frac{\sigma^2 N}{\left(\sum_{i=1}^{N} x_i^2\right) N - \left(\sum_{i=1}^{N} x_i\right)^2}$$

Similarly for the intercept, $\partial b/\partial y_j$ is

(%i10)   `dbdyj  :  B,  sum(x[i]*y[i],  i,  1,  N)  =  x[j],  sum(y[i],  i,`
         `↦  1,  N)  =  1;`

(%o10)
$$\frac{\left(\sum_{i=1}^{N} x_i\right) x_j - \sum_{i=1}^{N} x_i^2}{\left(\sum_{i=1}^{N} x_i^2\right) N - \left(\sum_{i=1}^{N} x_i\right)^2}$$

and the variance in the slope σ_b^2 is

(%i11)   `sigma^2*sum(dbdyj^2,j,1,N);`
(%i12)   **changevar(%, j=i, i, j), ratsimp;**

(%o11)
$$\frac{\left(\sum_{j=1}^{N} \left(\left(\sum_{i=1}^{N} x_i\right) x_j - \sum_{i=1}^{N} x_i^2\right)^2\right) \sigma^2}{\left(\left(\sum_{i=1}^{N} x_i^2\right) N - \left(\sum_{i=1}^{N} x_i\right)^2\right)^2}$$

(%o12)
$$\frac{\left(\sum_{i=1}^{N} x_i^2\right) \sigma^2}{\left(\sum_{i=1}^{N} x_i^2\right) N - \left(\sum_{i=1}^{N} x_i\right)^2}$$

We must have some estimate of the unknown error variance σ^2 to use these equations. Remember that the residuals include both experimental error and errors due to the model's lack-of-fit, while the error variance includes only experimental error. We can use the residual variance s^2 as an estimate of the error variance *if we assume that our model is correct.*

$$s^2 = \frac{SSE}{N - p} \qquad \text{Estimated variance of} \atop \text{an ordinary least-squares fit} \tag{9.13}$$

where we have $N - p$ degrees of freedom with p fit parameters.[1]

The residual variance has two contributions: one is variance in the experimental data and the other is variance due to flaws in the model. We call these contributions the *error variance* and the *lack-of-fit variance*, respectively.

In "good" models, the residual variance s^2 is close to the error variance σ^2. We can have $s^2 > \sigma^2$ if our model is underfitting the data, because the lack-of-fit error will increase s^2 (see Figure 9.9). On the other hand, if our model artificially (and incorrectly) erases some of the experimental error, we can have $s^2 < \sigma^2$. This is one symptom of overfitting. The difference between s^2 and σ^2 is the basis for statistical tests for lack-of-fit, as we'll see in Section 9.1.6.

The square root of s^2 is variously called the standard error of the regression, the *residual standard error, standard error of the fit*, and the *standard error of estimation*.[2] It estimates the standard deviation of the residuals around a mean residual of zero. If s is zero, we have a perfect fit, with all points lying exactly on the line. If s is large, at least some of the points lie far away from the line – and the linear model is less useful for predicting individual data points.

Maxima's `linear_regression` function displays the variance of the fit as `v_estimation`.

If we assume that we have a good model, we can replace σ^2 with s^2 in the expressions we derived above for σ_m^2 and σ_b^2 to estimate the standard errors of m and b:

$$s_m = \sqrt{\frac{s^2 N}{N \sum_{i=1}^{N} x_i^2 - \left(\sum_{i=1}^{N} x_i\right)^2}} \qquad \text{Standard error of the slope} \tag{9.14}$$

$$s_b = \sqrt{\frac{s^2 \sum_{i=1}^{N} x_i^2}{N \sum_{i=1}^{N} x_i^2 - \left(\sum_{i=1}^{N} x_i\right)^2}} \qquad \text{Standard error of the intercept} \tag{9.15}$$

These standard errors can be converted to 95% confidence intervals using the methods we applied in Chapter 8. Multiplying the standard error by the t-value with a quantile of 0.975 and $N - 2$ degrees of freedom gives the half-width of the 95% confidence interval[3]

$$m \pm t s_m \qquad \text{Confidence interval of the slope} \tag{9.16}$$

$$b \pm t s_b \qquad \text{Confidence interval of the intercept} \tag{9.17}$$

Maxima can calculate t as `quantile_student_t(0.975, N-2)`.

---

1 To see why there are $N - p$ degrees of freedom, imagine a linear fit with only two data points. Two points determine a line, so the residuals are both forced to be zero. They have zero degrees of freedom; the degrees of freedom are two less than the number of points. If we had three parameters (say a parabola), three points would perfectly fit the curve, and we have $N - 3$ degrees of freedom.

2 It isn't quite right to refer to s as the standard deviation of the fit. Recall that a sample standard deviation measures the scatter of data, while a sample standard error measures the variation of a parameter across many theoretical samples. In this case, the parameter is the variance of the residuals. If we collect a different set of data, the slope and the intercept will change, so the residuals will change too. For the entire population, the standard errors of the fit, the slope, and the intercept are zero, if some correct linear model actually exists.

3 Recall that the quantile for a two-sided test is $(1 - \alpha/2)$, where α is $1 - 0.95 = 0.05$ for 95% confidence.

The `linear_regression` function computes the confidence intervals for the fit parameters as `b_conf_int`; type `take_inference('b_conf_int, result)` to see them (if the output from `linear_regression` is stored as `result`).

 Worksheet 9.1: Fitting Data to a Line with Ordinary Least Squares (OLS)

In this worksheet, we'll see how to calculate m and b, along with their error estimates. We'll also take a first look at how residuals can be used to critically evaluate the models we'll build.

9.1.3 The Determination Coefficient

Computing the fit parameters and estimating their errors is only a first step. Small values of s, s_m, and s_b mean a closer fit of the data, but we cannot use these standard errors alone to report the quality of a fit, because their derivation implicitly assumes a good fit [56]. We need some more objective measure of how well the model fits the data.

In Figure 9.1 we partitioned the variation of a single y value around its mean into an unexplained variation (the residual) and an explained variation (the difference between the y value and the mean y). The total sum of squared deviations from the mean (SST) can also be partitioned into explained and unexplained parts, which we call SSR and SSE, respectively:

$$\text{SSR} = \sum_{i=1}^{N} (\hat{y}_i - \bar{y})^2 \qquad \text{A measure of explained variation in } y \tag{9.18}$$

$$\text{SSE} = \sum_{i=1}^{N} (y_i - \hat{y}_i)^2 \qquad \text{A measure of unexplained variation in } y \tag{9.19}$$

$$\text{SST} = \text{SSR} + \text{SSE} = \sum_{i=1}^{N} (y_i - \bar{y})^2 \qquad \text{A measure of total variation in } y \tag{9.20}$$

When all points fall exactly on the fit line, SSE = 0 and SST = SSR. For completely random data, SSR = 0 and SST = SSE. Between these extremes, random error blurs the linear relationship to some degree.

The determination coefficient (R^2) quantitatively shows where a fit lies between the extremes. R^2 is the ratio of the explained variation to the total variation:

$$R^2 \equiv \frac{\text{SSR}}{\text{SST}} = 1 - \frac{\text{SSE}}{\text{SST}} \qquad \text{The determination coefficient} \tag{9.21}$$

R^2 is zero when the data is completely random; the line explains none of the variation. R^2 can be equal to one only when all residuals are zero, so that all points lie exactly on the line (Figure 9.2).

Why is the determination coefficient called R^2? The name comes from the fact that it is equal to the square of Pearson's r, which we have already seen in Section 8.5.3. It is given by

$$r = \frac{\sum_{i=1}^{N} (x_i - \bar{x})(y_i - \bar{y})}{\sqrt{\sum_{i=1}^{N} (x_i - \bar{x})^2 \sum_{i=1}^{N} (y_i - \bar{y})^2}} \qquad \begin{array}{l}\text{The linear correlation} \\ \text{coefficient between } x \text{ and } y\end{array} \tag{9.22}$$

This expression gives us some insight into what R^2 (and r) can and cannot tell us:

- If we know that a linear correlation does in fact exist between x and y, r and R^2 indicate the strength of that correlation.

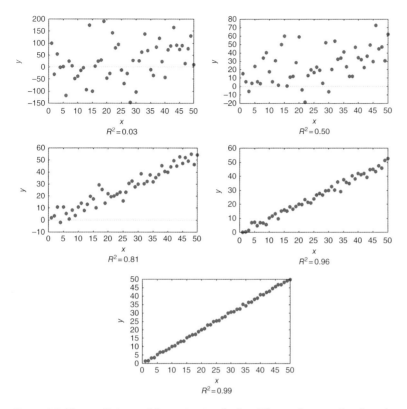

Figure 9.2 The coefficients of determination for five different datasets, fitted to a line.

- The equation for r contains no information about the distribution of x and y, so unless we make specific assumptions about those distributions, we have no general way to use r or R^2 to test hypotheses [56]. Without such tests, we can't decide whether the correlation is statistically significant, and we can't confidently state whether one correlation is better than another using r or R^2 values alone.
- Notice that r (and so, R^2) can be computed without information about the fit parameters. R^2 and r aren't measures of how good the fit is; they only indicate the strength of the linear correlation between x and y, if a linear correlation exists, which isn't the same thing.

Remember the following when you're tempted to report an R^2 value as evidence of goodness-of-fit, or to use R^2 to compare two fits.

- Fitting a line to a curve can yield a high R^2. If the relationship is *roughly* linear overall, the least-squares procedure will do its best to compensate for missing or incorrectly specified independent variables by distorting the fit parameters. As a result R^2 cannot "see" small but obvious departures from linearity, especially over small ranges of x. To convince yourself of this, try fitting a line to a gentle curve, or an oscillating function with small amplitude. The R^2 value will often be high.
- Simply changing the range of the x data can change R^2 because R^2 depends strongly on the variance in the independent variable (see Figure 9.3). It is unwise to compare fits with different datasets simply by comparing their R^2 values, particularly if their ranges or variances differ. We'll see a classic example of this in Worksheet 9.3.

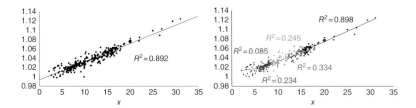

Figure 9.3 R^2 depends strongly on the variance of the independent variable. Partitioning the 300-point fit on the left into five 60-point subsets yields five R^2 values that aren't equal to each other or to the R^2 of the overall fit, even though the actual data and its underlying linearity haven't changed at all.

- When we have more than one independent variable, R^2 is misleading. R^2 rises when adding new variables – even when those variables are completely spurious. Fitting absorbance to concentration and the current wind speed gives a better R^2 than fitting absorbance to concentration alone.

  As is, R^2 isn't useful for comparing models, or for deciding whether a variable should be included in the model. But it can be adjusted somewhat to compensate for this effect. The adjusted R^2 (also called an adjusted determination coefficient, or ADC) tries to penalize the R^2 for additional parameters:

$$\text{ADC} = 1 - \frac{N-1}{N-p}\left(1 - R^2\right) \qquad \text{The ADC} \qquad (9.23)$$

  The ADC can provide some guidance in screening potential new variables to add to a model. If the ADC rises when the new variable is added, the new variable can be considered for inclusion. If the ADC drops, the variable is unnecessary.

  The ADC is close to the value of the unadjusted R^2 – unless the number of parameters is large relative to the number of data points. Both statistics share the weaknesses listed above.
- A high R^2 can accompany fits that are severely perturbed by outliers, especially when the dataset is large.
- Even a good fit can have a low R^2 value if the y values are much smaller than the x values.

We saw in Section 8.5.3 that r is easily computed using the `cor` function. Several other functions also compute r and R^2. Maxima's `plsquares` function displays the R^2 value, and `linear_regression` displays the ADC as `adc`. The `list_correlations` function will compute both R^2 and Pearson's r (among other things) from a data matrix.

Like any statistic, R^2 must be interpreted in the context of the data. Much of that context can be provided by scatter plots of the data and the residuals, as we'll see in the next section.

9.1.4 Residual Analysis

The residuals are a rich source of information about the quality of a fit. They can reveal outliers in the data and missing or incorrectly specified variables in the model. With precise and reliable data, they can also tell us something about the model's accuracy.

Residuals are easy to calculate, and they are also provided by many built-in functions. Maxima's `linear_regression` output includes them under the name `residuals`. If you're using `lsquares_estimates`, use `lsquares_residuals` to find the residuals.

Always present least-squares results with scatter plots of the data and the residuals. Routinely examine the following plots, and include any that are particularly informative when reporting the fit.

- Plot y against each x variable.
- Plot residuals against time or run number.
- Plot residuals against $\hat{y}$ (not y),[4] particularly when there is more than one x variable.
- Plot residuals against each x variable.
- Plot residuals against any other experimental variables that might be included in the model later.

Look for three things in these plots: the overall scale of the residuals; the appearance of trends, patterns, or "structure" in the residuals; and the presence of outliers (points that are distant from the others, or points with extreme x or y values).

- The residuals ought to be roughly the same size as the experimental uncertainty in the y_i data. For example, if the y_i values are pressures that are reproducible to ± 0.1 torr, the raw residuals should be about that size for a good fit.
- The residuals should scatter randomly around zero, and the amount of scatter should be consistent everywhere. Small residuals will be more common than larger ones, and the sign of the residuals should vary randomly. *Standardizing* the residuals (dividing them all by the standard error of the fit) helps clarify their distribution. Standardized residuals are in units of standard deviation, so roughly 2/3 should fall between ± 1, and about 95% should fall between ± 2.
- Outliers must be considered individually; we'll deal with them in Section 9.2.4. Points that have standardized residuals outside the range ± 3 are potential outliers, though we'd expect about 1% of the residuals to be about that size just from natural variation.

Any structure in residuals plots is cause for concern. Nonrandom patterns may indicate that some of the assumptions that the least-squares method makes about the data don't apply. Let's examine these assumptions, and their effect on the residuals:

1) *Errors for different points are independent and random.* As we saw in Chapter 8, error is often *not* entirely random. The least-squares method completely neglects nonrandom error or systematic error. Systematic error can sometimes correlate errors for different points. In this case the residuals cluster or follow regular trends. For example, plotting the residuals against the run number can reveal the presence of instrument drift (Figure 9.4).
2) *A correct linear model exists.* Fitting an infinite number of points should give the correct model. But the correct model (if it exists at all) might not be linear. Fitting a line to a curve yields residual plots that show trends (see Figure 9.5). An x, y scatter plot alone might *look* linear, if the curvature in the data is subtle, but a residual plot can make the trends clear.

Figure 9.4 Systematic error (like instrument drift) can cause trends or clusters in the residuals when they are plotted against the run number.

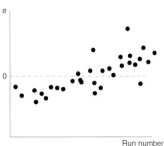

---

4 Plotting the residuals e against $\hat{y}$ is more useful than plotting them against y, because y and e are correlated $(y = \hat{y} + e)$, but $\hat{y}$ and e are independent.

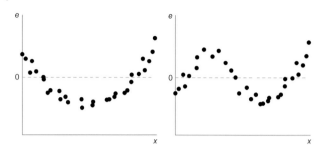

Figure 9.5 Trends in residuals sometimes show flaws in a model equation. There may be a missing variable, or a missing higher order term or interaction for variables already included.

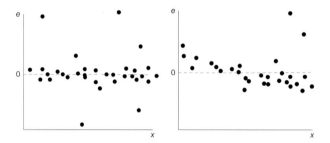

Figure 9.6 A non-normal distribution can lead to multiple outliers on a residuals plot. If the outliers aren't symmetrically distributed, they will strongly distort the fit, as in the plot on the right.

Also plot the residuals against variables that you've measured along with x and y, but haven't included in your model. If trends appear in one of those plots, your model should include that variable.

3) *The errors are normally distributed* around the correct model. In Chapter 8, we saw that real data doesn't always follow a normal distribution. One symptom of a non-normal distribution can be a large number of outliers (Figure 9.6). The problem can be especially severe if the distribution isn't symmetric; multiple outliers on just one side of the distribution invalidate the fit. Ordinary least-squares may still work for data with small to moderate deviations from normality, provided the other assumptions are met. If you aren't seeing many outliers, your data is probably normal enough. Otherwise, use the Shapiro–Wilk test discussed in Chapter 8 or the *rankit plot* (Worksheet 8.4) to check normality.

4) *The error variance is the same for all points.*[5] Real data often has an error variance that changes from point to point. The fit should treat imprecise points as less important than precise points. OLS gives all points equal contributions to the fit, so it produces fits that favor imprecise points more than it should. It may drag the fit line away from more precise points to do so.

Nonconstant error variance can create interesting patterns in the residuals (Figure 9.7), especially on a plot of the residuals against $\hat{y}$. The mean of the residuals also may not be close to zero. The fit parameters themselves aren't affected, but their error estimates (and all statistics computed from them) will be incorrect. When residual plots alone do not clearly reveal unequal variances, statistical tests can be applied to detect them.

The plot on the right in Figure 10.5 is typical for linear fits of the Clausius–Clapeyron equation (Equation 10.5) to vapor pressure and temperature data. Not only is a linear model inadequate (as shown by the curve in the residuals), the error in the data is distorted

---

5 Statisticians refer to this assumption as *homoskedasticity*. When the standard deviations differ from point to point, the residuals are said to be *heteroskedastic*.

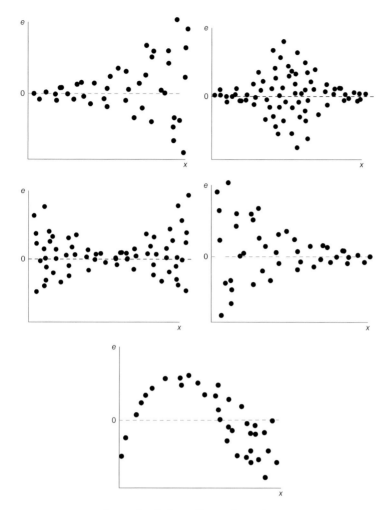

Figure 9.7 Megaphones, footballs, and butterflies: when the standard deviation isn't the same for all points, structures like these can appear on residual plots. The plot on the far right has a second problem; a line is being fit to curved data.

because the y variable is the logarithm of the vapor pressure. We'll see that the weighted least-squares method can solve the second problem. The first must be solved by choosing a better fit function.

5) *All of the error is in the y data.* If you have significant error in the x values, the fit needs to account for it. Ordinary least-squares can't do this (unless the error in y is relatively much larger than the error in x). You need to use total least-squares (TLS) when you have significant error in both x and y (see Section 9.4).

9.1.5 Testing the Fit Parameters

Often we would like to compare the slope with some "known" value, m_0. If the residuals are normally distributed, *and* the error variances are the same for every point,[6] we can use the t-statistic

---

6 Beware: if you have unequal error variances, the standard errors for the fit, slope, and intercept as computed above aren't valid – and the results of the t tests outlined in this section will be meaningless.

$$t = \frac{m - m_0}{s_m} \qquad \text{\textit{t}-Statistic for testing the null hypothesis } m = m_0 \qquad (9.24)$$

The p-value (the chance that you'll be wrong if you reject the null hypothesis) is calculated as follows:

$$p\text{-value} = 2(1 - \texttt{cdf\_student\_t}(t, N - 2)) \qquad \text{Probability of incorrectly rejecting the null hypothesis}$$

$$(9.25)$$

where $\texttt{cdf\_student\_t}(t, N - 2)$ is the value of cumulative t-distribution function at t with $N - 2$ degrees of freedom. In Maxima, the p-value can be computed by $\texttt{2*(1-cdf\_student\_t(t, N-2))}$.

A p-value of less than 0.05 is usually taken as sufficient grounds for rejecting the null hypothesis.[7] We can say that m and m_0 differ significantly with a confidence of $(1 - p\text{-value}) \times 100\%$.

The special case $m_0 = 0$ is particularly useful. If we have a p-value of less than 0.05 for this hypothesis, we can be 95% confident that *no linear relationship exists between x and y*.

A similar t-statistic can be used to compare the intercept b with some constant value b_0:

$$t = \frac{b - b_0}{s_b} \qquad \text{\textit{t}-Statistic for testing the null hypothesis } b = b_0 \qquad (9.26)$$

Maxima's $\texttt{linear\_regression}$ function computes the t-statistics for you with the null hypotheses $b_0 = 0$ and $m_0 = 0$. You'll find them under the name $\texttt{b\_statistics}$. It also computes the p-values under the name $\texttt{b\_p\_values}$.

 Worksheet 9.2: Why the determination coefficient is not enough

In this worksheet, you'll apply the ordinary least-squares method to the classic *Anscombe quartet*, four datasets that are designed to show the danger of blind reliance on regression statistics [57]. We'll also use ordinary least-squares to find the activation energy, reaction order, and rate constants in chemical reactions, and to find the dipole moment of a molecule.

9.1.6 Testing for Lack-of-Fit

Unfortunately, many of the effects we discussed in Section 9.1.4 may be superimposed in residual plots for real experimental data. For example, consider the fit and residuals plots in Figure 9.8. There does appear to be a slight curved trend in the residuals, but random error obscures it.

Humans have an amazing talent for recognizing patterns – even when they're not actually there. When you aren't sure whether the pattern you're seeing in the residuals is real, there are a number of objective statistical tests that can be applied. None of these tests can provide the depth of information found in a residual plot, so use them to supplement ambiguous plots, not replace them.

---

7 Remember, the p-value isn't the probability that the null hypothesis is correct; it's the probability that you'll be wrong if you reject the null hypothesis. It tells you how strong the evidence *against* the null hypothesis is. For example, if the p-value is 0.05, you can say that you are 95% confident that the null hypothesis is incorrect, but you can *not* say that you're 5% confident that the null hypothesis is correct. The distinction is subtle, but very important!

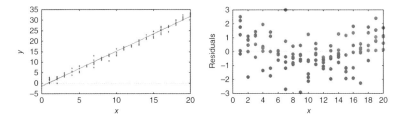

Figure 9.8 Is the dragonfly pattern in these residuals real, or is it just the result of random chance? Random error can obscure trends in the residuals, making interpretation of a scatter plot difficult. Objective statistical lack-of-fit tests can be helpful in cases like this.

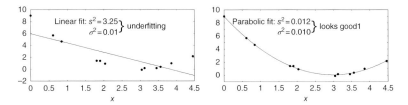

Figure 9.9 When the model isn't flexible enough to fit the data, or if it has the wrong form, the residual variance s^2 will be larger than the error variance σ^2.

As we saw in Section 9.1.2, the residual variance won't be a good estimate of the error variance if we have a bad fit (Figure 9.9). That suggests that we can build a statistic that compares s^2 with σ^2 to test for lack-of-fit. We'll need to either know σ^2 or have some way of estimating it that is independent of our model.

In chemistry, we sometimes know the error variance in advance. Manufacturers may provide expected standard errors for instrument readings; standard analytical protocols list them, too. Sometimes we can measure them ourselves by collecting replicate data.[8] If we know σ^2, we can compute the statistic

$$\chi^2 = (N - p)\frac{s^2}{\sigma^2} \qquad \text{Chi-squared statistic to test lack-of-fit} \atop \text{with known } \sigma^2 \qquad\qquad (9.27)$$

which follows the well-known χ^2 (chi-squared) distribution. For the null hypothesis that $s^2 = \sigma^2$, the p-value is computed as `1 - cdf_chi2(`χ^2`,N-p)` in Maxima. If the p-value is less than 0.05, we can conclude with 95% confidence that the model is inadequately fitting the data. In that case we need to find a better model.

What if we don't know σ^2? One convenient test is the lack-of-fit sum-of-squares test (LOFSS test) [58]. To apply the test, you *must* have replicate measurements for at least some of the y_i. The test compares the variation in the residuals with the variation in y_i without the model. If the two variations are the same, the fit is adequate. If the residual variation is higher than the experimental variation in y_i, though, the model can be rejected at some confidence level.

To test for lack-of-fit we separate the SSE into model-dependent and model-independent parts. If y_{ij} is the jth replicate measurement of y at x_i, and $\bar{y}_i$ is the average value of the n_i replicate y measurements at x_i, we can write

---

8 If you go this route, be sure that you've collected replicate data for several different samples. With a single sample, you'll get only the error variance for that sample, which may or may not be a good estimate of σ^2. If you find that the error variance is different for different samples you should do a weighted least-squares fit; see Section 9.3.

$$\text{SSE} = \sum_{i=1}^{N} \sum_{j=1}^{n_i} (y_{ij} - \hat{y}_i)^2$$

Partitioning SSE into
model-independent and
model-dependent terms
(9.28)

$$= \sum_{i=1}^{n_x} \sum_{j=1}^{n_i} (y_{ij} - \bar{y}_i)^2 + \sum_{i=1}^{n_x} \sum_{j=1}^{n_i} (\bar{y}_i - \hat{y}_i)^2$$

where n_x is the number of unique x_i values. The first term is the sum of squared model-independent errors; the second is the sum of squares from lack of fit. These terms can be used to build model-independent and model-dependent variance estimates:

$$s_{\text{ind}}^2 = \frac{\sum_{i=1}^{n_x} \sum_{j=1}^{n_i} (y_{ij} - \bar{y}_i)^2}{N - n_x}$$

Model-independent estimate of σ^2
(9.29)

$$s_{\text{dep}}^2 = \frac{\sum_{i=1}^{n_x} n_i (\bar{y}_i - \hat{y}_i)^2}{n_x - p}$$

Model-dependent estimate of σ^2
(9.30)

where p is the number of fit parameters (two for a line). The model-independent estimate of the variance will be a good estimate of σ^2 whether or not the model is valid. The model-dependent estimate will be accurate only if the model is correct; if not, it will be greater than the model-independent estimate.

To see whether the model-dependent estimate of variance is significantly larger than the model-independent estimate, perform an F-test, as described in Section 8.4.2. The F statistic is

$$F = \frac{s_{\text{dep}}^2}{s_{\text{ind}}^2}$$

F-Statistic for the LOFSS test
(9.31)

The p-value for the hypothesis that the two estimates are equal is computed by $1 - \text{cdf\_f}(F, n_x - p, N - n_x)$ in Maxima. If the p-value is less than 0.05, we can be 95% confident that the fit is inadequate.

The LOFSS test is quite general. We can use it in multiple linear regression and weighted least-squares (WMS) (discussed later in this chapter) and for nonlinear least-squares, introduced in chapter 10.

 Worksheet 9.3: Testing for lack-of-fit

In this worksheet, we'll analyze two large datasets containing replicate y data, and apply the LOFSS test to them.

9.2   Multiple Linear Regression

Models with a single independent variable can't always explain all of the variation in the data. We can build more flexible models by introducing additional independent variables. The `linear_regression` function can perform *multiple linear regression* with these additional variables if you add columns for the new variables to the data matrix. For example, let's fit the equation $y = a + bx_1 + cx_2$ to a series of (x_1, x_2, y) points:

```
(%i1)   load(stats)$
(%i2)   x1_data: [1, 2, 3, 4, 5, 6]$
(%i3)   x2_data: [1, 4, 9, 16, 25, 36]$
```

```
(%i4)   y_data: [10.5, 33.0, 65.4, 111.4, 167.8, 235.9]$
(%i5)   data: transpose(matrix(x1_data, x2_data, y_data));
(%i6)   linear_regression(data);
```

$$(\%o5) \quad \begin{pmatrix} 1 & 1 & 10.5 \\ 2 & 4 & 33.0 \\ 3 & 9 & 65.4 \\ 4 & 16 & 111.4 \\ 5 & 25 & 167.8 \\ 6 & 36 & 235.9 \end{pmatrix}$$

$(\%o6)$

LINEAR REGRESSION MODEL

$b\_estimation = [0.25999999999976, 4.568571428571886, 5.785714285714221]$

$b\_statistics = [0.29597917660812, 7.949495169612727, 71.98933570311807]$

$b\_p\_values = [0.78655151646895, 0.0041519190297185, 5.906968398194934\ 10^{-6}]$

$b\_distribution = [student\_t, 3]$

$v\_estimation = 0.24114285714285$

$v\_conf\_int = [0.07738525228849, 3.352383623183303]$

$v\_distribution = [chi2, 3]$

$adc = 0.99996723248096$

Both the fit parameters (`b_estimation`) and their 95% confidence intervals (`b_conf_int`) are available.

9.2.1   Matrix Form of Multiple Linear Regression

If we need more flexibility (for example, we want to fit a linear equation without an intercept) we can generalize the normal equations and solve them to find the fit parameters as before. With each additional variable, though, we add additional complexity. Fitting data to the model equation $y = mx + b$ requires 2 parameters, and solution of 2 normal equations. We get 2 equations for the fit parameters, and we have to work out 2 more equations for their standard errors. When we fit a model with p parameters, all of those 2's become p's. Writing out all of the equations and applying them individually is a lot of work. It is much easier to write the system of equations in matrix notation, as we saw in Section 7.2.6.

Suppose we have p parameters and N data points. Then the fit equation takes the form:

$$y_i = \alpha_1 + \alpha_2 x_{i2} + \alpha_3 x_{i3} + \cdots + \alpha_m x_{im} + \epsilon_i \quad i = 1, \ldots, N$$

Model function for multiple linear regression

(9.32)

where α_j is the jth parameter fitted to the (x_{ij}, y_i) data and ϵ_i is the error in y_i. The indices are $j = 1, \ldots, p$ and $i = 1, \ldots, N$; the i index numbers the data point and the j index numbers the parameter. Note that for the first parameter, the y-intercept, $x_{i1} = 1$.

To write the least-squares problem in matrix form, we define the following matrices and arrays:

$$\mathbf{y} = \begin{bmatrix} y_1 \\ y_2 \\ \vdots \\ y_N \end{bmatrix} \quad \mathbf{X} = \begin{bmatrix} 1 & x_{12} & x_{13} & \cdots & x_{1p} \\ 1 & x_{22} & x_{23} & \cdots & x_{2p} \\ \vdots & \vdots & \vdots & \ddots & \vdots \\ 1 & x_{N2} & x_{N3} & \cdots & x_{Np} \end{bmatrix} \quad \boldsymbol{\alpha} = \begin{bmatrix} \alpha_1 \\ \alpha_2 \\ \vdots \\ \alpha_p \end{bmatrix} \quad \boldsymbol{\epsilon} = \begin{bmatrix} \epsilon_1 \\ \epsilon_2 \\ \vdots \\ \epsilon_N \end{bmatrix} \quad \mathbf{e} = \begin{bmatrix} e_1 \\ e_2 \\ \vdots \\ e_N \end{bmatrix}$$

where

- $\mathbf{y}$ is an N-length column vector of the y data.
- $\mathbf{X}$ is the $N \times p$ *design matrix*, with elements x_{ij}. Each row corresponds to a data point. The column of ones corresponds to the intercept, followed by columns of data for each independent variable to be multiplied by a fit parameter in the fit equation. If you don't want to include a constant intercept in the model, omit α_1 from $\boldsymbol{\alpha}$ and don't put the first column of ones in $\mathbf{X}$.
- α is a p-length column vector of the fit parameters, with the intercept as the first element. In the fit equation, each parameter will be multiplied by a column in the design matrix.
- ϵ is an N-length column vector of experimental errors (which aren't known).
- $\mathbf{e}$ is a N-length column vector containing the residuals.

In matrix notation, the fit equation is

$$\mathbf{y} = \mathbf{X}\boldsymbol{\alpha}_{\text{true}} + \epsilon \qquad \text{Model function for Multiple Linear Regression (MLR) in matrix form} \qquad (9.33)$$

where $\boldsymbol{\alpha}_{\text{true}}$ gives the fit parameters of the "true" model. If instead we use the *estimated* fit parameters, the fit equation can be written in terms of the residuals as

$$\mathbf{e} = \mathbf{y} - \mathbf{X}\alpha \qquad \text{Residuals for MLR} \qquad (9.34)$$

The sum of squared residuals SSE is

$$\text{SSE} = \sum_i^N e_i^2 = \mathbf{e}^{\mathsf{T}}\mathbf{e} = (\mathbf{y} - \mathbf{X}\alpha)^{\mathsf{T}}(\mathbf{y} - \mathbf{X}\alpha) \qquad (9.35)$$

and the normal equations take the form

$$\frac{\partial \text{SSE}}{\partial \alpha} = 0 \qquad (9.36)$$

which is the derivative of a scalar with respect to an vector (see Section 7.3.3). These equations simplify to

$$\mathbf{X}^{\mathsf{T}}\mathbf{X}\alpha = \mathbf{X}^{\mathsf{T}}\mathbf{y} \qquad \text{Normal equations in matrix form} \qquad (9.37)$$

Solving the normal equations for α gives the parameters as a function of the design matrix and the y vector[9]

$$\boldsymbol{\alpha} = \left(\mathbf{X}^{\mathsf{T}}\mathbf{X}\right)^{-1}\mathbf{X}^{\mathsf{T}}\mathbf{y} \qquad \text{Matrix-form least-squares fit parameters} \qquad (9.38)$$

$\mathbf{X}$ is an $N \times p$ matrix, so $\mathbf{X}^{\mathsf{T}}\mathbf{X}$ is a $p \times p$ matrix. If we have fewer points than parameters, the inverse of $\mathbf{X}^{\mathsf{T}}\mathbf{X}$ won't exist, because the rank of $\mathbf{X}^{\mathsf{T}}\mathbf{X}$ will be less than p. It isn't a surprise that we must have more points than parameters – you can fit a line to a pair of points, but you cannot fit a unique parabola to them!

---

9 See Appendix A in Worksheet 9.5 for the derivation of Equations (9.38) and (9.38). In Appendix B of the same worksheet, we show that this equation gives solutions that are identical to the usual (non-matrix) solutions of the normal equations.

9.2.2 Estimating the Errors in the Fit Parameters in MLR

The matrix form of the least-squares equations makes it easy to write expressions for the standard errors in the estimated parameters α. The covariance matrix $\mathbf{C}$ contains the information we need. If all of the residuals have equal variances, the covariance matrix can be estimated from the design matrix $\mathbf{X}$ and the estimated variance of the regression s^2 as

$$\mathbf{C} = s^2 \left(\mathbf{X}^{\mathrm{T}}\mathbf{X}\right)^{-1} \qquad \begin{array}{l}\text{Estimated covariance matrix}\\ \text{(equal residual variances)}\end{array} \qquad (9.39)$$

$\mathbf{C}$ is a $p \times p$ matrix, with each row and column corresponding to the index of a particular parameter. The elements C_{ij} are the covariances between two parameters (see Section 8.5.3). Diagonal elements C_{ii} give the variance in parameter α_i, so the corresponding standard error is

$$s_i = \sqrt{C_{ii}} \qquad \begin{array}{l}\text{The estimated standard error}\\ \text{of the } i\text{th fit parameter}\end{array} \qquad (9.40)$$

These standard errors can be used to compute confidence intervals for each parameter.

In Maxima, the covariance matrix is available from the output of `linear_regression` as `b_covariances`.

What happens when we have unequal residual variances? Equation (9.39) can't be used to estimate the covariance matrix, because s^2 does not represent the variance of individual residuals. In this case, a better estimate of the covariance matrix is:

$$\mathbf{C} = \left(\mathbf{X}^{\mathrm{T}}\mathbf{X}\right)^{-1}\mathbf{X}^{\mathrm{T}}\boldsymbol{\Phi}\mathbf{X}\left(\mathbf{X}^{\mathrm{T}}\mathbf{X}\right)^{-1} \qquad \begin{array}{l}\text{Estimated covariance matrix}\\ \text{(unequal residual variances)}\end{array} \qquad (9.41)$$

where $\boldsymbol{\Phi}$ is a diagonal matrix with $\Phi ii = e_i^2 \frac{N}{N-p}$. This estimate of $\mathbf{C}$ is sometimes called the *sandwich estimator* because $\left(\mathbf{X}^{\mathrm{T}}\mathbf{X}\right)^{-1}$ sandwiches the $\mathbf{X}^{\mathrm{T}}\boldsymbol{\Phi}\mathbf{X}$ matrix.[10] The standard errors for the fit parameters are then computed by Equation (9.40), as before.

Which estimate is better? The sandwich estimator takes into account unequal residual variances, and the previous estimate doesn't. When you have equal variances, the sandwich estimator will give you nearly the same results as the previous estimate, so you can safely use it as your default method for computing the covariance matrix in OLS. There are a few caveats you should keep in mind when using the sandwich estimator:

- The sandwich estimator is most accurate when the residual variance changes slightly from point to point, and when the residuals are not too large.
- If you're seeing a strong pattern in the residuals like those in Figure 9.7, it's better to do a weighted least-squares fit (Section 9.3) than to use OLS with the sandwich estimator.
- You cannot do statistics based on the SSE when you're using the sandwich estimator. For example, s^2 is still the square root of the mean square residual, but it doesn't estimate the variance of the residuals any more.

What about the off-diagonal elements of C? They usually are *not* zero, which indicates that the fit parameters aren't independent. As we'll see later, these covariances will be essential in propagating error in further calculations that involve the fit parameters. C_{ij} is a measure of the

---

10 Statisticians whimsically call it the *heteroskedasticity consistent covariance matrix* (*HCCM*) or *White's estimator*. The standard errors computed from the HCCM are called *robust standard errors*, or sometimes *White's standard errors*. See Ref. [59] for more about the scope, limitations, and reasoning behind the sandwich estimator.

correlation of errors in different fit parameters. A positive C_{ij} means that errors in α_i and α_j increase together; a negative value means that an increase in error in α_i tends to decrease the error in α_j.

The magnitudes of the covariances are hard to interpret unless we normalize them. Dividing each element C_{ij} by $s_i s_j$ gives a normalized covariance matrix called the correlation matrix:

$$ r_{ij} = \frac{C_{ij}}{s_i s_j} \qquad \text{The correlation matrix} \tag{9.42} $$

The correlation matrix is symmetric, with ones along the diagonal. The off-diagonal elements r_{ij} lie between 1 and -1. A positive value means that errors in ith and jth parameters tend to have the same sign; if r_{ij} is negative, the errors will likely have opposite signs. An absolute value of 1 indicates that the parameters are perfectly correlated, with errors that are proportional to each other. A value of zero indicates that the parameters are completely independent.

A convenient matrix calculation for the determination coefficient is [60]

$$ R^2 = 1 - \frac{\mathbf{e}^{\mathsf{T}}\mathbf{e}}{\mathbf{Y}^{\mathsf{T}}\mathbf{Y} - N\overline{Y}^2} \qquad R^2 \text{ in matrix form} \tag{9.43} $$

The numerator measures unexplained variation; the denominator measures the total variation of y around the mean.

9.2.3 Example: Microwave Rotational Spectrum of HCl

Let's apply matrix multiple linear least-squares to some real data.

Linear molecules with a permanent dipole moment (like HCl) have simple microwave spectra that contain valuable information about the structure of the molecule. The wavenumber of each peak in the microwave spectrum corresponds to a transition between allowed rotational energy states. Each state is labeled by a rotational quantum number, J. For linear molecules without any unpaired electrons, only transitions between states with $\Delta J = \pm 1$ are allowed, and the wavenumbers can be modeled using the equation [61]

$$ \tilde{v}_{J \to J+1} = 2\tilde{B}(J+1) - 4\tilde{D}(J+1)^3 \qquad \begin{array}{l}\text{Rotational wavenumbers for}\\\text{a nonrigid rotor}\end{array} \tag{9.44} $$

where $\tilde{v}_{J \to J+1}$ is the wavenumber for the transition between energy levels with rotational quantum number J and $J + 1$, in cm^{-1}. The rotational constant $\tilde{B}$ can be used to estimate the molecule's moment of inertia (and bond length), and the centrifugal distortion constant $\tilde{D}$ accounts for the slight stretching of the bond that occurs as the molecule rotates.

We'll obtain $\tilde{B}$ and $\tilde{D}$ by fitting Equation (9.44) to the microwave wavenumbers for H^{35}Cl, published by Herzberg [62]. The fit equation isn't linear in J, but that doesn't matter; it is linear in the fit parameters. The fit equation can be written more compactly as $y = \tilde{B}x_1 + \tilde{D}x_2$, where $y = \tilde{v}_{J \to J+1}$, $x_1 = 2(J+1)$ and $x_2 = -4(J+1)^3$.

We will also compute the standard errors and confidence intervals for the rotational constants, analyze the residuals, and compute the covariance and correlation matrices.

We begin by setting up a column vector $\mathtt{J}$ to hold J values corresponding to the wavenumbers for the $J \to J+1$ transition in $\mathtt{Y}$. The design matrix $\mathbf{X}$ is built by appending columns containing the variables $2(J + 1)$ and $-4(J + 1)^3$. Notice that we do not include a column of ones in the design matrix here, because there is no constant intercept in this fit!

We use `transpose` to build column vectors from a list of values. The design matrix is built by adding columns with `addcol`.

```
(%i1)   J: transpose([3,4,5,6,7,8,9,10])$
(%i2)   Y: transpose([83.03, 104.1, 124.30, 145.03, 165.51,
            ➡ 185.86, 206.38, 226.50]);
(%i3)   X: addcol(2*(J+1), -4*(J+1)^3);
```

(%o2)

$$
\begin{pmatrix}
83.03 \\
104.1 \\
124.3 \\
145.03 \\
165.51 \\
185.86 \\
206.38 \\
226.5
\end{pmatrix}
$$

(%o3)

$$
\begin{pmatrix}
8 & -256 \\
10 & -500 \\
12 & -864 \\
14 & -1372 \\
16 & -2048 \\
18 & -2916 \\
20 & -4000 \\
22 & -5324
\end{pmatrix}
$$

Now estimate the fit parameters α, using Equation (9.38), and compute the residuals e using Equation (9.34).

```
(%i4)   alpha: invert(transpose(X) . X) . transpose(X) . Y;
(%i5)   e : Y - X . alpha;
(%i6)   wxplot2d([discrete,  list_matrix_entries(Y),
            ➡ list_matrix_entries(e)],
                [style,points],
                [ylabel, "residuals (cm^-1)"],
                [xlabel, "experimental wavenumber (cm^-1)"]
            )$
```

(%o4)

$$
\begin{pmatrix}
10.40285333002893 \\
4.4298372923856633 \; 10^{-4}
\end{pmatrix}
$$

(%o5)

$$
\begin{pmatrix}
-0.079422805546386 \\
0.29295856432995 \\
-0.15150201828507 \\
-0.0021729438897466 \\
-0.028422602982346 \\
-0.099619386061107 \\
0.094868316375596 \\
-0.0043278861703868
\end{pmatrix}
$$

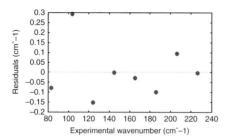

There is no discernable pattern in the residuals. They are all about 0.15 cm^{-1} or under, except for the residual for the second point at 104.1 cm^{-1}, which is about 0.30 cm^{-1}. This is a possible concern because most of the wavenumbers are reported out to the hundredths place.

Now estimate the variance of the regression, and the standard errors and confidence intervals for the parameters:

Calculate the variance of the regression with the matrix form of Equation (9.13).

```
(%i7)    s2 : transpose(e) . e / (N-2);
```

```
(%o7)    0.022473480029142
```

Estimate the covariance matrix **C** with Equation (9.39).

```
(%i8)    C: s2 * invert(transpose(X) . X);
```

$$
(\%o8) \quad \begin{pmatrix} 8.2768028662354856\ 10^{-5} & 4.4015824173178264\ 10^{-7} \\ 4.4015824173178264\ 10^{-7} & 2.715390898435614\ 10^{-9} \end{pmatrix}
$$

The standard errors of the parameters se are computed as the square roots of the diagonal elements:

```
(%i9)    se : transpose( makelist(sqrt(C[i][i]), i, 1, 2));
```

$$
(\%o9) \quad \begin{pmatrix} 0.0090976935902653 \\ 5.2109412762337038\ 10^{-5} \end{pmatrix}
$$

```
(%i10)   load(distrib)$
(%i11)   t : quantile_student_t(0.975,N-2)$
(%i12)   CI_low : alpha - t*se;
(%i13)   CI_high : alpha + t*se;
```

$$
(\%o11) \quad \begin{pmatrix} 10.38059207576483 \\ 3.1547658959420373\ 10^{-4} \end{pmatrix}
$$

$$
(\%o12) \quad \begin{pmatrix} 10.42511458429304 \\ 5.7049086888292893\ 10^{-4} \end{pmatrix}
$$

Finally, let's compute the correlation matrix from the covariance matrix to see just how strongly the fitted $\tilde{B}$ and $\tilde{D}$ parameters are correlated:

```
(%i14)   r[i,j] := C[i][j]/sqrt(C[i][i]*C[j][j]);
(%i15)   R : genmatrix(r, 2,2);
```

$$(\%\text{o}13) \quad r_{i,j} := \frac{\left(C_i\right)_j}{\sqrt{\left(C_i\right)_i \left(C_j\right)_j}}$$

$$(\%\text{o}14) \quad \begin{pmatrix} 1.0 & 0.92845605634425 \\ 0.92845605634425 & 1.0 \end{pmatrix}$$

The parameters are very strongly correlated, which tells us that any error propagation we do on calculations with the fitted $\tilde{B}$ and $\tilde{D}$ must use the covariance matrix, and not just the standard errors of the parameters.

 Worksheet 9.4: Multiple linear least-squares using matrices

In this worksheet, we'll apply the matrix form of ordinary least-squares to do fits with more than one independent variable.

9.2.4   Detecting and Dealing with Outliers

Sometimes a single discordant point can drag the fit line away from the rest of the data. If the fit parameters are largely determined by a single point, we want to know about it – and we want to identify the point, so we can recheck it or perhaps collect more data around it.

Residual plots are most commonly used to detect outliers. Outliers are points that lie far away from the others, or they are points with one or more coordinates that differ greatly from those of the other points. We can ask:

- *How far are the outlier's x values from those of other points?* A point with x values that are far from the average x values has the potential to strongly influence the fit. Such a point is said to have high leverage, because it can tip the fit line like a see-saw. The line swings around a fulcrum near the average values of x and y. Low-leverage points are closer to the fulcrum; they have to have large residuals to influence the fit much from those positions (see Figure 9.10).
- *How far is the outlier from the fit line?* We can look at the absolute size of the residual for a rough answer. To compare residuals appropriately, though, it is useful to divide them by their estimated standard deviations. Such residuals are said to be *standardized*. For example, a residual of 5 units is large when the residual's standard deviation is 1, but it's very small when the standard deviation is 50.
- *How much influence does the outlier have on the fit?* Don't worry about outliers that don't strongly affect the fit. But when a few points are far more influential than the rest of the data, we need to investigate them.

  We *could* assess the influence of an outlier by performing the fit with and without the outlier, but it's much easier to just look at the leverage and the residual size (see Figure 9.10). A large residual in a high-leverage position strongly influences the fit; the same residual in a low leverage position won't have as much influence.

When there is only one x variable, it's easy to spot high-leverage points (they are simply points with x values that are much different from the average x value). But in multiple linear regression, there may be many different x variables. It isn't as easy to pick out high-leverage points from multidimensional plots or from plots of the residuals against the individual x variables. In that situation, a numerical estimate of the leverage is helpful.

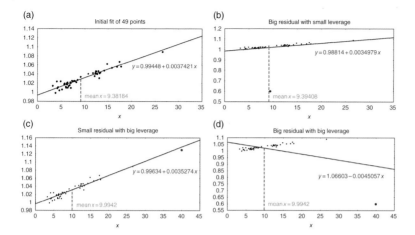

Figure 9.10 Both leverage and residual sizes determine the influence of a point on a fit. In (a), 49 points are fitted to a line. In (b) a 50-th point (red) with a large residual but low leverage is added; it changes the slope and intercept slightly. In (c), a point with a small residual but high leverage is added; again, the influence of the point on the fit is relatively small. In (d) a point with a large residual with high leverage is added. The fit is severely distorted.

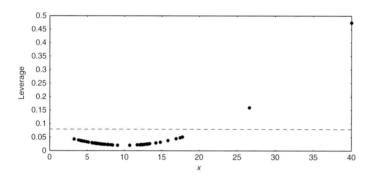

Figure 9.11 Leverages for the data in Figure 9.10d. Two points are above the $2p/N$ cutoff (the red dashed line).

Leverages can be estimated using the hat matrix, $\mathbf{H}$, which maps the observed values of y onto their predicted values $\hat{\mathbf{y}}$:

$$\hat{\mathbf{y}} = \mathbf{H}y \qquad \text{Definition of the hat matrix} \tag{9.45}$$

The hat matrix "puts a hat" on y. Combining the definition of $\hat{\mathbf{y}}$ with Equations (9.38) and (9.45) gives the hat matrix in terms the design matrix $\mathbf{X}$ as

$$\mathbf{H} = \mathbf{X}\left(\mathbf{X}^{\mathsf{T}}\mathbf{X}\right)^{-1}\mathbf{X}^{\mathsf{T}} \qquad \text{Calculation of the hat matrix} \tag{9.46}$$

The diagonal elements of the hat matrix $\mathbf{H}_{ii}$ estimate the leverage for the ith data point. The $\mathbf{H}_{ii}$ are never less than zero or greater than one. They all sum to p, the number of parameters. For example, $\mathbf{H}_{ii} = 0.475$ for the outlier in Figure 9.10d; for all but one of the other points, $\mathbf{H}_{ii}$ is less than about 0.05 (see Figure 9.11).

But what do these numbers mean? The point with the largest $\mathbf{H}_{ii}$ has the most extreme x values, while those with the smallest $\mathbf{H}_{ii}$ are those with average x values. A general rule is that if a leverage is greater than $2p/N$, the corresponding point *may* strongly influence the fit [63]. Such points should be carefully rechecked.[11]

Figure 9.10 shows that both the size of an outlier's residual and its leverage determine its influence on a fit.

Equation (9.46) must be used directly to compute leverages in Maxima.

Comparing raw or standardized residuals is tricky because of the way the least-squares method handles high-leverage points. Even when the error variance is constant, the residual variances will *not* all be equal. The least-squares method constrains the residuals so that the SSE are as close to zero as possible. This constraint forces individual residual variances σ_{e_i} to decrease with increasing leverage[12]:

$$\sigma_{e_i}^2 = \sigma^2(1 - \mathbf{H}_{ii}) \qquad \text{Variance of the } i\text{th residual} \tag{9.47}$$

where as before σ^2 is the error variance, which often is unknown. This equation quantitatively shows how the least-squares method places undue weight on high-leverage points. It suppresses the residuals for those points as much as it can, even if it must rock the fit line away from points with lower leverage to do so.

Residuals are much easier to compare if we scale them for this effect. Equation (9.47) suggests a way to do this. Dividing a residual by its leverage-corrected standard deviation converts it into a studentized residual:

$$\frac{e_i}{\sigma_{e_i}} = \frac{e_i}{\sigma\sqrt{1 - \mathbf{H}_{ii}}} \qquad \text{A "studentized" residual} \tag{9.48}$$

We need some estimate of σ to use studentized residuals. Two different approximations can be used. If we use $s \approx \sigma$ (using s from Equation 9.13), we obtain an *internally studentized residual* (ISR). ISRs account for the suppression of the raw residuals for high-leverage points, so they are more useful than raw or standardized residuals for detecting influential outliers.

To see whether or not a point has an undue influence on the fit, it is more useful to estimate σ so that it excludes the point in question:

$$\sigma_{(i)}^2 \approx \frac{1}{N-p-1} \sum_{j \neq i}^{N} e_j^2 = \frac{\text{SSE} - e_i^2}{N-p-1} \qquad \begin{array}{l} \text{Estimated error variance} \\ \text{without the } i\text{th point} \end{array} \tag{9.49}$$

Using this estimate in Equation (9.48) gives an *externally studentized residual* (ESR):

$$\text{ESR}_i = \frac{e_i}{\sqrt{\sigma_{(i)}^2(1 - \mathbf{H}_{ii})}} \qquad \text{An ESR} \tag{9.50}$$

Except for the scale on the y axis, ESR, ISR, and raw residual plots don't look much different – *unless the data includes outliers*. ESRs can resolve outliers that the raw residuals and ISRs don't reveal. In Figure 9.12, the outlier's raw residual is in the range of the other residuals, but its ESR is clearly different from the ESRs of other points. Its ESR is also around three. It's easy enough to see that the point is an outlier from any of the scatter plots in this case, but when you have more than one independent variable, outliers may not be clearly resolved on any of the residual plots.

---

11 You cannot delete the point based on this criterion!

12 We'll need estimates for the residual variances later when we do WLS, and it's important to realize that Equation (9.47) is *not* giving us such estimates. It merely shows us the effect of the OLS constraints on the residual variances. Also, the equation applies when the residuals all have the same variance σ^2, ignoring the leverage. It won't apply when you have a residual plot like those in Figure 9.8

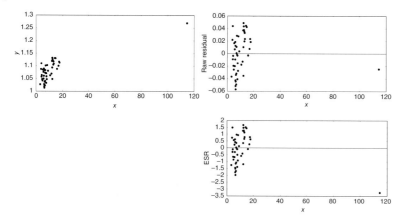

Figure 9.12 ESRs can identify outliers that the raw residuals miss. In the *xy* data on the top left plot, a plot of the raw residuals against *x* (top right) shows that the residual of the outlier isn't much different from the residuals of the other points. Plotting the ESRs against *x* (bottom) clearly picks out the outlier.

If the errors are normally distributed, a further advantage of the ESRs is that they follow the *t*-distribution, with $N - p - 1$ degrees of freedom. With a large sample, an ESR with an absolute value of 2 would correspond to a residual that is 2 standard deviations from the mean residual size. Residuals of this size or larger should occur roughly 5% of the time. An ESR with an absolute value of 3 or larger will occur only about 1% of the time. For the last two points in Figure 9.10d, the ESRs are about 2.9 and −9.8, respectively. We can be much more than 99% confident that the last point is an outlier, and better than 95% confident that the next-to-last point is an outlier, too. They may simply reflect natural variability in the data, but such points need to be rechecked. Were they recorded correctly? Was there anything different about the procedure used to measure them? Were they really sampled from the same population as the other points?

Except for the scale on the *y* axis, ESR, ISR, and raw residual plots don't look much different – *unless the data includes outliers*. ESRs can resolve outliers that the raw residuals and ISRs don't reveal. In Figure 9.12, the outlier's raw residual is in the range of the other residuals, but its ESR is clearly different from the ESRs of other points. Its ESR is also around three. It's easy enough to see that the point is an outlier from any of the scatter plots in this case, but when you have more than one independent variable, outliers may not be clearly resolved on any of the residual plots.

With smaller samples, we can use a *t* test to decide whether or not we have a single outlier at a given level of confidence. Our null hypothesis is that the *i*th point is *not* an outlier. If

$$|\text{ESR}_i| > t_{1-\alpha/2, N-p-1} \qquad \text{Criterion for selecting the } i\text{th point as an outlier} \qquad (9.51)$$

then we reject the null hypothesis at the $(1 - \alpha) \times 100\%$ confidence level.

Don't worry about outliers that don't strongly affect the fit. Influential outliers will have high leverage and large ESR's, and you can easily identify them by plotting the ESRs against the leverage (see Figure 9.13).

After we've identified an influential outlier, can we simply delete it? The answer is *no*. No statistical test will justify rejecting an outlier simply because it doesn't fit the model! If the cause of the outlier cannot be identified, it must be kept. Our choices are few:

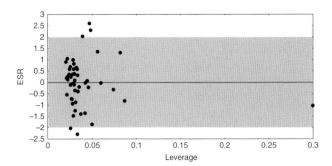

Figure 9.13 Plot of ESRs against leverage for the data in Figure 9.10a. There are a few outliers with high ESRs, but their leverage is low that they won't influence the fit. There is a lone point with high leverage and a moderate ESR that should be checked; it will have a much stronger influence on the fit than the other points.

- Recheck the point. If an outlier is the result of an experimental blunder or a data entry error, fix it.
- If we have reason to believe that a data point doesn't represent the population we're trying to sample, we can justify not including it. Did you apply a different method to get that point? Were experimental conditions different? Does that point represent a compound or solution that is different in some important way from the others in your study?
- Report the fit with and without the outlier. Clearly note that data was deleted in the second case.
- If the outlier occurs at an extreme value of x or y, perhaps the model applies to a smaller range. For example, suppose the model describes the kinetics of an enzymatic reaction over a wide temperature range. At the upper end of the temperature range, the enzyme began to denature. It's ok to omit the high-temperature data and report that the model doesn't apply at those temperatures.
- Collect more data in the region of the outlier. If the outlier really is a gross error, or simply a part of the natural variability of the data, collecting new data will decrease its leverage. But if the new data reproduces the outlier, you've made a discovery – and you need to build a better model!

 Worksheet 9.5: Finding outliers

In this worksheet, we'll look at how leverages, studentized residuals, and t-tests can be used to identify influential outliers in a fit.

9.3 WLS

The error in real data often varies from point to point. The experimental error may be directly related to the size of the measurement. For example, in high-performance liquid chromatography (HPLC), the error variance in instrument readings has a constant component (from the error in the injection volume) and proportional components (which scale as the square of the peak areas) [64]. Even if the error in an instrument reading is constant, data transformation can create a relationship between the error and the transformed measurement. For example, vapor pressure data collected at several different temperatures may have a constant error variance, but the error in the *log* of the vapor pressure of a liquid will decrease exponentially with increasing temperature.

OLS implicitly assumes that all points have equal error variance, so it doesn't perform well in situations like this. It has no way to tell a precise point from an imprecise point, so it treats both as equally important. It will drag the fit line towards imprecise points and away from precise points more than it should. This is particularly troublesome when the imprecise points have large residuals with high leverage. It will also overestimate the error in its fit parameters.

We need some way to weight the points so that the fit pays less attention to highly uncertain points. If each y_i is normally distributed with its own standard deviation σ_i, we can modify the figure of merit to account for unequal data variances as follows:

$$\text{Choose } \alpha \text{ to minimize } \chi^2 \equiv \sum_{i=1}^{N} \left(\frac{y_i - \hat{y}_i}{\sigma_i} \right)^2 \quad \begin{array}{l}\text{The weighted least-}\\\text{squares method}\end{array} \tag{9.52}$$

This new method is called weighted least-squares because each residual is now weighted by a factor of $1/\sigma_i^2$. For example, suppose we have a dataset that contains two points with $y_i = 3.0 \pm 0.1$ and $y_j = 4.0 \pm 0.2$. The weights of the points will be 100 and 25, respectively, so the contribution of the second point is only one-fourth of the contribution of the first point. Weighting makes points with larger uncertainties relatively less important in the fit; in ordinary least-squares, both points are equally important.

The figure of merit in Equation (9.52) is called chi-squared ("chi squared"). A major advantage of χ^2 over the SSE used in ordinary least-squares (Equation 9.10) is that it has a well-known probability distribution function. As we'll see below, we can use χ^2 directly to test for goodness-of-fit. This cannot be done with the SSE.

9.3.1 The Fit Parameters in WLS

Simple least-squares is a special case of WLS, with all of the weights set equal to one. The WLS fit parameters and their error estimates are calculated from expressions that are extensions of those we've already seen.

We can write χ^2 in matrix form as

$$\chi^2 = (\mathbf{Y} - \mathbf{X}\alpha)^{\mathsf{T}} \mathbf{W} (\mathbf{Y} - \mathbf{X}\alpha) \quad \chi^2 \text{ in matrix form} \tag{9.53}$$

where the $N \times N$ weight matrix $\mathbf{W}$ has the weights along its diagonal and zeros elsewhere:

$$\mathbf{W} = \begin{bmatrix} \frac{1}{\sigma_1^2} & 0 & \cdots & 0 \\ 0 & \frac{1}{\sigma_2^2} & \cdots & 0 \\ \vdots & \vdots & \ddots & \vdots \\ 0 & 0 & \cdots & \frac{1}{\sigma_N^2} \end{bmatrix}$$

The normal equations take the form

$$(\mathbf{X}^{\mathsf{T}} \mathbf{W} \mathbf{X})\alpha = \mathbf{X}^{\mathsf{T}} \mathbf{W} \mathbf{Y} \quad \begin{array}{l}\text{WLS normal equations}\\\text{in matrix form}\end{array} \tag{9.54}$$

and their solution is

$$\alpha = (\mathbf{X}^{\mathsf{T}} \mathbf{W} \mathbf{X})^{-1} \mathbf{X}^{\mathsf{T}} \mathbf{W} \mathbf{Y} \quad \text{WLS fit parameters} \tag{9.55}$$

9.3.2 Error Estimates for the WLS Fit Parameters

The WLS fit parameters are often quite close to those given by OLS. But the WLS error estimates can be much smaller, because weighting the fit makes it much more efficient when the variances of the residuals are unequal.

When the weights are known exactly, the covariance matrix in WLS can be estimated as

$$\mathbf{C} = (\mathbf{X}^T \mathbf{W} \mathbf{X})^{-1} \qquad \text{Covariance matrix for WLS} \tag{9.56}$$

This equation reduces to Equation (9.39) if all of the σ_i's are set to one, and we use the approximation $s^2 \approx \sigma^2$ (that is, we assume that we have a good fit). The diagonal elements of the covariance matrix give the estimated variances in the fit parameters as before, so the standard error in the ith parameter is $\sqrt{C_{ii}}$.

9.3.3   Finding the Weights

A weighted least-squares fit isn't any more difficult to compute than an ordinary least-squares fit – *once you have an accurate set of weights*. Unfortunately, we seldom know the error variances σ_i^2 that we need to compute the weights.

The weights usually have to be estimated. The approach we take depends on the structure of the data and the information we have. We use error propagation to find weights for transformed data, variance modeling to find data with residuals that vary with x, group variances for grouped data, or direct estimation for data with a large number of replicates for each point.

1) **Error propagation**. When the raw y data have been transformed to Y and then fitted, the error in y will propagate into the error in Y. If the variances in the raw data ($\sigma_{y_i}^2$) are known, the variance in Y_i (σ_i^2) is

$$\sigma_i^2 = \left(\frac{dY}{dy} \right)^2 \sigma_{y_i}^2 \qquad \begin{array}{l} \text{Propagation of error} \\ \text{in a transformation from } y \text{ to } Y \end{array} \tag{9.57}$$

2) **Variance modeling**. In a large dataset with relatively small residuals, the squared residuals are a rough estimate of the point's error variance. If the variance changes with one or more of the x variables, try fitting the squared residuals (or the absolute value of the residuals) to those variables, their squares, or their cross-products. If we find a good fit, we have a workable *variance model* that can then be used to compute the weights. If the fit of the squared residuals to x is $g(x)$, then $\sigma_i^2 \approx g(x)\frac{N}{N-p}$, where the factor $\frac{N}{N-p}$ improves the estimate for small samples.
   Variance models based on fits of the residuals are only approximately valid if the residuals are relatively small [65].
3) **Variance of grouped data**. Sometimes the points we want to fit are means for groups of data measured at different levels of x. For example, suppose we are studying a compound's ability to suppress a biochemical reaction in cells. A dose of 1 unit was applied to 12 cell cultures; a dose of 2 units was applied to 15 cell cultures, and so on. We'd like to fit some measure of the average suppression observed for each of the groups to the dose. If we assume that each group has the same error variance σ^2, the weights can be estimated as n_i/σ^2, where n_i is the number of cell cultures tested for dose level i.
4) **Direct variance calculation**. Collect replicate data for each point and use it to estimate each point's error variance directly. This is the simplest but least practical way to get the weights, because the variances won't be accurate unless we've collected an enormous amount of data over the entire range of x.

In Worksheet 9.7 , we'll estimate the weights for a WLS fit on the temperature dependence of vapor pressure using error propagation and variance modeling, and compare the results.

9.3.4   Residual Analysis in WLS

A residual tells us how much an individual point contributes to the figure of merit. In ordinary least-squares, the residuals $\mathbf{e} = \mathbf{Y} - \mathbf{X}\boldsymbol{\alpha}$ contribute to the sum of squares $\mathbf{e}^T\mathbf{e}$. In WLS, the residuals must include the weights, or they won't accurately reflect a point's contribution to χ^2.

To see how the weights should be included in the residuals, it's instructive to treat weighting as a data transformation. After transforming the variables, we can do a weighted least-squares fit simply by applying simple least-squares to the transformed model [66].

Our basic model is $\mathbf{Y} = \mathbf{X}\boldsymbol{\alpha} + \mathbf{e}$. Multiplying both sides of this equation by $\mathbf{W}^{1/2}$ gives

$$\mathbf{W}^{1/2}\mathbf{Y} = \mathbf{W}^{1/2}\mathbf{X}\boldsymbol{\alpha} + \mathbf{W}^{1/2}\mathbf{e} \qquad \text{Weighting as a data transformation} \tag{9.58}$$

Making the substitutions

$$\begin{aligned} \mathbf{Y}_w &= \mathbf{W}^{1/2}\mathbf{Y} \\ \mathbf{X}_w &= \mathbf{W}^{1/2}\mathbf{X} \\ \mathbf{e}_w &= \mathbf{W}^{1/2}\mathbf{e} \end{aligned} \tag{9.59}$$

the transformed model is

$$\mathbf{Y}_w = \mathbf{X}_w\boldsymbol{\alpha} + \mathbf{e}_w \qquad \begin{array}{l} \text{A WLS model that} \\ \text{can be fitted by} \\ \text{ordinary least-squares} \end{array} \tag{9.60}$$

After the transformation, the variances of the residuals $\mathbf{e}_w$ are all the same, so that requirement of ordinary least-squares has been met. We can perform an ordinary least-squares on the model in Equation (9.60) to accomplish a weighted least-squares fit.

Residuals in WLS are defined by Equation (9.59). They are called *weighted residuals*, or *Pearson residuals*. Unlike ordinary (unweighted) residuals, the weighted residuals are dimensionless. When the weights are reciprocal variances, the weighted residuals tell us how many standard deviations a point is above or below the fit line.[13] They reflect variances that change from point to point, while unweighted residuals implicitly assume that all points have the same variance. If your residual plots still show patterns like those in Figure 9.8 after doing a weighted least-squares fit with accurate weights, you've probably forgotten to use weighted residuals [68].

Leverages are calculated as before from the diagonal elements of the hat matrix. In WLS, the hat matrix must include the weights:

$$\mathbf{H} = \mathbf{W}^{1/2}\mathbf{X}\left(\mathbf{X}^T\mathbf{W}\mathbf{X}\right)^{-1}\mathbf{X}^T\mathbf{W}^{1/2} \qquad \text{The hat matrix in WLS} \tag{9.61}$$

9.3.5   Evaluating Goodness-of-Fit

We've seen that R^2 values are easily misinterpreted in OLS. This is even more so in WLS, where weighting introduces additional complications. Calculating R^2 for Equation (9.60) gives the determination coefficient in $\mathbf{X}_w$ and $\mathbf{Y}_w$, not in $\mathbf{X}$ and $\mathbf{Y}$. If the weights have been chosen properly, you'll get a higher R^2 in these transformed variables than you would have obtained by applying ordinary least-squares directly to the unweighted data. But the increase in R^2 doesn't necessarily represent an improved fit; it also reflects how successfully the weights have equalized the variance of the points. You can see this by computing R^2 using the original unweighted data with the WLS fit coefficients. The increase in R^2 in this case is much more modest; it may even drop slightly, implying that WLS degrades the fit relative to OLS [69].

---

13 If the weights aren't accurate reciprocal variances, you cannot interpret the weighted residuals this way, and your estimates of the parameter uncertainties will be unreliable [67].

There is a simpler and more accurate statistic to assess goodness-of-fit in WLS. It is the χ^2 value itself. For a reasonable fit, χ^2 should approach $N - p$ with larger sample sizes. When it is greater than $N - p$, it means that either the model is inadequate or the residual variances have been underestimated. It may also mean that the measurement errors aren't normally distributed. When χ^2 is less than $N - p$, the residual variances have been overestimated, or the model is overfitting the data.

We can do even better than this. The probability Q that the observed chi-squared will be at least as large as the fit's χ^2 when the model is correct is

$$Q = 1 - \text{cdf}(\chi^2, N - p) \qquad \text{Goodness-of-fit measure for WLS} \qquad (9.62)$$

where cdf is the cumulative probability distribution function. In Maxima, Q is calculated as `1 - cdf_chi2(`χ^2`, N-p)`; the function is loaded in the `stats` package. A Q of one would be a perfect fit. A Q of less than 0.05 indicates potential trouble. Press [56] says that

> "...reasonable experimenters are often rather tolerant of low probabilities Q. It is not uncommon to deem acceptable on equal terms any models with, say, $Q > 0.001$. This is not as sloppy as it sounds: Truly *wrong* models will often be rejected with vastly smaller values of Q, 10^{-18}, say. However, if day-in and day-out you find yourself accepting models with $Q \approx 10^{-3}$, you really should track down the cause."

For example, a two-parameter fit of 45 points had a χ^2 value of 41.83. Maxima gives Q as 0.522, meaning that the chance of observing a χ^2 of 41.83 or larger when the model is correct is 52.2%. Since Q is higher than the usual criterion for statistical significance (0.05), the model fits the data well.

No statistic is a magic bullet, and this measure of goodness-of-fit is no exception. When the errors aren't normally distributed, there may be many outliers, and this decreases Q. Underestimating or overestimating the weights also strongly affects Q. A high Q can't assure us that we have the best possible model, either; there may be another that would fit the data even better.

 Worksheet 9.6: Fitting data to a line with WLS

In this worksheet, we'll use weighted and ordinary least-squares to find the enthalpy of vaporization for an organic liquid from vapor pressure and temperature measurements.

9.4 Fitting Data to a Line with Errors in Both *X* and *Y*

Until now, we have assumed that experimental error is exclusively confined to the y measurements. Both ordinary and weighted least-squares are built on the assumption that the independent variables are exact.

But we can't always ignore the errors in the x measurements. We aren't always sure which variable we should choose as the dependent variable. In OLS and WLS, that choice is critical; fitting y to x implicitly assumes that all of the error is in y, and it gives a different fit than we'd get by fitting x to y, and much different error estimates. When both x and y contain significant error, we shouldn't use OLS or WLS.

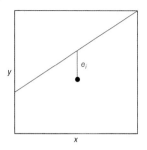 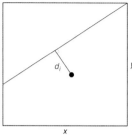

Figure 9.14 In ordinary least-squares, we minimize the sum of the squared y residuals, $\sum e_i^2$ (left). When the error in x isn't negligible, we have to minimize the sum of squared point-to-line distances, $\sum d_i^2$.

Fitting data to a line now becomes much harder because the fit must do more than just minimize the vertical distances between the points and the fit line (the y residuals). It must minimize the perpendicular distances between the points and the fit line (Figure 9.14). We need to build a new figure of merit $\sum d_i^2$. We can start with the χ^2 figure of merit used in WLS, and propagate the variances in both x and y into the error variances used to build the weights. Writing the variances in x_i and y_i as $\sigma_{x_i}^2$ and $\sigma_{y_i}^2$, propagation of error gives the variance σ_i^2 in the errors $E_i = y_i - mx_i - b$ as

$$\sigma_i^2 = \left(\frac{\partial E_i}{\partial x_i}\right)^2 \sigma_{x_i}^2 + \left(\frac{\partial E_i}{\partial y_i}\right)^2 \sigma_{y_i}^2$$
$$= m^2\sigma_{x_i}^2 + \sigma_{y_i}^2 \tag{9.63}$$

and the figure of merit in Equation (9.52) becomes

$$\chi^2 = \sum_{i=1}^{N} \frac{(y_i - (mx_i + b))^2}{\sigma_{y_i}^2 + m^2\sigma_{x_i}^2} \qquad \begin{array}{l}\text{Figure of merit for a linear fit}\\\text{with errors in both } y \text{ and } x\end{array} \tag{9.64}$$

Each term in the sum is the square of the perpendicular distance d_i in Figure 9.14. Choosing m and b to minimize this new figure of merit is called the total least-squares (TLS) method.

TLS is a weighted least-squares method with weights that now depend on the fit parameters:

$$w_i = \frac{1}{(\sigma_{y_i}^2 + m^2\sigma_{x_i}^2)} \qquad \text{weights in TLS} \tag{9.65}$$

In fact, we can just use WLS directly if $\sigma_i^2 \approx \sigma_{y_i}^2$, or if the ratio $\sigma_{x_i}^2/\sigma_{y_i}^2$ is the same for all points [70]. If we find that σ_i^2 is the same for all points, we can just use ordinary least-squares.

But if none of these conditions apply, we'll have to try something new.

9.4.1 Finding Fit Parameters in TLS

We *could* find the fit parameters by solving the normal equations $d\chi^2/dm = 0$ and $d\chi^2/db = 0$ for m and b, as we did with OLS and WLS. But unlike the OLS and WLS normal equations, the TLS equations are nonlinear in the fit parameters. That makes them much more difficult to solve, as we saw in Section 5.4.

It's easier to use Maxima to minimize χ^2 by brute force, using the numerical function minimizers we applied in Section 6.4.2. To take this approach, we must have an initial guess for m and b; we can start with the OLS slope and intercept. We then use a function minimizer to find the value of m and b where χ^2 is at its minimum value.

We'll use a classic test dataset from Pearson to illustrate this method [71]:

```
(%i1)   X : [0,.9,1.8,2.6,3.3,4.4,5.2,6.1,6.5,7.4]$
(%i2)   N: length(X)$
```

```
(%i3)  Y : [5.9, 5.4, 4.4, 4.6, 3.5, 3.7, 2.8, 2.8, 2.4, 1.5]$
(%i4)  plot2d([discrete, X, Y], [style, points])$
```

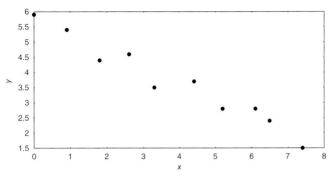

York [72] used the following weights with Pearson's data to test his TLS calculations:

```
(%i5)  wxi : [1000,1000,500,800,200,80,60,20,1.8,1.0]$
(%i6)  wyi : [1,1.8,4,8,20,20,70,70,100,500]$
(%i7)  sigma_y : sqrt(1/wyi)$
(%i8)  sigma_x : sqrt(1/wxi)$
```

We need an initial guess for the slope and intercept of the line. We can get them by doing an ordinary least-squares fit, without the weights.

```
(%i9)   data: apply(matrix, makelist([X[i], Y[i]], i, 1, N))$
(%i10)  load(stats)$
(%i11)  OLS: linear_regression(data);
(%i12)  guess : take_inference ( b_estimation, OLS);
```

(%o11)

$$\begin{array}{c} \text{LINEAR REGRESSION MODEL} \\ b\_estimation = [5.761185190439042, -0.53957727498404] \\ b\_statistics = [30.40440791394536, -12.80848528119326] \\ b\_p\_values = [1.4868004605261831\ 10^{-9}, 1.3024677540940388\ 10^{-6}] \\ b\_distribution = [student\_t, 8] \\ v\_estimation = 0.10008294027945 \\ v\_conf\_int = [0.045662061388214, 0.36732221318839] \\ v\_distribution = [chi2, 8] \\ adc = 0.94769184305952 \end{array}$$

(%o12) $[5.761185190439042, -0.53957727498404]$

Now find m and b that minimize χ^2. As we saw in Section 6.4.2, we can numerically minimize a function with lbfgs. First, we define an expression for χ^2:

```
(%i13)  chi2 : sum((Y[i] - m*X[i] - b)^2/(sigma_y[i]^2 + m^2*
        ➥ sigma_x[i]^2), i, 1, N)$
(%i14)  load(lbfgs)$
(%i15)  result: lbfgs(chi2, '[b,m], guess, 1e-8, [1,0]);
```

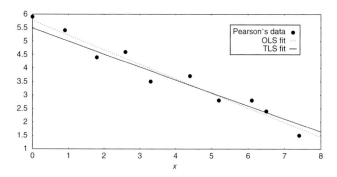

Figure 9.15 Comparison of TLS and OLS fits with Pearson's data, with Young's weights. The TLS line minimizes both the x and y residuals, subject to the weights assigned in both the x and y directions; the OLS line minimizes only the y residuals, and treats all points as equally important.

```
***************************************************
  N=     2   NUMBER OF CORRECTIONS=25
INITIAL VALUES F=  1.284678744378317D+01   GNORM=  3.675985762090254D+01
***************************************************
I   NFN    FUNC                 GNORM                  STEPLENGTH
1    3     1.279437209142716D+01  2.426863327535518D+01  4.678741983642800D-05
2    4     1.274971598151280D+01  6.194200622172588D+00  1.000000000000000D+00

⋮

11  14     1.186635319406145D+01  1.408269221024853D-05  1.000000000000000D+00
12  15     1.186635319406144D+01  4.017035402458462D-08  1.000000000000000D+00
THE MINIMIZATION TERMINATED WITHOUT DETECTING ERRORS. IFLAG = 0
```

(%o15) $[b = 5.479910224141367, m = -0.48053340747164]$

Both fits are shown in Figure 9.15. The TLS line has a somewhat smaller slope than the OLS line. TLS drags the fit line in the x direction towards the first few points (which have high x weights) and in the y direction towards the last few points (which have high y weights).

9.4.2   Error Estimates for the TLS Fit Parameters

We'll need to compute a covariance matrix $\mathbf{C}$ to find the standard errors σ_m and σ_b for the slope and intercept. It's much more difficult to calculate $\mathbf{C}$ efficiently in TLS than it was for OLS and WLS. Fortunately, we can use a brute-force approach here, too.

The χ^2 value and its second derivatives $\partial^2 \chi^2 / \partial b^2$, $\partial^2 \chi^2 / \partial m^2$, and $\partial^2 \chi^2 / \partial b \partial m$ contain all information about the error in the fit parameters. These derivatives are conveniently arranged in the Hessian matrix:

$$\mathcal{H} = \begin{bmatrix} \dfrac{\partial^2 \chi^2}{\partial b^2} & \dfrac{\partial^2 \chi^2}{\partial b \partial m} \\[2mm] \dfrac{\partial^2 \chi^2}{\partial m \partial b} & \dfrac{\partial^2 \chi^2}{\partial m^2} \end{bmatrix} \qquad \text{The Hessian matrix for TLS} \qquad (9.66)$$

which can be computed with the `hessian` function (see Section 6.4.1).

The covariance matrix is computed from the Hessian matrix as

$$\mathbf{C} = \frac{\chi^2}{N - p} \mathcal{H}^{-1} \qquad \text{The covariance matrix from the inverse Hessian} \qquad (9.67)$$

As usual, the diagonal of the covariance matrix lists the variances in the fit parameters.

Using the expression for `chi2` and the `result` computed by the minimization above, the covariance matrix is

```
(%i16)   Hessian : hessian(chi2,[b,m])$
(%i17)   C: (chi2/(N-2))*invert(Hessian), result;
```

$$(\%o17) \quad \begin{pmatrix} 0.063396796063625 & -0.012014460004753 \\ -0.012014460004753 & 0.0024581902679705 \end{pmatrix}$$

The standard errors in the intercept and slope are, respectively,

```
(%i18)   sqrt(C[1,1]);
(%i19)   sqrt(C[2,2]);
```

```
(%o18)   0.25178720393147
(%o19)   0.049580139854285
```

Notice that `chi2` is a symbolic expression that includes m's and b's; the `hessian` function will of course fail if you give it a numerical value for χ^2. In the calculation of C, we need numerical values, so the comma'd `result` substitutes the b and m values found previously in the minimization.

9.4.3   Assessing Goodness-of-Fit in TLS

The use of weights makes interpretation of R^2 for the fit quite complicated, because R^2 is looking at the weight-transformed variables. But we don't need R^2; we have a χ^2 value, which is much better. We can use it to compute the quantitative goodness-of-fit measure Q and interpret it just as we did in WLS. As before, the value of Q hinges on the accuracy of our weights. Underestimating the errors we use to compute the weights will give us a low Q; overestimating them will inflate Q.

9.4.4   Multiple Linear Regression with TLS

Like OLS and WLS, TLS can handle more than one independent variable. Adding more x variables to the model adds more terms to the error variance. For example, to fit $y = a + bx_1 + cx_2$, we use the following expression for χ^2:

$$\chi^2 = \sum_{i=1}^{N} \frac{\left(y_i - (a + bx_{1i} + cx_{2i})\right)^2}{\sigma_{y_i}^2 + b^2\sigma_{x_{1i}}^2 + c^2\sigma_{x_{2i}}^2} \qquad \begin{array}{l} \text{The figure of merit} \\ \text{for TLS with two independent} \\ \text{variables} \end{array} \qquad (9.68)$$

We can minimize χ^2 and compute the covariance matrix from the Hessian exactly as before, but now we need estimates for the variances in y, x_1, and x_2 for each data point. If some of those variances don't change from point to point, or if some are negligible compared with the others, simplifications are possible.

The methods used in this section to find the fit parameters and their errors are "brute force" because they are computationally intensive and inefficient for large datasets. They are also vulnerable to numerical problems. But they're not much work for *us* – they can be written in a few lines of code. They are also quite general: they work for OLS, MLR, and WLS as well. More importantly, they apply to fit functions that are more complicated than lines.

In Chapter 10, we'll see several numerical methods for estimating errors in the fit parameters. They work for all of the fits described in this chapter, and they are usually a better choice than computing the covariance matrix from the Hessian.

 Worksheet 9.7: Fitting data to a line with errors in both x and y

In this worksheet, we'll use ordinary, weighted, and total least-squares to fit reaction kinetics data to a line when there are errors in both x and y.

9.5 Calibration and Standard Additions

One of the most common applications of fitting in chemistry is the prediction of unobserved values of either y or x. Predicting a value of y within the range of the x data is interpolation; predicting y outside that range is extrapolation. Predicting x from observed y values is *inverse* interpolation or extrapolation.

In analytical chemistry, linear fits are routinely used to relate some concentration-dependent quantity y with the amount of analyte in a sample. Typically we measure y for a series of samples with known concentrations x. The measurements are fit to the concentrations, often with a proportionality or a straight line. The result is a calibration curve that can be used to convert measurements into concentrations for further samples by inverse interpolation.

In physical chemistry, calibration is often used to convert raw experimental measurements into units that we're more interested in. For example, the time it takes for a solution to flow through a capillary tube is related to its viscosity. The times depend on the bore of the capillary and other properties of the apparatus; the viscosities do not. Fitting the known viscosities of various liquids to measured times provides a way to convert time measurements into viscosities.

Calibration can also be used to correct instrument readings for systematic error. For example, a spectrometer can be calibrated by measuring the wavelengths of the emission lines in the mercury spectrum and fitting them to their known values. Wavelengths corrected for instrument bias can then be read from the fit.

9.5.1 Error Estimates for Calibrated Values

Suppose that we have a linear calibration curve $y = mx + b$, and we want to take a measurement of y for an unknown sample and compute x. Obviously, the calibrated value is $x = (y - b)/m$, but how can we estimate its error? How does the error in y and the fit parameters propagate into the error in x?

If we have several measurements of y for the unknown, we can directly estimate its standard deviation s_y. If we have only one measurement, the OLS standard error of the fit serves as an estimate of standard error in y (provided that we've measured y for all samples by the same procedure, under identical conditions). If the error variance isn't constant (as in HPLC calibrations, for example, see Ref. [73]) we might use a variance model to accurately predict s_y.

The errors from y and the fit parameters propagate into the error in the predicted value of x. In general, for a fit of p parameters $\alpha_1, \alpha_2, \ldots, \alpha_p$ that has a covariance matrix $\mathbf{C}$, the estimated variance in x is

$$s_x^2 = \left(\frac{\partial F}{\partial y}\right)^2 s_y^2 + \sum_i^p \sum_j^p \left(\frac{\partial F}{\partial \alpha_i}\right)\left(\frac{\partial F}{\partial \alpha_j}\right) C_{ij} \qquad \text{Variance in } x \text{ read from a multiparameter fit against } y$$

$$(9.69)$$

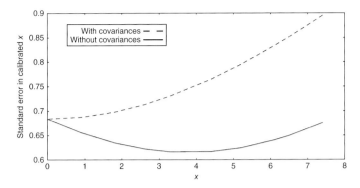

Figure 9.16 Plot of the estimated standard error in calibrated values of x for an OLS fit of Pearson's data, with and without the covariance term. Neglecting the covariance seriously overestimates the error with this data, especially for high values of x.

where x is computed from the fit as $F(y)$. For a linear fit $y = mx + b$, the equation simplifies to

$$s_x^2 = \frac{1}{m^2}\left(s_y^2 + x^2 s_m^2 + 2x C_{12} + s_b^2\right) \qquad \begin{array}{l}\text{Variance in } x \\ \text{read from a fitted line}\end{array} \qquad (9.70)$$

How important is the covariance in Equation (9.70)? It depends on the data, but ignoring the C_{12} term leads to wildly inaccurate error estimates. The slope and intercept are nearly always correlated, and sometimes quite strongly. For example, we compare the estimated standard error in calibrated x as a function of x in Figure 9.16 with and without the covariance term. Since the covariance term scales with x, s_x becomes increasingly inaccurate as x increases when covariances are neglected. For this data, C_{12} is negative, and neglecting covariances severely overestimates s_x.

 Always report a covariance matrix with your fit. Without it, others can look at your fit but cannot accurately estimate errors in calculations of their own that use your fit parameters. Presenting a fit without a covariance matrix is like serving a sundae without a spoon.

It is sometimes helpful to show s_x directly on your calibration curve using imprecision contours. Plot lines with points $(x - s_x, \hat{y})$ and $(x + s_x, \hat{y})$ along with your fit line. The lines show the size of the error in values of x read from the fit line (see Figure 9.17). You can also plot 95% confidence contours by scaling s_x with t, where we compute t as `quantile_student_t(0.05,N-2)`.

9.5.2 Standard Additions

Absolute calibration is not always practical in chemical analysis. Sometimes it isn't possible to prepare solutions of known concentration with chemical environments that are identical to the unknown samples we'd like to analyze. For example, suppose we want to determine the concentration of a compound in blood. The instrumental technique we use to determine the compound's concentration may have a much different precision and accuracy when applied to a blood sample than it has with an aqueous solution of the compound.

One alternative method is to spike the blood sample with known amounts of the compound, and see what effect that has on instrument readings. For example, suppose the (unknown)

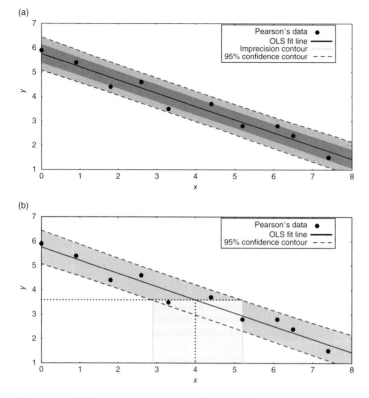

Figure 9.17 (a) Imprecision contours and 95% confidence contours for reading x from an OLS fit (using Pearson's data). The contours are very gently curved. (b) Reading the confidence contours. If $y = 3.6$, the value of x read from the fit line is 4, with a 95% confidence interval that stretches from about 2.8 to 5.2.

concentration of the compound in a blood sample is C_{sample}. Suppose further that our instrument gives us a reading of A_{sample} for the sample, which is proportional to C_{sample}:

$$A_{sample} = kC_{sample} \tag{9.71}$$

We then add V_{std} milliliters of a solution with a known concentration of C_{std}. The new instrument reading is $A_{sample+std}$:

$$A_{sample + std} = k \left(\frac{C_{sample} V_{sample} + C_{std} V_{std}}{V_{sample} + V_{std}} \right) \tag{9.72}$$

where the original volume of the sample is V_{sample}. We have two equations in two unknowns that we can solve for k and C_{sample}.

We could try this again and again with different volumes of standard. In fact, we can linearize Equation (9.72) if we dilute all solutions (including the original sample) to a volume of V before taking our readings. The new equation is

$$A = k \left(\frac{C_{sample} V_{sample} + C_{std} V_{std}}{V} \right) \tag{9.73}$$

A plot of A against V_{std} has a slope $m = kC_{std}/V$ and an intercept $b = kC_{sample} V_{sample}/V$. The concentration of the compound in the blood sample is $C_{sample} = (b/m)(C_{std}/V_{sample})$.

To find an error estimate for C_{sample}, we propagate the errors in b and m (assuming that C_{std} and V_{sample} have errors that are negligible in comparison with those contributed by the fit parameters):

$$
\begin{aligned}
s_{C_{\text{sample}}}^2 &= \left(\frac{\partial C_{\text{sample}}}{\partial b} \right)^2 s_b^2 + \left(\frac{\partial C_{\text{sample}}}{\partial m} \right)^2 s_m^2 + 2 \left(\frac{\partial C_{\text{sample}}}{\partial m} \right) \left(\frac{\partial C_{\text{sample}}}{\partial b} \right) cov(b, m) \\
&= C_{\text{sample}}^2 \left(\frac{C_{11}}{b^2} + \frac{C_{22}}{m^2} - 2 \frac{C_{12}}{mb} \right)
\end{aligned}
$$

$$(9.74)$$

The second equation is written in terms of covariance matrix elements C_{11}, C_{22}, and C_{12}, obtained from a fit of Equation (9.73).

10

Fitting Data to a Curve

With four parameters I can fit an elephant, and with five I can make him wiggle his trunk.
— John von Neumann

In Chapter 9, we fit linear models to experimental data. A linear model of y as a function of x can be generally written as

$$y = \sum_i c_i f_i(x) \qquad \text{General form of a linear model} \qquad (10.1)$$

where c_i is an adjustable fit parameter and $f_i(x)$ is a function of x that does not depend on any of the fit parameters. If any of the functions f_i depends on fit parameters, the model is *nonlinear*.

Notice that by this definition polynomial fits are considered linear. For example, a parabolic fit $y = c_1 + c_2 x + c_3 x^2$ has $f_1(x) = 1, f_2(x) = x,$ and $f_3(x) = x^2$, which don't depend at all on the fit parameters $c_1, c_2,$ and c_3. We can fit polynomials using linear regression, as we'll see in Section 10.2.

Nonlinear models are often encountered in chemistry [74]. Before software for nonlinear curve fitting was in routine use, chemists transformed data to cast nonlinear models into linear forms (Section 10.1). This cannot always be done, and even when it can, the residuals from the resulting fit may no longer be uniform and random. We saw in Chapter 9 that data with unequal variances will yield distorted fit parameters and error estimates in ordinary least-squares.

The nonlinear fit parameters can be derived as before by minimizing the sum of squared errors (*SSE*) for the model equation. Unfortunately, the expression for the SSE can sometimes have multiple minima. For this reason, nonlinear least squares problems are usually solved numerically and iteratively, starting with some guess for the parameters. Each iteration refines the parameters, often by using linear extrapolation.

Estimating the precision of a nonlinear fit and its parameters also presents a problem. Many different kinds of fitting functions are possible, so deriving general statistical formulas for the errors in the fit parameters as we did in Chapter 9 is difficult and inconvenient, as is estimating parameter errors with the Hessian matrix (Section 9.4.2). In this chapter we'll look at two efficient numerical procedures for estimating error in fit parameters: jackknifing (Section 10.4.1) and bootstrapping (Section 10.4.2).

10.1 Transforming Data to a Linear Form

One simple way to handle nonlinear data is to nonlinearly transform the x or y variables to new variables X and Y that have a linear relationship. We can then apply the linear least-squares methods outlined in Chapter 9 to perform the fit.

Table 10.1 Nonlinear models can sometimes be cast into linear form using transformations.

Model	Transformation	Linear fit equation
$y = \exp(b + mx)$	$Y = \ln y$	$Y = b + mx$
$y = (b + mx)^n$	$Y = y^{1/n}$	$Y = b + mx$
$y = \frac{1}{b+mx}$	$Y = 1/y$	$Y = b + mx$
$y = b + m \ln x$	$X = \ln x$	$y = b + mX$
$y = bx^m$	$Y = \ln y, X = \ln x, B = \ln b$	$Y = B + mX$

The transformation involves using the $f_i(x)$ functions in Equation (10.1) as independent variables, rather than using the x data directly. Sometimes the dependent variable data is also transformed. Table 10.1 lists some common transformations for linearizing data.

For example, consider linearizing a parabolic fit $y = ax + bx^2$ by substituting $Y = y/x$ to obtain $Y = a + bx$. Let's use the following generated data (with $a = -0.2$, $b = 0.1$, and random normally distributed noise in y):

```
(%i1)  xdata: [1,2,3,4,5,6,7,8,9,10,11,12,13,14,15,16,17,18,19,
            20,21,22,23,24,25,26,27,28,29,30,31,32,33,34,35,36,
         ➡ 37,38,39,40,41,42,43,44,45,46,47,48,49,50]$
(%i2)  ydata:
         ➡ [3.59,6.35,5.05,12.2,-3.92,-7.35,-1.92,4.91,-5.04,
        11.9,-3.31,18.06,11.98,17.6,8.81,11.11,29.8,10.61,15.7,
        14.2,47.49,24.99,43.1,60.5,55.75,53.23,74.28,75.54,72.44,
         ➡
        70.68,83.13,71.71,81.99,92.56,97.76,131.73,135.62,
        110.19,130.02,144.41,139.23,156.5,172.98,156.18,159.32,
        192.31,194.25,183.08,226.71,218.91]$
(%i3)  plot2d([discrete,xdata,ydata], [style, points])$
```

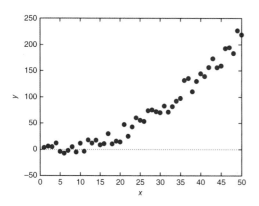

Now transform the *y* data, and fit and plot the transformed data.

```
(%i4)   ytrans: ydata/xdata$
(%i5)   data: transpose(matrix(xdata,ytrans))$
(%i6)   load(stats)$
(%i7)   fit: linear_regression(data);
(%i8)   [a,b] : take_inference( b_estimation, fit);
(%i9)   plot2d([[discrete, xdata, ytrans], [discrete, xdata, a+b
          ➥ *xdata]],
          [style, points, lines],
            [legend, "transformed data", "linear fit"],
            [gnuplot_preamble, "set key outside"]
        )$
```

$$
\begin{array}{c}
\text{\textit{LINEAR REGRESSION MODEL}} \\
b_{estimation} = [0.2802900054738764, 0.07593594043146895] \\
b_{statistics} = [1.004040792163626, 7.970054134412728] \\
b\_p\_values = [0.3203938157101862, 2.443953928121810^{-10}] \\
b_{distribution} = [student_t, 48] \\
v_{estimation} = 0.9452076407146764 \\
v\_conf\_int = [0.6573205891265982, 1.475229912099587] \\
v_{distribution} = [chi2, 48] \\
adc = 0.5606238753458869
\end{array}
$$

(%o7)

(%o8) $[0.2802900054738764, 0.07593594043146895]$

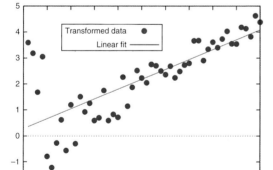

The adjusted determination coefficient *adc* is not close to one, which suggests a poor fit. Let's look at the residuals plot:

```
(%i10)  plot2d([discrete, xdata, take_inference('residuals, fit)
          ➥ ], [style, points], [title, "raw residuals"])$
```

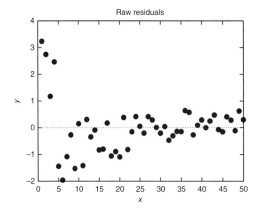

The variance of the residuals appears to decrease (and become slightly more positive) as x increases, again showing that the fit is poor. In this case, the residuals have been severely distorted by the transformation when the value of x is small.

 In general, uniform error in the data may be nonuniform after transformation. This forces an incorrect weighting on the data, resulting in inaccurate fit error estimates and parameters. It is much better to fit the curve directly without a linear transformation, as outlined in the following sections.

Transformations sometimes also have numerical difficulties. For example, a negative data point can lead to an imaginary number in a logarithmic transformation; a zero value in a denominator can cause an undefined point in the transformed data.

10.2 Polynomial Least-Squares Fitting

The `linear_regression` function can fit polynomials as well as lines. To fit an n-degree polynomial $c_0 + c_1 x + c_2 x^2 + \cdots + c_n x^n$, build a data matrix with the first n columns corresponding to $x_i, x_i^2, \ldots, x_i^n$, with the y data in column $n + 1$.

Let's use the data in the last section to build a matrix with a column for each power of x, followed by a column for the y data:

```
(%i11)   data: transpose(matrix( xdata, xdata 2 , ydata))$
```

Now perform the fit and extract the parameters.

```
(%i12)   fit: linear_regression(data);
```

$$\text{LINEAR REGRESSION MODEL}$$
$$b_{estimation} = [0.9019000000000688, -0.4660376996953062, 0.09705353679933548]$$
$$b_{statistics} = [0.1955395514366298, -1.117017365076006, 12.23731697644457]$$
$$b\_p\_values = [0.845813507769831, 0.2696647248680719, 2.22044604925031310^{-16}]$$

(%o12)
$$b_{distribution} = [student_t, 47]$$
$$v_{estimation} = 108.9829728190026$$
$$v\_conf\_int = [75.52566880823863, 170.989660265371]$$
$$v_{distribution} = [chi2, 47]$$
$$adc = 0.9768593897111947$$

The adjusted determination coefficient is close to 1, suggesting a much better fit than we obtained from the linear transformation. But notice the p values for the parameters: there is an 85% chance and 27 intervals:

```
(%i13)   take_inference ( b_conf_int, fit);
```

$$(\%o13)\quad \begin{pmatrix} [[-8.376983765470236, 10.18078376547037], \\ [-1.305368358926832, 0.3732929595362194], \\ [0.08109852538990822, 0.1130085482087627]] \end{pmatrix}$$

The confidence intervals for both c_0 and c_1 include zero. In future fits, we might try to simplify the model equation by excluding those parameters.

Plot the fit together with the data to visually verify the fit:

```
(%i14)   [c_0,c_1,c_2] : take_inference ( b_estimation, fit)$
(%i15)   plot2d([[discrete, xdata, ydata], [discrete, xdata, c_0+
         ➥ c_1*xdata+c_2*xdata^2]],
            [style, points, lines],
            [legend, "untransformed data", "parabolic fit"],
            [gnuplot_preamble, "set key outside"]
         )$
```

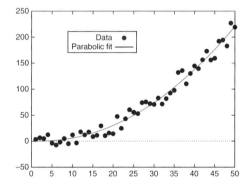

Problems with the fit (like polynomials that oscillate between data points) are often made apparent by the residuals plot. For this fit,

```
(%i16)   plot2d([discrete, xdata, take_inference ( residuals, fit)
         ➥ ], [style, points], [title, "raw residuals"])$
```

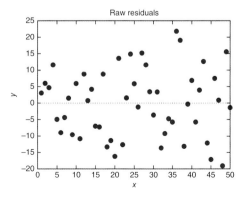

The residuals with the direct polynomial fit are distributed around zero, with apparently uniform variance. Compare this residuals plot for the linear transformation fit in the previous section. The standard deviation of the residuals is

```
(%i17)   variance: take_inference('v_estimation, fit)$
(%i18)   stdev: sqrt(variance);
```

```
(%o18)   10.43949102298587
```

and the residuals plot does show that about two-thirds of the residuals fall within ± 10 of zero.

10.2.1   How Many Fit Parameters Are Needed?

We saw in Chapter 9 that the variance of the residuals included both experimental error and error due to lack of fit. Adding more parameters to a model reduces the residual variance. This becomes too much of a good thing when the number of parameters begins to approach the number of data points. In Figure 10.1, four points are fit with a line (two parameters), a parabola (three parameters), and a cubic polynomial (four parameters). The cubic polynomial is an example of overfitting. The function oscillates so that it can pass exactly through all of the data points, but in the process it becomes inaccurate for interpolation and, especially, extrapolation.

 To avoid overfitting, the number of data points must greatly exceed the number of parameters.

In building a model, one usually starts with the fewest parameters possible and then adds new parameters one at a time. Each added parameter can be checked statistically to determine whether it results in a significant reduction in the variance of the fit.

Our null hypothesis is that adding the parameter has no significant effect on the variance of the fit. We can do an F-test (Section 8.4.2) to compare the variances for a fit with and without the added parameter. The F statistic is the ratio of the variances for the the two fits:

$$F = \frac{\text{variance of fit 1}}{\text{variance of fit 2}} \tag{10.2}$$

where the fraction is inverted if $F < 1$. The variances we need are listed in the `linear_regression` output as `v_estimation`.

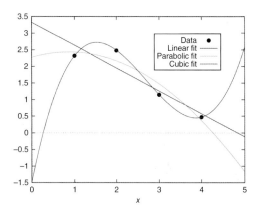

Figure 10.1 Overfitting a set of four points.

As we saw in Section 8.4.2, the *p*-value is computed as

$$P = 2 * (1 - \text{cdf\_f}(F, v_1, v_2)) \qquad \begin{array}{l} \textit{p-value for the} \\ \textit{F-test} \end{array} \qquad (10.3)$$

where v_1 and v_2 are the degrees of freedom for the two fits. In this case the degrees of freedom for a fit is the number of data points minus the number of fit parameters. If the *p*-value is less than 0.05, we can reject the null hypothesis with 95% confidence or better, and add the new parameter to our model. If the *p*-value is greater than 0.05, we can't objectively justify adding the parameter.

The `plsquares` function introduced in Chapter 9 is useful for quickly screening a series of fitting polynomials:

> `plsquares(`*data matrix, variable list, dependent variables, max variable exponent, max degree*`)`
> Fits a polynomial in one or more variables to the data in *data matrix*. Each row in the data matrix is a point; each column corresponds to a variable in *variable list*. A polynomial is fitted for each dependent variable listed in *dependent variables*. Optional arguments are *max variable exponent* (the maximum exponent allowed in each independent variable (1 by default)) and *max degree* (the maximum polynomial degree). When `max degree` is zero, there is no limit on the maximum degree of the polynomial. You must `load(plsquares)` before using this function.

For example, let's quickly look at polynomial fits of degree 1 through 4 to the following data:

```
(%i1)   xdata: [1,2,3,4,5,6,7,8]$
(%i2)   ydata: [0.2,8.9,27.9,60.0,108.1,175.0,264.1,378.0]$
(%i3)   data: transpose(matrix(xdata,ydata))$
(%i4)   load(plsquares)$
(%i5)   plsquares(data, [x,y], y);
(%i6)   plsquares(data, [x,y], y, 2);
(%i7)   plsquares(data, [x,y], y, 3);
(%i8)   plsquares(data, [x,y], y, 4);
```

Determination Coefficient for y = 0.89885062424549
(%o4) $y = 52.5\,x - 108.475$

Determination Coefficient for y = 0.99888603918471
(%o5) $y = 8.75714285714286\,x^2 - 26.31428571428574\,x + 22.88214285714292$

Determination Coefficient for y = 0.99999982420385
(%o6) $y = 0.49141414141413\,x^3 + 2.12305194805201\,x^2 - 1.006457431457685\,x - 1.44285714285688$

Determination Coefficient for y = 0.99999990941288
(%o7) $y = 0.0024621212121295\,x^4 + 0.4470959595958\,x^3 + 2.39071969697066\,x^2 - 1.620580808083131\,x - 1.024999999998319$

Although the determination coefficient decreases as we add parameters, we must do *F*-tests to see whether adding successive parameters results in a significant reduction in the fit variance, as we'll see in Worksheet 10.0.

Worksheet 10.0: Polynomial Curve Fitting

In this worksheet, we'll fit polynomials to experimental data using `linear_regression`. We'll determine the optimum degree of the polynomial to fit using *F*-tests and by examining changes in residual plots as parameters are added.

10.3 Nonlinear Least-Squares Models

The `lsquares` package provides several tools for fitting truly nonlinear models. The general routine `lsquares_estimates` can fit both explicit and implicit equations to data. It returns a list of solutions, each of which is a list of parameter equations:

> `lsquares_estimates(`*data matrix, variable list, model equation, parameter list*`)`
>
> `lsquares_estimates(`*data matrix, variable list, model equation, fit parameters,* `initial=`*initial parameter guess list,* `tol=`*tolerance*`)`
>
> Find the best fit of *model equation* with variables in *variable list* to the data in *data matrix*. The fit parameters appearing in the model equation are listed in *parameter list*. The data matrix contains one "point" per row. The *variable list* gives a name for each column in the data matrix; not all of the variables need to appear in the model equation. Exact solutions are sought first; if they cannot be found, iterative numerical solutions are sought. The second form of `lsquares_estimates` can be used to specify initial estimates for the fit parameters and a stopping tolerance for searching for numerical solutions. The default stopping tolerance is 1×10^{-3}; setting it to a smaller number can result in a more accurate fit.

The `lsquares_estimates` function tries to find an exact solution, and if that fails, it tries to find an approximate numerical solution with `lbfgs` (Section 6.4.2). The search for an exact solution can be extremely slow. When it fails, try the following:

- Convert the data matrix to numeric form before fitting unless you need symbolic fit parameters. Nonlinear curve-fitting is computationally intensive. Symbolic data makes it even more so.
- When `lsquares_estimates` hangs, seek approximate solutions directly with the functions `lsquares_mse` and `lsquares_estimates_approximate`. Replace

```
(%i1)  fit: lsquares_estimates(dataMatrix, variableList,
         ➡ modelEquation, parameterList);
```

with

```
(%i1)  mse: lsquares_mse(dataMatrix, variableList,
         ➡ modelEquation)$
(%i2)  fit: lsquares_estimates_approximate(mse, parameterList
         ➡ );
```

We'll see examples of this approach in Worksheet 10.2.

- Provide the tolerance option `tol=tolerance` to prevent premature convergence.
- Provide an initial guess of parameter values `initial=list`. This is especially helpful when adding new parmeters one at a time – use the results of successful fits with fewer parameters as part of your initial guess.
- The SSE surface for nonlinear fits may have multiple minima. Repeat the fit with several different initial guesses to help detect convergence to local rather than global minima.
- Some functions (such as exponential and hyperbolic functions) may cause numerical problems during fitting calculations. If they cause the fit to fail, rewrite the fitting function to eliminate them. For example, take the logarithm of both sides of the fitting equation to eliminate `exp`.

To see how `lsquares_estimates` works, let's fit the simple model equation

$$\ln P = A - \frac{B}{T} \qquad \text{The August equation for temperature dependence of vapor pressure} \qquad (10.4)$$

to the experimental vapor pressure of water P and the temperature T. This is a simplified version of the *Clausius-Clapeyron equation*

$$\ln \frac{P}{P^\circ} = \frac{\Delta H_{vap}}{R} \left(\frac{1}{T^\circ} - \frac{1}{T} \right) \qquad \text{The Clausius-Clapeyron equation} \qquad (10.5)$$

where P° and T° are a reference vapor pressure and temperature, respectively, and ΔH_{vap} is the enthalpy of vaporization, which is taken as a constant over the temperature range from T to T°. R is the ideal gas law constant (8.314 J mol^{-1} K^{-1}). We expect the parameters in the model equation to be

$$A \approx \frac{\Delta H_{vap}}{RT^\circ} + \ln(P^\circ)$$
$$B \approx \frac{\Delta H_{vap}}{R} \qquad (10.6)$$

Water has a vapor pressure $P^\circ = 1$ atm at $T^\circ = 373$ K, and an enthalpy of vaporization of 40.7 kJ mol^{-1}, so we expect $A \approx 13.1$ and $B \approx 4900$ if the Clausius-Clapeyron equation holds.

The raw vapor pressure data for water is in degrees Celsius and torr [31]. In Maxima, converting the data to kelvins and atmospheres is easily done using `transform_sample`, which we saw in Section 2.4.7.

```
(%i1)   raw_data: matrix([0,4.579], [5,6.543], [10, 9.209],
          ➥ [15,12.788], [20,17.535],[25,23.756], [30,31.824],
          ➥ [35, 42.175], [40,55.324], [45, 71.88],
          ➥ [50,92.51], [55,118.04], [60,149.38], [65,187.54],
          ➥ [70, 233.7], [75,289.1], [80,355.1], [85,433.6],
          ➥ [90,525.76], [95,633.9],[100,760] )$
(%i2)   load(stats)$
(%i3)   data : transform_sample( raw_data, [T,P], [T+273.15, P
          ➥ /760] )$
```

We now have a matrix with kelvin temperatures in the first column and pressures in atmospheres in the second. Now let's fit the data to Equation (10.4).

```
(%i4)   numer : true /* We want numerical results, not symbolic
        ➠ */$
(%i5)   august: lsquares_estimates( data, [T,P], P = exp(A-B/T),
        ➠ [A,B], tol=1e-10);
```

```
*****************************************************
  N=    2    NUMBER OF CORRECTIONS=25
          INITIAL VALUES
  F=  6.114368430667324D+00    GNORM=  1.330907165321096D+01
*****************************************************
I NFN    FUNC                    GNORM                   STEPLENGTH
1   2    6.355789448545657D-01   1.480927584156316D+00   7.513672073128697D-02
2   3    4.750622503200111D-01   1.100272585611286D+00   1.000000000000000D+00
3   4    2.122987572118746D-01   4.390143477419503D-01   1.000000000000000D+00
4   5    1.356461906626626D-01   2.192275377454690D-01   1.000000000000000D+00

  ⋮

34  38   2.947616282606039D-06   5.277227315255748D-04   1.000000000000000D+00
35  39   2.464437988015836D-06   8.158737379612421D-05   1.000000000000000D+00
36  40   2.451350080781738D-06   2.108853895551714D-06   1.000000000000000D+00
37  41   2.451297969843650D-06   7.572658347247666D-08   1.000000000000000D+00
 THE MINIMIZATION TERMINATED WITHOUT DETECTING ERRORS.
 IFLAG = 0
```

(%o5) $[[A = 13.98093570320805, B = 5204.830970615888]]$

Here FUNC is the SSE with the trial parameters at each step of the iteration, and GNORM is the norm of the gradient on the SSE surface. Both should approach small values if the iteration is converging successfully. It took 37 steps to converge to the fit parameters with an SSE that was changing by a tolerance of $tol=10^{-10}$. If you don't want to see output in this detail, add the option $iprint$=[-1,0] to the arguments of either lsquares_estimates or lsquares_estimates_approximate.

The fit parameters are roughly the values we expected. Let's plot the fit and the data together.

```
(%i6)   xdata: makelist(data[i,1], i, 1, length(data))$
(%i7)   ydata: makelist(data[i,2], i, 1, length(data))$
(%i8)   plot2d([[discrete, xdata, ydata], [discrete, xdata, exp(
        ➠ A-B/xdata)]],
          [style, points, lines],
          [legend, "data", "fit"],
          [xlabel, "T / K"],
          [ylabel, "P / atm"]
        ), first(august)$
```

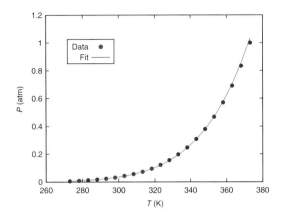

We can easily evaluate the residuals by hand, but there is a Maxima function that makes this easier:

lsquares_residuals(*data matrix, variable list, model equation, solution*)
Computes residuals for a fit of the *model equation* to the *data matrix* with columns labeled by *variable list*. The *solution* is a list of equations that specify the fit parameters in the *model equation* (usually found by a previous call to lsquares_estimates).

Again choosing the first and only solution for the parameters found, we have

```
(%i9)    residuals: lsquares_residuals(data, [T,P], log(P)=A-B/T,
     ➥   first(august))$
(%i10)   plot2d([discrete, xdata, residuals], [style, points], [
     ➥   ylabel, "raw residuals"], [xlabel, "T / K"]);
```

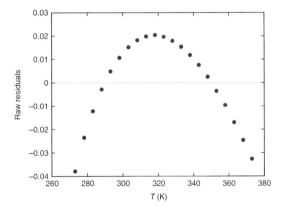

The residual size and pattern shows that Equation (10.4) can't adequately reproduce the variation in the data. In Worksheet 10.2, we'll see how to build a better model.

To compare model equations using an *F*-test, we'll need to calculate the variance of the fit (Equation 9.13). The lsquares package includes a function for computing the residual mean square error (MSE):

> lsquares_residual_mse(*data matrix, variable list,*
> *model equation, solution*)
> Computes the residual MSE for the *model equation* with parameters specified by *solution* fit to the *data matrix*. The MSE is defined as

$$\frac{1}{N} \sum_{i=1}^{N} e_i^2$$

> where e_i is the difference between the left- and right-hand sides of the *model equation*.

We can get the variance of the fit from the MSE by

$$v_{\text{fit}} = \frac{1}{N-p} \sum_{i=1}^{N} e_i^2 = \frac{N}{N-p} \text{MSE} \qquad \text{Variance of the fit from the MSE} \qquad (10.7)$$

multiplying the MSE by the number of data points N, and then dividing by the number of degrees of freedom ($N - p$, where there are p fit parameters). For this fit,

```
(%i11)   lsquares_residual_mse(data, [T,P], log(P)=A-B/T, first(
         ➥ august));
(%i12)   v_estimation: length(data)*%/(length(data)-2);
(%i13)   s_estimation: sqrt(%);
```

```
(%o11)   3.264057895610227 10^-4
(%o12)   3.607642937253409 10^-4
(%o13)   0.018993796190476
```

The standard error of the fit (also known as the standard error of the regression) is consistent with the residual plot above; about two-thirds of the data fall within 0.019 of zero.

 Worksheet 10.1: Nonlinear Least-Squares Curve Fitting

In this worksheet, we'll successively add parameters to the simple vapor pressure fit, using *F*-tests and residual plots with each addition to ensure that the added parameters significantly improve the model. We'll also fit integrated rate laws to kinetic data, Lorentzian and Gaussian lineshapes to spectral peaks, and expressions relating concentrations and equilibrium constants to titration data.

10.4 Estimating Error in Nonlinear Fit Parameters

We need error estimates and confidence intervals for the fit parameters. It is possible to compute the variance of the fit parameters by computing the Hessian matrix around the minimum in the SSE surface, following the brute-force procedure outlined in Section 9.4.2. The inverse of the Hessian matrix along with χ^2 can be used to obtain the covariance matrix $\mathbf{C}$ (Equation 9.67), and the diagonal elements of $\mathbf{C}$ give the variances in each fit parameter. In practice, though, this procedure is computationally inefficient and is quite vulnerable to numerical problems.

It is much easier to compute error estimates using numerical techniques. Many techniques are available, but we'll look at only two: jackknifing and bootstrapping.

10.4.1 Estimating Parameter Errors with the Jackknife Method

In jackknifing, we recompute the fit parameters with one data point removed from the dataset [75, 76]. Repeating the procedure for each data points generates a list of "jackknifed" fit parameters. The mean of all of the jackknifed fit parameters should be the fit parameters for the original dataset, and the standard deviation of the jackknifed fit parameters gives us the standard error of the original fit parameters.

The following example estimates the standard error and 95% confidence interval for each parameter using jackknifing for the fit of Equation (10.4) to the temperature/pressure `data` matrix set up in the previous section.

An empty matrix `jackknife` is created to hold the perturbed parameters:

```
(%i14)   jackknife : matrix()$
```

For each row *i* in the `data` matrix, do the following[1]:

1) Build a submatrix with the *i*th row deleted, using `submatrix(i, data)`.
2) Compute the fit parameters using this submatrix.
3) Add the fit parameters as a row to the `jackknife` matrix, using `addrow`.

```
(%i15)   N: length(data)$
(%i16)   for i : 1 thru N do ev(
                jackknife: addrow(jackknife,[A,B]),
                lsquares_estimates(submatrix(i,data), [T,P], log
            ➡   (P) = A-B/T, [A,B])
         )$
```

Here we use an evaluation environment(ev) rather than the usual `block` statement for the body of the loop because we want the `addrow` to use the values of *A* and *B* computed by `lsquares_estimates`.

Each row in the `jackknife` matrix now contains a perturbed fit; each column holds the values of one fit parameter.

```
(%i17)   jackknife;
```

---

1 The `submatrix` and `addrow` functions were introduced in Section 2.4.5.

$$\begin{pmatrix} 13.93943336727351 & 5190.809644620807 \\ 13.9594748444081 & 5197.524846018335 \\ 13.9717482689542 & 5201.67240886454 \\ 13.97919915397329 & 5204.226152058393 \\ 13.98327844838948 & 5205.661788111166 \\ 13.98494451447145 & 5206.291258832624 \\ 13.98502264096295 & 5206.386641364393 \\ 13.98404358685203 & 5206.119600701381 \\ 13.98247197904075 & 5205.641690394245 \\ 13.98068176891621 & 5205.076740776241 \\ 13.9789875632177 & 5204.521187957116 \\ 13.97763967716424 & 5204.062904849869 \\ 13.97683740010337 & 5203.765789610169 \\ 13.97680121082263 & 5203.698929344257 \\ 13.97765781540364 & 5203.905642504309 \\ 13.97965678806581 & 5204.462619132454 \\ 13.98305975694693 & 5205.45149162019 \\ 13.98747323757219 & 5206.761169270886 \\ 13.99395231202381 & 5208.705719063188 \\ 14.00195025592008 & 5211.127494344794 \\ 14.01174279462322 & 5214.110879821542 \end{pmatrix}$$

(%o8)

Compute the standard error s_E of each fit parameter by multiplying the standard deviations for each column by $(N - 1)/\sqrt{N}$, where N is the number of data points (rows in the `data` matrix).[2]

```
(%i18)   s_E: (N-1)/sqrt(N) * std1(jackknife);
```

(%o9)   [0.061632309963245, 19.92412613465668]

Compute the 95% confidence interval for each fit parameter from the standard errors as ts_E, obtaining the t statistic using the procedure we used in Section 8.3.2.

```
(%i19)   CI_95 : quantile_student_t(0.975, N-1)*s_E;
```

(%o10)   [0.12856274575388, 41.56099883560644]

The final jackknifed 95% confidence intervals for the fit parameters are $A = 13.98 \pm 0.13$ and $B = 5204 \pm 42$, with $N = 21$.

In practice, `lsquares_estimates` will often fail to converge correctly. Our code will perform much better by using `lsquares_estimates_approximate` with an initial guess. Let's repeat the calculation with these changes. Here, `fit` holds the initial guess, which is just the fit parameters obtained with the full dataset.

```
(%i20)   jackknife: matrix()$
(%i21)   mse: lsquares_mse(data, [T,P], log(P) = A - B/T)$
(%i22)   fit: [A,B], lsquares_estimates_approximate(mse, [A,B],
```

---

2 Recall that taking the sample standard deviation `std1`(D) where D is a matrix returns a list of standard deviations for each column.

```
              initial=[14, 5200], tol=1e-15
           )$
(%i23)     for i : 1 thru N do block(
                 datai: submatrix(i, data),
                 mse: lsquares_mse(datai, [T,P], log(P) = A-B/T),
                 ev(
                         jackknife : addrow(jackknife, [A,B]),
                         lsquares_estimates_approximate(mse, [A,B
                             ➥ ], initial=fit, tol=1e-12)
                     )
           )$
(%i24)     s_E: (N-1)/sqrt(N) * std1(jackknife)$
(%i25)     fit;
(%i26)     CI_95 : quantile_student_t(0.975, N-1)*s_E;
```

(%o16)   [13.98093570320803, 5204.830970615883]
(%o17)   [0.1285627457538864, 41.56099883560807]

Once again, the final jackknifed 95% confidence intervals for the fit parameters are $A = 13.98 \pm 0.13$ and $B = 5204 \pm 42$, with $N = 21$.

10.4.2   Estimating Parameter Errors with the Bootstrap Method

Bootstrapping approximates a distribution for the fit parameters by randomly generating new samples from the original data. If the original data was representative of the entire population, fit parameters computed for the new samples should be distributed as they are for the entire population. We can then compute error estimates from the list of fit parameters for the new samples.

Each of the samples is constructed by drawing points at random from the original dataset until we have a sample that is the same size as the original. For example, if we had five data points $[p_1, p_2, p_3, p_4, p_5]$ in the original set, the bootstrap samples could be $[p_2, p_3, p_3, p_5, p_1]$, $[p_1, p_3, p_4, p_5, p_1]$, and so on. Note that some points in the original set may be repeated in the samples. We generate a very large number of samples $N_{samples}$, and compute and save fit parameters for each sample in a matrix bootstrap.

Bootstrapping requires a fairly large dataset, and it is more computationally intensive than jackknifing. Since we can have a much larger number of samples in bootstrapping, it can potentially yield superior error estimates (in fact, jackknifing is a just special case of bootstrapping, with a smaller number of samples).

The following example estimates the standard error and 95% confidence interval for each fit parameter using 1000 bootstrap samples.

First, an empty matrix bootstrap is created to hold the list of perturbed parameters.

```
(%i27)    bootstrap : matrix()$
```

For each sample, repeat the following steps:

1) Create an empty matrix sample to hold the resampled data.
2) For each point in the original dataset, add one randomly chosen point to sample.
3) Fit the data in sample and store the fit parameters as a new row in bootstrap.

Using `lsquares_estimates_approximate`, the bootstrap calculation is

```
(%i28)    for j : 1 thru 1000 do block(
                 sample : matrix(),
                 for i : 1 thru N do
                       sample : addrow(sample, row(data,
                             ➥ random_discrete_uniform(N))),
                 mse: lsquares_mse(sample, [T,P], log(P)=A-B/T),
                 bootstrap : ev(
                       addrow(bootstrap, [A, B]),
                       lsquares_estimates_approximate(mse, [A,B
                             ➥ ], initial=fit, tol=1e-12,
                             iprint=[-1,0])
                 )
          )$
```

Compute the standard error of each fit parameter; it will be equal to the standard deviation of the corresponding column in `bootstrap`.

```
(%i29)    s_E: std1(bootstrap);
```

```
(%o19)    [0.05844155721511786, 18.92558627661487]
```

Compute the 95% confidence interval for each fit parameter from the standard errors using the procedure we used in Section 8.3.2.

```
(%i30)    CI_95 : quantile_student_t(0.975, length(data)-1)*s_E;
```

```
(%o20)    [0.12297828158511, 39.76784736270468]
```

The final bootstrapped 95% confidence intervals for the fit parameters are $A = 13.98 \pm 0.12$ and $B = 5204 \pm 40$, with $N = 21$. The confidence intervals are slightly smaller than those we obtained by jackknifing. Because the bootstrap samples are assembled randomly, you may see slightly different confidence intervals.

The covariance matrix for the fit parameters can easily be computed from the bootstrap matrix using `cov1`

```
(%i31)    cov1(bootstrap);
```

$$(\%o21) \quad \begin{pmatrix} 0.003354474161940177 & 1.079159970824687 \\ 1.079159970824687 & 349.1073556773894 \end{pmatrix}$$

However, the bootstrap procedure can directly provide error estimates and even probability distributions for any result calculated from experimental data, without the need for computing the covariance matrix or Jacobian.

For example, consider the following experimental x, y, and z data:

```
(%i1)    X: [5.18, 4.22, 4.69, 2.57, 5.77, 3.98, 4.9, 5.82, 6.05,
              ➥  4.51, 4.99, 3.68]$
```

```
(%i2)   Y: [4.96, 5.12, 4.2, 4.62, 4.53, 2.47, 5.65, 5.01, 5.02,
        ➡   3.83, 5.7, 4.17]$
(%i3)   Z: [-3.64, -2.22, 2.04, -5.86, 2.71, 0.95, -1.22, -1.52,
        ➡   -0.76, 1.27, -1.15, 0.99]$
```

Suppose we want to find the mean and standard deviation of the following function:

```
(%i4)   Lbar(x,y,z) := mean(sqrt(x^2 + y^2 + z^2));
```

$$(\%o4) \quad \text{Lbar}\,(x, y, z) := \text{mean}\left(\sqrt{x^2 + y^2 + z^2}\right)$$

Set up a data matrix:

```
(%i5)   data: transpose(matrix(X, Y, Z))$
(%i6)   N: length(data)$
```

Now apply the bootstrap method, computing `Lbar` for a large number of samples taken from the data.

```
(%i7)   bootstrap : matrix()$
(%i8)   for j : 1 thru 1000 do block(
               sample : matrix(),
               for i : 1 thru N do
                    sample : addrow(sample, row(data,
                        ➡ random_discrete_uniform(N))),
               bootstrap :  addrow(bootstrap, Lbar(col(sample
                    ➡ ,1), col(sample,2), col(sample,3)))
        )$
```

The mean of `Lbar` over all the samples is

```
(%i9)   mean(bootstrap);
```

```
(%o9)   [7.064491946132205]
```

and its standard deviation is

```
(%i10)  std1(bootstrap);
```

```
(%o10)  [0.3066091617488531]
```

We can do the same for any function that computes one or more parameters from the data.

 Worksheet 10.2: Parameter Errors in Nonlinear Curve Fitting

In this worksheet, we'll use jackknifing and bootstrapping to estimate the errors in nonlinear fit parameters.

11

Differential Equations

True laws can only be expressed in differential equations.

– Bertrand Russell [77]

A differential equation relates an independent variable with derivatives of a dependent variable. It has the general form:

$$f\left(x, y, \frac{dy}{dx}, \frac{d^2y}{dx^2}, \ldots\right) = 0 \qquad \text{A differential equation} \tag{11.1}$$

where f explicitly depends on at least one of the derivatives of y. Such equations model many (if not most) physical processes, and solving them is a central task in reaction kinetics and quantum theory.

Consider an elementary reaction A → products. The rate of the first-order reaction $A \xrightarrow{k}$ products is related to the concentration of A at time t by

$$\frac{dA}{dt} = -kA \qquad \text{First-order rate law} \tag{11.2}$$

where A is the concentration of A as an unknown function of time t. An ordinary (not partial) derivative appears in the equation, so the equation is called an ordinary differential equation, or ODE.

The solution of a differential equation is a function, not the value of a variable. In this case, solving the equation gives us A as a function of t. In Section 1.2.4 we saw that the `ode2` function can find solutions to ODEs:

```
(%i1)    diff(A(t),t) = -k*A(t);
(%i2)    ode2(%, A(t), t);
```

$$(\%o1) \quad \frac{d}{dt} A(t) = -k\,A(t)$$

$$(\%o2) \quad A(t) = \%c\,e^{-kt}$$

The first line tells Maxima that A depends on t (see Section 6.3.1). The second line specifies the differential equation. The `ode2` function gives a general solution in terms of `%c`, which is some unknown constant. The general solution represents an entire family of solutions with different values of `%c` (Figure 11.5). In Section 11.1, we'll see how to find specific values for the constant by imposing initial conditions or boundary conditions on the general solution.

Equation (11.2) directly gives dA/dt, which is the slope of the tangent lines for the solutions $A(t)$. This lets us visualize the equation geometrically as a vector field, which is easily plotted

Symbolic Mathematics for Chemists: A Guide for Maxima Users, First Edition. Fred Senese.
© 2019 John Wiley & Sons Ltd. Published 2019 by John Wiley & Sons Ltd.
Companion website: http://booksupport.wiley.com

in Maxima. In Section 11.3, we'll see how solutions are represented as trajectories through the vector field.

The highest derivative that appears in the equation determines the *order* of the differential equation . A first-order equation like Equation (11.2) can be written in the form:

$$\frac{dy}{dx} = f(x, y) \qquad \text{General form of a first-order ODE} \tag{11.3}$$

A second-order equation has the form:

$$\frac{d^2y}{dx^2} = f\left(x, y, \frac{dy}{dx}\right) \qquad \text{General form of a second-order ODE} \tag{11.4}$$

where the expression on the right-hand side may or may not include first derivatives of y. For example, the wave equation that describes an electron trapped in a one-dimensional "box" with potential energy V is

$$-\frac{h^2}{8\pi^2 m}\frac{d^2\psi}{dx^2} + V\psi = E\psi \qquad \text{The one-dimensional Schrödinger equation} \tag{11.5}$$

is an example of a second-order ODE. Here, solution of the equation involves not only finding the unknown wavefunction ψ (which depends on x) but also its associated energy E.

When the unknown function in the equation depends on more than one variable, partial derivatives appear. For example, if the temperature distribution in a two-dimensional plate is $\theta(x, y)$, the steady-state equation for heat conduction in the plate is [78]

$$\frac{\partial^2\theta}{\partial x^2} + \frac{\partial^2\theta}{\partial y^2} = 0 \qquad \text{Laplace's equation} \tag{11.6}$$

This is a partial differential equation (PDE). The additional variables involved in PDEs often make them much more difficult to solve than ODEs. In chemistry we can often separate the independent variables and convert the PDE into a system of ODEs, as we'll see in Section 11.6.

11.1   Symbolic Solutions of ODEs

The `ode2` function tries to find symbolic solutions for both first- and second-order ODEs. You can access it through the menu with $\boxed{\text{Equations}} \rangle\!\rangle \boxed{\text{Solve ODE...}}$, through the General Math panel with the $\boxed{\text{Solve ODE...}}$ button, or by directly typing

> `ode2(diffeqn, y, x)`
> Solve a first- or second-order ODE `diffeqn`. It returns `false` if it cannot find a solution.

The differential equation must be typed carefully. We can't simply type Equation (11.2) as `diff(A,t) = -k*A`, because Maxima doesn't know that `A` depends on `t`, so it will evaluate `diff(A,t)` as zero. There are three ways to keep this from happening:

1) Use `depends(A,t)` to show the dependence (as we did in our first example in this chapter).
2) Use the single quote operator to prevent evaluation of the derivative,

```
(%i1)   'diff(A, t) = -k*A;
(%i2)   ode2(%, A, t);
```

$$(\%o1) \qquad \frac{d}{dt}A = -kA$$

$$(\%o2) \qquad A = \%c\, e^{-kt}$$

3) Explicitly show the dependence of A on t by consistently typing it as A(t).

```
(%i1)    diff(A(t),t) = -k*A(t);
(%i2)    ode2(%, A(t), t);
```

$$(\%o1) \quad \frac{d}{dt}\,\mathrm{A}(t) = -k\,\mathrm{A}(t)$$

$$(\%o2) \quad \mathrm{A}(t) = \%c\,e^{-kt}$$

We can verify the solution by substituting it back into the differential equation:

```
(%i1)    A(t) := %c*exp(-k*t);
(%i2)    diff(A(t), t) = -k*A(t);
```

$$(\%o1) \quad \mathrm{A}(t) := \%c\,\exp((-k)\,t)$$

$$(\%o2) \quad -\%c\,k\,e^{-kt} = -\%c\,k\,e^{-kt}$$

The `ode2` function tries to find a solution by indefinite integration; the constant `%c` is the integration constant. In this case, we obtain an explicit solution with the form A(t) = f(t). When there is more than one explicit solution `ode2` will write the solution as an implicit equation. For example,

```
(%i1)    depends(y,x)$
(%i2)    y^3*diff(y, x) - x*y^2 = 0;
(%i3)    ode2(%, y, x);
```

$$(\%o2) \quad y^3\left(\frac{d}{dx}\,y\right) - x\,y^2 = 0$$

$$(\%o3) \quad \frac{y^2}{2} = \frac{x^2}{2} + \%c$$

The explicit solutions are

```
(%i4)    solve(%, y)
```

$$(\%o4) \quad [y = -\sqrt{x^2 + 2\,\%c}, y = \sqrt{x^2 + 2\,\%c}]$$

Now let's solve a second-order ODE. One common type of differential equation encountered in physical chemistry is

$$\frac{d^2 y}{dx^2} + A\frac{dy}{dx} + By = 0 \qquad \text{A second-order ODE with constant coefficients } A \text{ and } B$$

$$(11.7)$$

The form of the general solution depends on the relationship between the constants A and B.

```
(%i1)    eqn: 'diff(y,x,2)+A*'diff(y,x)+ B*y=0;
```

$$(\%o1) \quad y\,B + \left(\frac{d}{dx}\,y\right)A + \frac{d^2}{dx^2}\,y = 0$$

```
(%i2)    ode2(eqn, y, x);
```

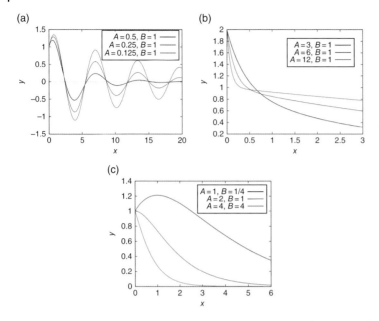

Figure 11.1 General solutions of Equation (11.7) when (a) $4B > A^2$, (b) $4B < A^2$, and (c) $4B = A^2$.

Is $4B - A^2$ *positive, negative, or zero?* positive;

$$(\%o2) \quad y = e^{-\frac{xA}{2}}\left(\%k1 \sin\left(\frac{x\sqrt{4B - A^2}}{2}\right) + \%k2 \cos\left(\frac{x\sqrt{4B - A^2}}{2}\right)\right)$$

With positive A and B and $4B > A^2$, the solution is a damped wave (Figure 11.1a). When A is negative, the amplitude of the wave *increases* with x.

When $4B < A^2$, the solution is a combination of an exponential increase and an exponential decay (Figure 11.1b). The dominant term depends on the signs and relative values of A and B.

(%i3)   **ode2**(eqn, y, x);

Is $4B - A^2$ *positive, negative, or zero?* negative;

$$(\%o3) \quad y = \%k1\, e^{\frac{x\left(\sqrt{A^2 - 4B} - A\right)}{2}} + \%k2\, e^{\frac{x\left(-\sqrt{A^2 - 4B} - A\right)}{2}}$$

When $4B = A^2$, the solution has both linear and exponential factors, which sometimes leads to a curve with a point of inflection (Figure 11.1c).

(%i4)   **ode2**(eqn, y, x);

Is $4B - A^2$ *positive, negative, or zero?* zero;

$$(\%o3) \quad y = (\%k2\, x + \%k1)\, e^{-\frac{xA}{2}}$$

11.1.1   Initial Value Problems

Once we have a general solution for a differential equation, a specific solution can be found by applying an initial condition, which sets the required value of the solution at a single point.

The Maxima function `ic1` applies an initial condition for first-order differential equations. Use the menu dialog $\boxed{\text{Equations} \gg \text{Initial Value Problem(1)}}$, or type

```
ic1(solution, x=x_value, y=y_value)
```
Solve a first-order initial value problem. Here x and y are the names of the independent and dependent variables, respectively.

after calling `ode2`. For example, an initial condition for Equation (11.2) might be that the concentration of A is A_0 at time zero. The specific solution becomes

```
(%i1)   'diff(A,t) = -k*A;
(%i2)   ode2(%, A, t);
(%i3)   ic1(%, t=0, A=A_0);
```

$$(\%o1) \quad \frac{d}{dt}A = -k\,A$$

$$(\%o2) \quad A = \%c\,e^{-kt}$$

$$(\%o3) \quad A = e^{-kt}\,A\_0$$

The general solution of the equation includes the arbitrary constant `%c`, which is replaced by a specific value when the initial value $A = A_0$ at $t = 0$ is included.

General solutions of a second-order differential equation contain two constants, so two conditions are needed to eliminate them. The conditions are the value of the dependent variable *and its derivative* at some particular point. These can be specified with the `ic2` function, inserted by $\boxed{\text{Equations} \gg \text{Initial Value Problem(2)}}$, or by typing

```
ic2(solution, x=x_value, y=y_value,
    'diff(y,x)=diff_value)
```
Solve a second-order initial value problem. Here x is and y are the names of the independent and dependent variables, respectively.

after a call to `ode2`. For example, let's solve the classical harmonic oscillator equation

$$\frac{dx^2}{dt^2} = -\omega^2 x \qquad \text{Classical harmonic oscillator equation} \qquad (11.8)$$

for the position x of an oscillating particle as a function of time t, where ω is a positive constant. The position depends on the initial position and speed of the oscillating particle. Suppose that the particle is initially stationary at $x = A$ (where A is amplitude or maximum displacement of the particle). Then $dx/dt = 0$ at $t = 0$, and we have

```
(%i1)   assume(omega > 0)$
(%i2)   'diff(x,t,2) = -omega^2*x;
(%i3)   ode2(%, x, t);
(%i4)   ic2(%, t=0, x=A, 'diff(x,t)=0);
```

$$(\%o1) \quad \frac{d^2}{dt^2}x = -\omega^2 x$$

$$(\%o2) \quad x = \%k1\sin(\omega t) + \%k2\cos(\omega t)$$

$$(\%o3) \quad x = \cos(\omega t)\,A$$

where the initial conditions have replaced the arbitrary constants k_1 and k_2 in the general solution with 0 and A, respectively.

11.1.2 Boundary Value Problems

The two constants in the general solution of a second-order ODE can also be found by specifying two (x, y) points rather than a single point where x, y, and dy/dx are known. We can specify these boundary conditions using the bc2 function, inserted with Equations ⟩ Boundary Value Problem , or by entering

bc2(*solution*, x=*x_value_1*, y=*y_value_1*, x=*x_value_2*, y=*y_value_2*)
Solve a two-point boundary condition problem.

after a call to ode2. For example, suppose we have the equation $\frac{d^2y}{dx^2} + 2\left(\frac{dy}{dx}\right) + 2y = 0$ with boundary conditions $y(\pi) = 0$ and $y(\pi/2) = 1$:

```
(%i1)  'diff(y,x,2) +2*'diff(y,x) + 2*y = 0;
(%i2)  ode2(%,y,x);
(%i3)  bc2(%, x=%pi, y=0, x=%pi/2, y=1);
```

$$(\%o1) \quad \frac{d^2}{dx^2}y + 2\left(\frac{d}{dx}y\right) + 2y = 0$$

$$(\%o2) \quad y = e^{-x}\left(\%k1\sin(x) + \%k2\cos(x)\right)$$

$$(\%o3) \quad y = e^{\frac{\pi}{2}-x}\sin(x)$$

Let's verify that the particular solution is in fact a solution of the differential equation:

```
(%i4)  diff(y,x,2) +2*diff(y,x) + 2*y = 0, %;
(%i5)  radcan(%);
```

$$(\%o4) \quad 2\left(e^{\frac{\pi}{2}-x}\cos(x) - e^{\frac{\pi}{2}-x}\sin(x)\right) + 2e^{\frac{\pi}{2}-x}\sin(x) - 2e^{\frac{\pi}{2}-x}\cos(x) = 0$$

$$(\%o5) \quad 0 = 0$$

where we had to use the radcan function to simplify the left-hand side of the result (Section 1.2.2).

Unfortunately, bc2 will sometimes only give trivial solutions; in more complex boundary value problems, we'll have to apply the boundary conditions by hand. For example, let's look at a fundamental problem in quantum mechanics: the *particle in a box* problem.

Imagine a particle like an electron that moves in a straight line between two walls. The particle can't penetrate the walls. Its potential energy is zero between the walls, and infinite inside them. The x coordinate of the particle can range anywhere from zero to L, the length of the line between the walls (the "box").

Quantum mechanics postulates that all mechanical information about the particle is encoded in its *wavefunction*, Ψ. Ψ can be interpreted as a *probability amplitude*, that is, the probability of locating the particle between x and $x + dx$ is $\Psi^*\Psi dx$. This expression must integrate to one over the length of the box, that is, the wavefunction must be *normalized*:

$$\int_0^L \Psi^*\Psi dx = 1 \qquad \text{Normalization of } \Psi \tag{11.9}$$

Ψ can be obtained by solving the Schrödinger equation, which for this problem can be written as

$$-\frac{\hbar^2}{2m}\frac{d^2\Psi}{dx^2} = E\Psi \qquad \text{The Schrödinger equation for a particle in a box} \qquad (11.10)$$

with $\hbar = h/2\pi$. Our objective is to solve this second-order differential equation for *both E* and Ψ.

First, let's enter the Schrödinger equation. We'll also tell Maxima to assume that the mass, energy, box length *L*, and the reduced Planck constant $\hbar$ are positive, to keep it from asking about them later.

```
(%i1)   assume(m>0, E>0, L>0, hbar > 0);
(%i2)   Schrodinger: -(hbar^2/(2*m))*'diff(Psi,x,2) = E*Psi;
```

$$(\%o1) \quad [m > 0, E > 0, L > 0, hbar > 0]$$

$$(\%o2) \quad -\frac{hbar^2\left(\frac{d^2}{dx^2}\Psi\right)}{2\,m} = \Psi\,E$$

Now find the general solution using `ode2`:

```
(%i3)   solution: ode2(Schrodinger, Psi, x);
```

$$(\%o3) \quad \Psi = \%k1\sin\left(\frac{\sqrt{2}\,\sqrt{m}\,x\,\sqrt{E}}{hbar}\right) + \%k2\cos\left(\frac{\sqrt{2}\,\sqrt{m}\,x\,\sqrt{E}}{hbar}\right)$$

Now let's find a physically meaningful particular solution of the equation by imposing boundary conditions. If the electron is trapped in a box with length *L*, the position *x* of the electron is restricted to the interval $0 \le x \le L$. The wavefunction Ψ is required to be zero at the interval endpoints (at $x = 0$ and $x = L$).

Unfortunately, we only get a trivial solution $\Psi = 0$ if we try to impose the conditions with `bc2`:

```
(%i4)   bc2(solution, x=0, Psi=0, x=L, Psi=0);
```

$$(\%o4) \quad \Psi = 0$$

To find nontrivial solutions we must apply the boundary conditions by hand. First, simplify the general solution by substituting $k = \sqrt{2mE}/\hbar$:

```
(%i5)   solution1: ratsubst(k, sqrt(2*m*E)/hbar, solution);
```

$$(\%o5) \quad \Psi = \%k1\sin(k\,x) + \%k2\cos(k\,x)$$

Requiring that $\Psi(0) = 0$ means that $\%k2 = 0$, since the sine term is zero and the cosine term is one when $x = 0$:

```
(%i6)   solution1, x=0;
```

$$(\%o6) \quad \Psi = \%k2$$

so the solution simplifies to

(%i7)   `solution2: ratsubst(0, %k2, solution1);`

(%o7) $\Psi(x) = \%k1 \sin(kx)$

Applying the second boundary condition $\Psi(L) = 0$, we have

(%i8)   `solution2, x=L;`

(%o8) $\Psi = \%k1 \sin(kL)$

To have $\Psi(L) = 0$, we could have $\%k1 = 0$. But that would mean that ψ must be zero everywhere, which it isn't. The sine function is zero at integer multiples of π. We can have $\Psi(L) = 0$ if $kL = n\pi$, where n is an integer. Let's substitute this condition into the solution:

(%i9)   `solution3: solution2, k=n*%pi/L;`

(%o9) $\Psi = \%k1 \sin\left(\dfrac{\pi n x}{L}\right)$

We need an additional condition to find $\%k1$. We can apply Equation (11.9):

(%i10) `rhs(solution3);`
(%i11) `integrate(conjugate(%)*%, x, 0, L) = 1;`
(%i12) `solve(%, %k1);`

(%o10) $\%k1 \sin\left(\dfrac{\pi n x}{L}\right)$

(%o11) $\dfrac{\%k1^2 L}{2} = 1$

(%o12) $[\%k1 = -\dfrac{\sqrt{2}}{\sqrt{L}}, \%k1 = \dfrac{\sqrt{2}}{\sqrt{L}}]$

Choosing the positive solution for $\%k1$, we have

(%i13) `solution3, %[2];`

(%o13) $\Psi = \dfrac{\sqrt{2} \sin\left(\frac{\pi n x}{L}\right)}{\sqrt{L}}$

Substituting this back into the Schrödinger equation and solving for the energy, we have

(%i14) `Schrodinger, %$`
(%i15) `solve(%, E), nouns;`

(%o15) $[E = \dfrac{\pi^2 \, hbar^2 \, n^2}{2L^2 m}]$

> (M) **Worksheet 11.0: Symbolic Solution of ODEs**
>
> In this worksheet, we'll explore particular and general solutions for Equation (11.10), and impose boundary conditions on it.

11.2 Power Series Solution of ODEs

Some differential equations that are commonly encountered in quantum chemistry cannot be solved by `ode2`. Such equations can sometimes be solved by assuming that the solution y can be written as a power series in x [79]:

(%i1) `y(x) := sum(a[i]*x^i, i, 0, inf);`

(%o1) $y(x) := \sum\limits_{i=0}^{\infty} a_i x^i$

Consider the Hermite equation, which is related to the equation that describes a one-dimensional quantum mechanical harmonic oscillator:

$$\frac{d^2 y}{dx^2} - 2x\frac{dy}{dx} + 2ny = 0 \qquad \text{The Hermite equation} \qquad (11.11)$$

where n is usually a nonnegative integer. Substituting the power series into Equation (11.11), we have

(%i2) `HermiteEqn: diff(y(x),x,2) - 2*x*diff(y(x),x) + 2*n*y(x)`
       ➡ `= 0;`

(%o2) $2n\left(\sum\limits_{i=0}^{\infty} a_i x^i\right) - 2x\left(\sum\limits_{i=0}^{\infty} i\,a_i x^{i-1}\right) + \sum\limits_{i=0}^{\infty} (i-1)\,i\,a_i x^{i-2} = 0$

Solving the equation now becomes a matter of finding the coefficients in these sums. Since the equation must hold for any value of x, the coefficients of like powers of x must be zero. We need to collect the coefficients of like powers together and combine the sums, using the techniques we used in Section 5.1.2.

(%i3) `intosum(%);`

(%o3) $\left(\sum\limits_{i=0}^{\infty} 2a_i n x^i\right) + \left(\sum\limits_{i=0}^{\infty} -2i\,a_i x^i\right) + \sum\limits_{i=0}^{\infty} (i-1)\,i\,a_i x^{i-2} = 0$

The `intosum` function pulls factors under the summation sign. The index in the third sum must be shifted by 2 so that sum is also written in terms of x^i. First, use the `bashindices` function to give each sum a unique index:

(%i4) `bashindices(%);`

(%o4) $\left(\sum\limits_{j3=0}^{\infty} 2a_{j3} n x^{j3}\right) + \left(\sum\limits_{j2=0}^{\infty} -2j2\,a_{j2} x^{j2}\right) + \sum\limits_{j1=0}^{\infty} (j1-1)\,j1\,a_{j1} x^{j1-2} = 0$

Now use `changevar` to specifically target the sum over `j1` and shift its index by 2:

(%i5) **changevar(%, j1=j3+2, j3, j1);**

(%o5) $\left(\displaystyle\sum_{j3=0}^{\infty} 2a_{j3} n\, x^{j3} \right) + \left(\displaystyle\sum_{j3=-2}^{\infty} \left(j3^2 + 3j3 + 2 \right) a_{j3+2}\, x^{j3} \right) + \displaystyle\sum_{j2=0}^{\infty} -2j2\, a_{j2}\, x^{j2} = 0$

Apply `niceindices` to put all sums over the same index again.

(%i6) **niceindices(%);**

(%o6) $\left(\displaystyle\sum_{i=0}^{\infty} 2a_i n\, x^i \right) + \left(\displaystyle\sum_{i=-2}^{\infty} \left(i^2 + 3i + 2 \right) a_{i+2}\, x^i \right) + \displaystyle\sum_{i=0}^{\infty} -2i\, a_i\, x^i = 0$

Apply `sumcontract` to combine the sums:

(%i7) **sumcontract(%);**

(%o7) $\displaystyle\sum_{i=0}^{\infty} 2a_i n\, x^i + \left(i^2 + 3i + 2 \right) a_{i+2}\, x^i - 2i\, a_i\, x^i = 0$

Factor out x^i on the left-hand side of the equation:

(%i8) **map(factor, lhs(%))=0;**

(%o8) $\displaystyle\sum_{i=0}^{\infty} \left(2a_i n + i^2\, a_{i+2} + 3i\, a_{i+2} + 2a_{i+2} - 2i\, a_i \right) x^i = 0$

Pick out the coefficient of x^i:

(%i9)   **part(%, 1);**
(%i10) **part(%, 1);**
(%i11) **part(%, 1);**

(%o9) $\displaystyle\sum_{i=0}^{\infty} \left(2a_i n + i^2\, a_{i+2} + 3i\, a_{i+2} + 2a_{i+2} - 2i\, a_i \right) x^i$

(%o10) $\left(2a_i n + i^2\, a_{i+2} + 3i\, a_{i+2} + 2a_{i+2} - 2i\, a_i \right) x^i$

(%o11) $2a_i n + i^2\, a_{i+2} + 3i\, a_{i+2} + 2a_{i+2} - 2i\, a_i$

Each coefficient of a power of x must be zero. Solve for a_{i+2} to obtain a *recurrence relation* that relates coefficients a in the series.

(%i12) **solve(%=0, a[i+2]);**
(%i13) recurrence: **factor(first(%));**

(%o12) $\left[a_{i+2} = -\dfrac{2a_i n - 2i\, a_i}{i^2 + 3i + 2} \right]$

$$(\%o13) \quad a_{i+2} = -\frac{2a_i\,(n-i)}{(i+1)\,(i+2)}$$

We can now use this recurrence relation to write out the power series solutions for Equation (11.11). When n is odd, we can choose the initial coefficients $a_0 = 0$ and $a_1 = 1$ to obtain a specific solution. For example, with $n = 3$, all even coefficients $a_0 = a_2 = a_4 = \ldots = 0$, and all odd coefficients greater than a_n will be zero as well:

```
(%i14)   n: 3$
(%i15)   a[0]  :  0;
(%i16)   a[1]  :  1;
(%i17)   a[2]  :  rhs(recurrence),  i=0;
(%i18)   a[3]  :  rhs(recurrence),  i=1;
(%i19)   a[4]  :  rhs(recurrence),  i=2;
(%i20)   a[5]  :  rhs(recurrence),  i=3;
```

$$(\%o15) \quad 0$$
$$(\%o16) \quad 1$$
$$(\%o17) \quad 0$$
$$(\%o18) \quad -\frac{2}{3}$$
$$(\%o19) \quad 0$$
$$(\%o20) \quad 0$$

so our solution is

```
(%i21)   y = sum(a[i]*x^i,  i,  0,  n);
```

$$(\%o21) \quad y = x - \frac{2x^3}{3}$$

When n is even, choose the initial conditions $y(0) = a_0 = 1$ and $(dy(x)/dx)_{x=0} = a_1 = 0$. For example, with $n = 4$,

```
(%i22)   n: 4$
(%i23)   a[0]  :  1;
(%i24)   a[1]  :  0;
(%i25)   a[2]  :  rhs(recurrence),  i=0;
(%i26)   a[3]  :  rhs(recurrence),  i=1;
(%i27)   a[4]  :  rhs(recurrence),  i=2;
(%i28)   a[5]  :  rhs(recurrence),  i=3;
```

$$(\%o23) \quad 1$$
$$(\%o24) \quad 0$$
$$(\%o25) \quad -4$$
$$(\%o26) \quad 0$$
$$(\%o27) \quad \frac{4}{3}$$
$$(\%o28) \quad 0$$

and the solution is

```
(%i29)    y = sum(a[i]*x^i, i, 0, n);
```

$$(\%\text{o}29) \quad y = \frac{4x^4}{3} - 4x^2 + 1$$

The solutions are polynomials of degree n.

The choice of a_0 and a_1 is equivalent to choosing initial values, since $y(0) = a_0$ and $(dy/dx)_{x=0} = a_1$. Only choices that make even coefficients zero for odd values of n and make odd coefficients zero for even values of n will yield finite polynomials as solutions.

A special set of finite polynomial solutions called Hermite polynomials are obtained with the following initial conditions:

$$
\begin{aligned}
y(0) &= \begin{cases} (-2)^{\frac{n}{2}}(n-1)(n-3)(n-5)\ldots(1) & \text{for even} \quad n \\ 0 & \text{for odd} \quad n \end{cases} \\
y'(0) &= \begin{cases} 0 & \text{for even} \quad n \\ -(-2)^{\frac{n+1}{2}} n(n-2)(n-4)\ldots(1) & \text{for odd} \quad n \end{cases}
\end{aligned}
$$

Initial values for standard Hermite polynomials

$$(11.12)$$

For example,

```
(%i30)    n: 4$
(%i31)    a[0]  :  (-2)^(n/2)*3*1$
(%i32)    a[1]  :  0$
(%i33)    a[2]  :  rhs(recurrence),  i=0$
(%i34)    a[3]  :  rhs(recurrence),  i=1$
(%i35)    a[4]  :  rhs(recurrence),  i=2$
(%i36)    a[5]  :  rhs(recurrence),  i=3$
(%i37)    y = sum(a[i]*x^i, i, 0, n),  expand;
```

$$(\%\text{o}37) \quad y = 16x^4 - 48x^2 + 12$$

Maxima has a built-in function `hermite` for computing the Hermite polynomials directly. The first seven Hermite polynomials are

```
(%i1)    for n:0 thru 6 do
               print(H[n](x)  =  hermite(n,x));
```

$$H_0(x) = 1$$
$$H_1(x) = 2x$$
$$H_2(x) = -2\left(1 - 2x^2\right)$$
$$H_3(x) = -12x\left(1 - \frac{2x^2}{3}\right)$$
$$H_4(x) = 12\left(\frac{4x^4}{3} - 4x^2 + 1\right)$$

$$H_5(x) = 120x \left(\frac{4x^4}{15} - \frac{4x^2}{3} + 1 \right)$$

$$H_6(x) = -120 \left(-\frac{8x^6}{15} + 4x^4 - 6x^2 + 1 \right)$$

Let's demonstrate that Hermite polynomials are indeed solutions of Equation (11.11). With $n = 6$,

```
(%i1)   y(x) := hermite(6,x);
(%i2)   diff(y(x),x,2)-2*x*diff(y(x),x)+2*6*y(x)=0;
(%i3)   ratsimp(%);
```

(%o1) $y(x) := H_6(x)$

(%o2) $-1440 \left(-\dfrac{8x^6}{15} + 4x^4 - 6x^2 + 1 \right) + 240x \left(-\dfrac{16x^5}{5} + 16x^3 - 12x \right)$

$- 120 \left(-16x^4 + 48x^2 - 12 \right) = 0$

(%o3) $0 = 0$

> Ⓜ **Worksheet 11.1: Power Series Solution of ODEs**
>
> In this worksheet, we'll use power series to solve three well-known differential equations that occur in quantum chemistry.

11.3 Direction Fields

Let's look at Equation (11.2), which gives the derivative d[A]/dt as a function of [A]. The derivative is the slope of the tangent line on a plot of the solution [A] against t. By computing the value of this slope at various values of [A] and t, we can plot a field of vectors that are tangent to solutions of the equation. For example, with $k = 1$, the vector field for Equation (11.2) is shown in Figure 11.2. The graph is called a direction field; it is a geometric representation of the differential equation.

Figure 11.2 Direction field for the first-order ODE in Equation (11.2).

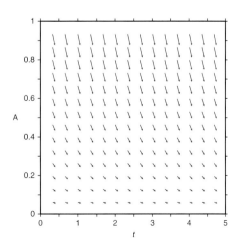

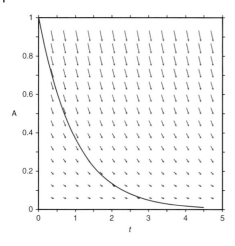

Figure 11.3 Direction field for the first-order ODE in Equation (11.2), with a particular solution that passes through the point with $t = 0$ and [A]=1 drawn in red.

Table 11.1 Options for `plotdf`.

Option	Description
`[trajectory_at, x_0, y_0]`	Plot a numerical solution of the ODE through the point with $x = x_0$ and $y = y_0$.
`[direction, option]`	Set the direction of the integration of the independent variable along a trajectory. *option* can be `forward`, `backward`, or `both` (the default).
`[x, x_min, x_max]`	Range for the x axis.
`[y, y_min, y_max]`	Range for the y axis.
`[parameters, "a=1,b=2"]`	Declare adjustable parameters a and b in the equation, and set their values.
`[sliders, "a=0:2,b=1:3"]`	Add sliders that control the values of a and b, with $0 \leq a \leq 2$ and $1 \leq b \leq 3$.

The direction field correctly shows that the solution must decrease (the tangent vectors all point down), and that the rate of decrease is lower as [A] decreases (the tangent vectors get shorter at lower values of [A]).

We can use the direction field to draw curves that are particular solutions of the differential equation. For example, suppose we know that [A] = 1 at $t = 0$. Starting at that point on the direction field, we can draw a curve that runs parallel to nearby arrows, and is tangent to any arrows it encounters, as shown in Figure 11.3.

Direction field plots for first-order differential equations can be simply generated in Maxima with the `plotdf` command:

`plotdf(f(x,y), [x,y], options)`

Creates a file containing an interactive plot of the vector field for the first-order differential equation $dy/dx = f(x, y)$, where the expression for `f(x, y)` is given as the first argument. The *options* are listed in Table 11.1. If you have an older version of Maxima, you may have to `load(plotdf)` before you can use this function.

Figure 11.4 The `plotdf` toolbar. From left to right, the icon functions are (1) close the window, (2) manipulate the ODE in use and plot new trajectories, (3) refresh the plot, (4) save the plot as a Postscript file, (5) zoom in, (6) zoom out, (7) plot the two variables in a pair of coupled autonomous equations against time (see Section 11.3.2), and (8) view help for `plotdf`.

Figure 11.5 Direction field for the first-order ODE in Equation (11.2), with a family of solution curves shown in red.

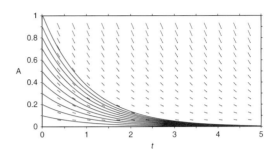

To generate the direction field in Figure 11.2, use the following code:

```
(%i1)   plotdf(-k*A, [t,A], [t, 0, 5], [A, 0, 1]), k=1;
```

```
(%o1)   C : /Users/Fred/maxout.xmaxima
```

The direction field plot is displayed in a pop-up window, and also saved in an external `.xmaxima` file in your home directory. The complete path to the file is shown in the output for the `plotdf` command. The toolbar across the top of the pop-up `plotdf` window lets you zoom, manipulate, and save the plot (Figure 11.4).

You can add plots of particular solutions to the direction field simply by clicking on the initial point for the solution. For example, clicking on the point at $t = 0$ and $A = 1$ generates Figure 11.3. You can also generate the figure with code

```
(%i1)   plotdf(-k*A, [t,A], [t, 0, 5], [A, 0, 1], [trajectory_at
         ➥ , 0, 1]), k=1;
```

```
(%o1)   C : /Users/Fred/maxout.xmaxima
```

Clicking on the tic marks on the A axis generates a family of solution curves with different starting concentrations of A, as shown in Figure 11.5.

11.3.1   Direction Fields with Adjustable Parameters

Let's look at the elementary reaction $A \underset{k_2}{\overset{k_1}{\rightleftharpoons}} B$. Let the initial concentrations of A and B be a and b, respectively. After time t, the concentration of A will be $a - x$ and the concentration of B will be $b + x$. The differential equation for the rate of reaction is

$$\frac{\mathrm{d}x}{\mathrm{d}t} = k_1(a - x) - k_2(b + x) \qquad \text{Rate law for opposing first-order reactions} \qquad (11.13)$$

where we have the initial condition $x = 0$ at $t = 0$. How will changing the parameters $k_1, k_2, a,$ and b change the solutions, which give x as a function of time?

We can use the `parameters` and `sliders` options for `plotdf` to make each parameter adjustable (see Table 11.1):

```
(%i1)   plotdf(k1*(a-x)-k2*(b+x), [t, x], [t, 0, 5], [x, 0, 1],
            [trajectory_at,0,0],
            [parameters, "k1=1,k2=0.5,a=1.75,b=2"],
            [sliders, "k1=0.1:2,k2=0.1:2,a=1.1:2,b=1.1:2"]);
```

(%o1) $C : /Users/Fred/maxout.xmaxima$

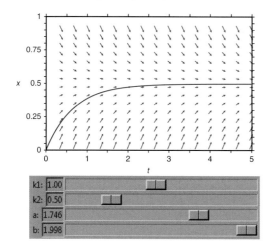

Moving the sliders for any of the four parameters will dynamically change the direction field, and the particular solution plotted in red. Notice that x will approach a constant value as time increases, whatever the initial value of x. At $t = \infty$, this value is x_∞, where

$$\frac{b + x_\infty}{a - x_\infty} = \frac{k_1}{k_2} = K \qquad \text{Equilibrium constant for opposing first-order reactions}$$

(11.14)

At the slider settings shown above, the concentration of B at $t = \infty$ is $b + x_\infty \approx 2.5$, and the concentration of A is $a - x_\infty \approx 1.25$. Since $K = [B]_\infty/[A]_\infty$, we obtain $K = 2$. This is consistent with the values of the rate constants, because k_1/k_2 also gives $K = 2$.

This example shows how direction fields are useful in connecting kinetics with equilibrium. More generally, direction fields are a valuable tool for exploring the long-term evolution of the solutions of a differential equation.

11.3.2 Direction Fields and Autonomous Equations

A differential equation is said to be autonomouss when its derivatives don't explicitly depend on the independent variable. For example, the equation $dy/dt = -kt$ is not autonomous because the derivative depends on the independent variable t. The equation $dy/dt = y^2 - 1$ *is* autonomous; t appears *only* in the derivative.

Autonomous equations have the general form:

$$\frac{dy}{dt} = f(y) \qquad \text{General form of an autonomous equation}$$

(11.15)

At points y_0 where the derivative is zero, $f(y_0) = 0$, and $y = y_0$ is a particular solution of the equation. In other words, autonomous equations can be solved simply by finding the roots of $f(y)$. These particular solutions are called equilibrium points of the autonomous equation.

For example, let's find the equilibrium points of the equation:

$$\frac{dy}{dt} = (y + 1)(y - 1) \tag{11.16}$$

If we try to use ode2 on this equation, we get a general solution that is an implicit equation:

```
(%i1)   'diff(y,x)  =  (y+1)*(y-1);
(%i2)   ode2(%, y, x);
```

$$(\%o1) \quad \frac{d}{dx} y = (y - 1)(y + 1)$$

$$(\%o2) \quad -\frac{\log(y + 1) - \log(y - 1)}{2} = x + \%c$$

It's difficult to solve this equation explicitly for y. But, because the equation is autonomous, the equilibrium points are easy to find. They occur when $(y + 1)(y - 1) = 0$, that is, when $y = -1$ and $y = 1$.

Direction fields are useful for visualizing and classifying the equilibrium points of autonomous differential equations. Let's look at the direction field for Equation (11.16):

```
(%i1)   plotdf((y+1)*(y-1),  [x, y],  [x, -1, 1],  [y, -2, 2], [
        ➥ trajectory_at,-1,1]);
```

$$(\%o1) \quad C:/Users/Fred/maxout.xmaxima$$

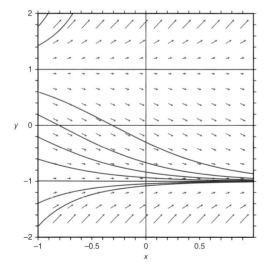

Curves for several particular solutions are plotted. Notice that the equilibrium point at $y = 1$ is *unstable*, because nearby solutions move away from it as x increases. Solutions that pass near the equilibrium point at $y = -1$ all move towards it, so it represents a *stable* equilibrium solution.[1]

Higher-order ODEs can sometimes be rewritten as systems of coupled first-order autonomous ODEs. For example, the classical harmonic oscillator equation (Equation 11.8) can be rewritten in terms of the velocity $v = dx/dt$:

$$\frac{dx}{dt} = v$$

Classical harmonic oscillator as two coupled autonomous equations　　　(11.17)

$$\frac{dv}{dt} = -\omega^2 x$$

The `plotdf` function can plot direction fields for these coupled equations in terms of v and x. The right-hand sides of each coupled autonomous equation are listed in the first argument to `plotdf`. In the following example, we provide a slider for the adjustable parameter ω (set initially at one, with a slider range from zero to two) and plot the solution that passes through $x = 5, v = 0$:

```
(%i1)  plotdf([v, -omega^2*x], [x, v], [parameters, "omega=1"],
            [sliders, "omega=0:2"], [trajectory_at, 5, 0]);
```

```
(%o1)  C:/Users/Fred/maxout.xmaxima
```

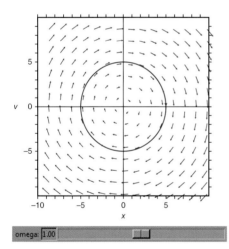

Move the slider back and forth to see how ω affects the direction field and the plotted trajectory. Note that there is a trivial equilibrium point at $x = 0, v = 0$ (when the particle is stationary); all solutions orbit around it.

You can also plot y and x as functions of time in `plotdf` for the plotted trajectory. Click on the icon that looks like a graph with two crossed curves (see Figure 11.4) to obtain this plot in a separate window:

---

1 It is possible to have an equilibrium point with nearby solutions that move towards it on one side, and away from it on the other. Such equilibrium points are called *semi-stable*.

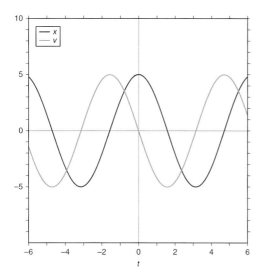

This plot gives $x(t)$ and $v(t)$ for the trajectory that passes through the point $x = 5, v = 0$.

 Worksheet 11.2: Direction Fields

In this worksheet, we'll plot direction fields for several chemically important differential equations with adjustable parameters. We'll find the equilibrium points of autonomous equations and classify them as stable, unstable, or semi-stable using their direction fields. We'll also see how to recast a second-order ODE as a pair of coupled autonomous equations.

11.4 Solving Systems of Linear Differential Equations

A linear differential equation has the form:

$$b + a_0 y + a_1 \frac{dy}{dt} + \dots + a_n \frac{d^n y}{dt^n} = 0 \qquad \text{A linear differential equation} \qquad (11.18)$$

where y is a function of t. The coefficients b and a_i may be zero, constants, or functions of t, but not functions of y. An equation with coefficients that are functions of y is called a *nonlinear differential equation*; these equations tend to be more difficult to solve than linear equations.

When $b = 0$ and both sides of Equation (11.18) are divided by a_n, the equation is said to be *reduced*. Reduced linear equations have an important property that is often exploited in quantum chemistry: linear combinations of the solutions are themselves solutions of the equation. For example, if $y = \phi_1(x)$ and $y = \phi_2(x)$ are both solutions of the reduced equation, then $y = c_1 \phi_1(x) + c_2 \phi_2(x)$ will also be a solution, where c_1 and c_2 are arbitrary constants.

A single linear equation can be solved using `ode2` (which recognizes linear equations as a special case). Systems of linear equations can be solved using the menu dialog $\boxed{\text{Equations}}$ $\boxed{\text{Solve ODE with Laplace}}$, which inserts a call to the `desolve` function:

```
desolve(diffeqn, y)
desolve([diffeqn_1, diffeqn_2, ...], [y_1, y_2, ...])
```
Solve a linear ODE, or solve a system of linear ODEs. The *y* values must be given in functional form, *e.g.* `y(x)`. If a solution cannot be found, `desolve` returns `false`.

The `desolve` function uses the method of Laplace transforms[2] to obtain general solutions for linear ODEs of any order. For example, to solve the first-order rate law in Equation (11.2),

```
(%i1)   diff(A(t),t) = -k*A(t);
(%i2)   desolve(%, A(t));
```

$$(\%o1) \quad \frac{d}{dt} A(t) = -k A(t)$$

$$(\%o2) \quad A(t) = A(0) e^{-kt}$$

The concentration must be written as `A(t)` consistently, *not* as simply `A` when working with `desolve`.

 Solution variables must be explicitly written in terms of their independent variables in differential equations to be solved by `desolve`. For example, writing the equation $(dy/dx) = \sin(x)$ as `'diff(y,x)=sin(x)` won't work; the equation must be written as `'diff(y(x),x) = sin(x)`.

Solutions obtained with `desolve` often are shown in terms of derivatives of the solution at an initial value of zero. For example, solving the classical harmonic oscillator equation (Equation 11.8) with `desolve` yields

```
(%i1)   desolve('diff(x(t),t,2)=-omega^2*x(t), x(t));
```

Is ω zero or nonzero? `nonzero;`

$$(\%o1) \quad x(t) = \frac{\sin(\omega t)\left(\frac{d}{dt} x(t)\Big|_{t=0}\right)}{\omega} + x(0)\cos(\omega t)$$

We can use the `at` function with `subst` or `ratsubst` to substitute values for the derivative in the general solution (Section 6.3.2). In general, though it is easier to specify initial conditions and boundary values, *before* calling `desolve` will usually result in simpler expressions. This can be done using the menu dialog Equations ⟩ At Value..., which inserts the code

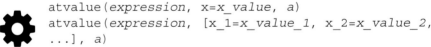

```
atvalue(expression, x=x_value, a)
atvalue(expression, [x_1=x_value_1, x_2=x_value_2,
...], a)
```

Sets initial or boundary conditions before calling `desolve` by setting *expression* to value *a* at the specified point `x=a`.

For example, if we know that the concentration in Equation (11.2) is 0.5 M at time zero,

```
(%i1)   eqn: diff(A(t),t) = -k*A(t);
(%i2)   atvalue(A(t), t=0, 0.5);
(%i3)   desolve(eqn, A(t));
(%i4)   kill(A)$
```

$$(\%o1) \quad \frac{d}{dt} A(t) = -k A(t)$$

---

2 Laplace transforms reduce linear ODEs to more easily solved algebraic forms. See Section 12.4 for more about Laplace transforms.

```
(%o2)    0.5
```

$$(\%o3) \quad A\,(t) = \frac{e^{-k\,t}}{2}$$

Notice that immediately after we solve the equation, we kill the assignment made with `atvalue` so that it won't cause confusing results in further calculations.

 Always use `kill` to remove assignments made with `atvalue` when you're finished with them.

The advantage of `desolve` over `ode2` is that it can solve systems of equations rather than just a single equation. This makes it indispensable for solving problems in reaction kinetics, as we'll see in Worksheet 11.3. For example, suppose we have the reaction $A \xrightarrow{k_1} B \xrightarrow{k_2} C$. The rate laws are three first-order ODEs:

$$\frac{dA}{dt} = -k_1 A$$
$$\frac{dB}{dt} = k_1 A - k_2 B \tag{11.19}$$
$$\frac{dC}{dt} = k_2 B$$

where A, B, and C are the concentrations of A, B, and C at time t. Suppose that $A = a$, $B = 0$, and $C = 0$ at $t = 0$. Then

```
(%i1)   eqn1: diff(A(t),t) = -k[1]*A(t);
(%i2)   eqn2: diff(B(t),t) = k[1]*A(t) - k[2]*B(t);
(%i3)   eqn3: diff(C(t),t) = k[2]*B(t);
(%i4)   atvalue(A(t), t=0, a)$
(%i5)   atvalue(B(t), t=0, 0)$
(%i6)   atvalue(C(t), t=0, 0)$
(%i7)   desolve([eqn1,eqn2,eqn3], [A(t),B(t),C(t)]);
```

$$(\%o1) \quad -\frac{d}{d\,t}\,A\,(t) = k_1\,A\,(t)$$

$$(\%o2) \quad -\frac{d}{d\,t}\,B\,(t) = k_2\,B\,(t) - k_1\,A\,(t)$$

$$(\%o3) \quad \frac{d}{d\,t}\,C\,(t) = k_2\,B\,(t)$$

$$(\%o7) \quad [A\,(t) = a\,e^{-k_1\,t}, B\,(t) = \frac{k_1\,a\,e^{-k_1\,t}}{k_2 - k_1} - \frac{k_1\,a\,e^{-k_2\,t}}{k_2 - k_1}, C\,(t) = \frac{k_1\,a\,e^{-k_2\,t}}{k_2 - k_1} - \frac{k_2\,a\,e^{-k_1\,t}}{k_2 - k_1} + a]$$

Let's plot the solutions when $k_1 = 1$, $k_2 = 2$, and $a = 1$:

```
(%i8)   %, k[1] = 1, k[2] = 2, a=1;
(%i9)   plot2d([A(t),B(t),C(t)], [t,0,6], [legend, "A(t)", "B(t)
        ➥ ", "C(t)"]), % $
(%i10)  kill(all)$
```

$$(\%o8) \quad [A\,(t) = e^{-t}, B\,(t) = e^{-t} - e^{-2\,t}, C\,(t) = -2\,e^{-t} + e^{-2\,t} + 1]$$

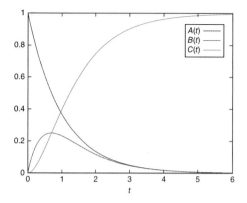

11.5   Numerical Solution of ODEs

Sometimes it isn't possible to obtain an analytical solution for an ODE or a system of ODEs with any of the methods we've looked at so far. In this case we can apply the Maxima function `rk` to obtain solutions numerically. The name "rk" comes from the method used (the fourth-order Runge–Kutta method [80]).

```
rk(dy/dx, y, y_0, [x, x_0, x_n, Δ x])
rk([dy_1/dx, ..., dy_n/dx], [y_1,...,y_n],
   [y_10,...,y_n0], [x, x_0, x_n, Δ x])
```

The first form solves a single first-order $dy/dx = f(x, y)$, where dy/dx is the expression on the right-hand side of the equation. Here y is the dependent variable with initial value y_0 and x is the independent variable which can take on the values x_0, $x\_0 + Δ\ x$, $x\_0 + 2Δ\ x$, ..., x_n. The second form solves a system of n first-order ODEs, with dependent variables y_1, ..., y_n and initial values y_10, ..., y_n0.

The solutions are returned as a list of points that can easily be plotted with `plot2d`. For example, let's solve the first-order rate law $dA/dt = -kA$ with $k = 1$ numerically, and plot the solution. The initial value of A is 0.1, and we'll compute points with $0 \le t \le 8$ every 0.01 time units:

```
(%i1)  data: rk(-k*A, A, 0.1, [t, 0, 8, 0.01]), k=1$
(%i2)  plot2d([discrete, data], [x, 0, 8],
           [xlabel, "t / s"],
           [ylabel, "concentration of A / (mol/L)"]
       )$
```

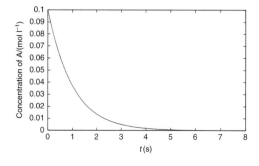

To numerically solve a second-order ODE we must rewrite it as a pair of coupled first-order ODEs. Let's repeat the solution of the classical harmonic oscillator equation outlined in the previous section with rk. Let $\omega = 1$, and let the initial conditions be $v = 0$ and $x = 5$ at $t = 0$. Computing points for $0 \leq t \leq 10$ every 0.1 time units, the call to rk is

```
(%i1)   txv: rk([v, -omega^2*x], [x,v], [5,0], [t, 0, 10, 0.1]),
     ➥   omega=1$
```

In this case the points returned in txv have the form [t, x, v], so we have to build lists of [t,x] and [t,v] points before we can plot x and v as a function of t:

```
(%i2)   tx: makelist([txv[i][1], txv[i][2]], i, 1, length(txv))$
(%i3)   tv: makelist([txv[i][1], txv[i][3]], i, 1, length(txv))$
(%i4)   plot2d([[discrete, tx], [discrete, tv]],
            [legend, "x", "v"],
            [xlabel, "t"]
        )$
```

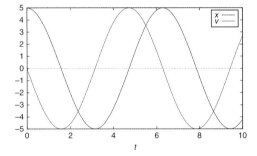

Finally, let's use rk to find the concentrations of A, B, C, and D as a function of time in the reaction network

$$
\begin{array}{ccc}
A & \overset{k_1}{\underset{k_2}{\rightleftharpoons}} & B \\
k_7 \uparrow\downarrow k_8 & & k_4 \uparrow\downarrow k_3 \\
D & \overset{k_5}{\underset{k_6}{\rightleftharpoons}} & C
\end{array}
$$

The rate laws are

```
(%i1)   dAdt : -k[1]*A+k[2]*B -k[8]*A+k[7]*D$
(%i2)   dBdt : k[1]*A-k[2]*B-k[3]*B+k[4]*C$
(%i3)   dCdt : k[3]*B-k[4]*C-k[5]*C+k[6]*D$
(%i4)   dDdt : k[5]*C-k[6]*D-k[7]*D+k[8]*A$
```

Now compute [t, A, B, C, D] points with $0 \leq t \leq 6$ s, with time increments of 0.01 s. Let the initial concentration of A be 1 M, with B, C, and D not present at $t = 0$, and use an evaluation environment to set numerical values for each of the eight rate constants:

```
(%i5)   tABCD : rk([dAdt, dBdt, dCdt, dDdt], [A,B,C,D],
     ➥   [1,0,0,0], [t, 0, 6, 0.01]),
            k = [0.1, 0.2, 0.4, 0.8, 1.6, 3.2, 6.4, 12.8]$
```

Finally, build lists of points so that the concentrations of each species can be plotted as a function of time:

```
(%i6)   N : length(tABCD)$
(%i7)   tA : makelist([tABCD[i][1], tABCD[i][2]], i, 1, N)$
(%i8)   tB : makelist([tABCD[i][1], tABCD[i][3]], i, 1, N)$
(%i9)   tC : makelist([tABCD[i][1], tABCD[i][4]], i, 1, N)$
(%i10)  tD : makelist([tABCD[i][1], tABCD[i][5]], i, 1, N)$
(%i11)  plot2d([[discrete,tA], [discrete,tB], [discrete,tC], [
        ➡ discrete, tD]],
          [legend, "A", "B", "C", "D"],
          [xlabel, "t"]
        )$
```

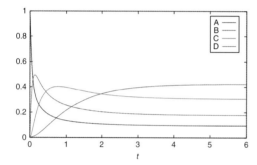

(M) **Worksheet 11.3: Solution of Systems of ODEs**

In this worksheet, we'll use the `desolve` and `rk` functions to plot the concentrations of reactants as a function of time for networks of coupled chemical reactions.

11.6   Solving Partial Differential Equations

A PDE is a differential equation that involves more than one independent variable. It is written in terms of partial derivatives. Examples are Equation (11.6) and the Schrödinger equation for a particle trapped in a three-dimensional box,

$$\frac{-\hbar^2}{2m}\nabla^2\psi - E\psi = 0 \qquad \text{Particle in a three-dimensional box} \tag{11.20}$$

where the Laplacian operator ∇^2 has been defined in Equation (7.40). E, m, and $\hbar$ are constants and ψ is a function of x, y, and z.

PDEs are in general much more difficult to solve than ODEs. Solving a many-variable PDE often involves rewriting it as a system of one-variable ODEs, a technique called separation of variables. Nearly all of the PDEs encountered in chemistry can be solved this way, although for some PDEs separation of variables is impossible.

To solve Equation (11.20) by separation of variables, we must assume that the solution ψ can be written as a product of functions that each depend on a single variable:

$$\psi(x, y, z) = X(x)Y(y)Z(z) \tag{11.21}$$

Writing out the Laplacian operator and making this substitution gives

```
(%i1)    -hbar^2/(2*m)*(diff(Psi(x,y,z),x,2)+diff(Psi(x,y,z),y,2)
     ➡   + diff(Psi(x,y,z),z,2)) - E*Psi(x,y,z) = 0;
(%i2)    subst(X(x)*Y(y)*Z(z), Psi(x,y,z), %);
(%i3)    %, nouns;
```

$$(\%o1) \quad -\Psi(x, y, z)\, E - \frac{hbar^2 \left(\frac{d^2}{dz^2}\Psi(x,y,z) + \frac{d^2}{dy^2}\Psi(x,y,z) + \frac{d^2}{dx^2}\Psi(x,y,z) \right)}{2\,m} = 0$$

$$(\%o2) \quad -X(x)\,Y(y)\,Z(z)\,E$$
$$-\frac{hbar^2 \left(\frac{d^2}{dz^2}(X(x)\,Y(y)\,Z(z)) + \frac{d^2}{dy^2}(X(x)\,Y(y)\,Z(z)) + \frac{d^2}{dx^2}(X(x)\,Y(y)\,Z(z)) \right)}{2\,m} = 0$$

$$(\%o3) \quad -X(x)\,Y(y)\,Z(z)\,E$$
$$-\frac{hbar^2 \left(X(x)\,Y(y) \left(\frac{d^2}{dz^2}Z(z) \right) + X(x) \left(\frac{d^2}{dy^2}Y(y) \right) Z(z) + \left(\frac{d^2}{dx^2}X(x) \right) Y(y)\,Z(z) \right)}{2\,m} = 0$$

where we've used *nouns* to force evaluation of the derivatives (Section 4.1). Solving for E and expanding, we have

```
(%i4)    soln: solve(%,E),expand;
```

$$(\%o4) \quad [E = -\frac{hbar^2 \left(\frac{d^2}{dz^2}Z(z) \right)}{2\,m\,Z(z)} - \frac{hbar^2 \left(\frac{d^2}{dy^2}Y(y) \right)}{2\,m\,Y(y)} - \frac{hbar^2 \left(\frac{d^2}{dx^2}X(x) \right)}{2\,m\,X(x)}]$$

Notice that the energy is a sum of three terms, and each term depends on a single variable. Naming the three terms $E_x, E_y,$ and E_z yields three single-variable ODEs:

```
(%i5)    E[x] = partition(rhs(soln),x)[2];
(%i6)    E[y] = partition(rhs(soln),y)[2];
(%i7)    E[z] = partition(rhs(soln),z)[2];
```

$$(\%o5) \quad E_x = -\frac{hbar^2 \left(\frac{d^2}{dx^2}X(x) \right)}{2\,m\,X(x)}$$

$$(\%o6) \quad E_y = -\frac{hbar^2 \left(\frac{d^2}{dy^2}Y(y) \right)}{2\,m\,Y(y)}$$

$$(\%o7) \quad E_z = -\frac{hbar \left(\frac{d^2}{dz^2}Z(z) \right)}{2\,m\,Z(z)}$$

where we have used the `partition` function to separate terms in each variable from the right-hand side of the solution (Section 5.3.3). Each of these equations can now be solved separately using the procedure outlined in Section 11.1.2.

When separation of variables is not possible, PDEs can sometimes be solved using Laplace transforms (Section 12.4). First, the PDE and its boundary conditions are transformed into an ODE with accompanying boundary conditions. The ODE is then solved, and the results are inverted to obtain the solution to the PDE [81].

12

Operators and Integral Transforms

All is waves.

– Erwin Schrödinger

If one has really technically penetrated a subject, things that previously seemed in complete contrast might now be seen as purely mathematical transformations of each other.

– John von Neumann

In quantum mechanics, a wavefunction is a mathematical function that encodes all mechanical information about a "particle," including its position, momentum, and energy. The expected values of these properties can be extracted from the wavefunction using operators. For example, to extract the energy E from a wavefunction Ψ, we apply the Hamiltonian operator, $\hat{H}$:

$$\hat{H}\Psi = E\Psi \qquad \text{The Schrödinger equation} \tag{12.1}$$

where $\hat{H}\Psi$ is *not* $\hat{H}$ times Ψ, but $\hat{H}$ operating on Ψ. Operators operate on functions to produce new functions; in this case, the new function is a constant (E) times the original function.

Maxima allows quantum mechanical expressions and equations to be written in operator notation. In Section 12.1, we'll see how abstract operators can be defined and assigned properties that determine their behavior.

Functions for waves can be expanded as a series of simpler functions called basis functions. For example, molecular orbitals Ψ are expanded in terms of a set of functions $\phi_1, \phi_2, \ldots$ by

$$\Psi = \sum_i c_i \phi_i \qquad \begin{array}{l}\text{Expansion of a wavefunction}\\\text{as a linear combination of}\\\text{basis functions}\end{array} \tag{12.2}$$

where the solution of Equation (12.1) provides the expansion coefficents c_i. Almost any periodic function can similarly be represented using a basis set of orthogonal sine and cosine functions (or complex exponential functions). We'll see how to compute the expansion coefficients for series like this in Section 12.2.

We can also map complicated functions onto simpler or more convenient ones using integral transforms, which are operators that use integration rather than summation to expand the function. Integral transforms take the form:

$$F(y) = \int K(x, y) f(x) \mathrm{d}x \qquad \text{General form of an integral transform} \tag{12.3}$$

The function $K(x, y)$ is called the *kernel* of the transform. The Fourier transform uses an imaginary exponential function (or equivalent sine and cosine expressions) as a kernel to map functions of time onto functions of frequency (Section 12.3). Instruments like FT-IR and FT-NMR

Symbolic Mathematics for Chemists: A Guide for Maxima Users, First Edition. Fred Senese.
© 2019 John Wiley & Sons Ltd. Published 2019 by John Wiley & Sons Ltd.
Companion website: http://booksupport.wiley.com

spectrometers use Fourier tranforms to generate spectra from time-dependent raw data. The Laplace transform uses a real exponential function as a kernel to map a function in one variable onto a rational expression in another variable (Section 12.4).

12.1 Defining Operators

The Hamiltonian operator for a particle of mass m with only kinetic energy moving in one dimension is

$$\hat{H} = -\frac{h^2}{8\pi^2 m}\frac{d^2}{dx^2} \qquad \text{1D Hamiltonian operator with zero potential energy} \qquad (12.4)$$

We can write the left-hand side of Equation (12.1) in Maxima by defining $\hat{H}$ as a function:

```
(%i1)   Hop(f)  := -h^2/(8*%pi^2*m)*diff(f,x,2);
```

$$(\%o1) \quad \text{Hop}(f) := \frac{-h^2}{8\,\pi^2\,m}\,\text{diff}(f,x,2)$$

However, Maxima supports operator notation so that we can type H Psi rather than H(Psi). The prefix function declares a function to be an operator with a single operand (which follows the operator after a space)[1]:

```
(%i2)   prefix("Hop");
```

$$(\%o2) \quad \textit{Hop}$$

 The double quotes around the argument for prefix are necessary.

Let's try applying the operator to the particle-in-a-box wavefunction $\Psi(x) = \sqrt{2/L}\sin(n\pi x/L)$, where L is the length of the box (a real and positive number) and n is an integer:

```
(%i3)   declare(n,integer)$
(%i4)   declare(L,real)$
(%i5)   assume(L>0)$
(%i6)   Psi(x)  := sqrt(2/L)*sin(n*%pi*x/L);
```

$$(\%o6) \quad \Psi(x) := \sqrt{\frac{2}{L}}\sin\left(\frac{n\,\pi\,x}{L}\right)$$

Now Equation (12.1) can be written as is, and solved[2]:

```
(%i7)   Hop Psi(x)  = E*Psi(x);
(%i8)   solve(%,E);
```

---

1 You can also define postfix operators (with arguments that come *before* the operator, like the factorial x!) and infix operators (which have arguments before and after the operator, like the multiplication sign in $a \times b$).
2 This approach works only when E is a constant. When Ψ is approximate, E is not a constant, but a function of the spatial coordinates. In that case, you can compute an "expected" value for E by multiplying both sides of the equation by conjugate(Psi(x)) and then integrate with respect to x before solving for E.

$$(\%o7) \quad \frac{h^2\,n^2 \sin\left(\frac{\pi n x}{L}\right)}{2^{\frac{5}{2}}\,m\,L^{\frac{5}{2}}} = \frac{\sqrt{2}\,E \sin\left(\frac{\pi n x}{L}\right)}{\sqrt{L}}$$

$$(\%o8) \quad [E = \frac{h^2\,n^2}{8\,m\,L^2}]$$

When applying the operator Ω to a function Ψ yields a constant ω times the original function, Ψ is called an eigenfunction of Ω, and ω is called an eigenvalue of Ω.

Let's show that e^{ikx} is an eigenfunction of the momentum operator, $\hat{p} = \frac{h}{2\pi i}\frac{d}{dx}$:

```
(%i1)   prefix("Pop")$
(%i2)   Pop f := (h/(2*%pi*%i))*diff(f,x);
(%i3)   Pop exp(%i*k*x);
```

$$(\%o2) \quad \text{Pop}\,(f) := \frac{h}{2\,\pi\,i}\,\text{diff}\,(f,x)$$

$$(\%o3) \quad \frac{h\,k\,e^{i\,k\,x}}{2\,\pi}$$

When $\hat{p}$ operates on $\exp(ikx)$, the result is a constant $hk/2\pi$ times the original function, so $hk/2\pi$ is an eigenvalue and $\exp(ikx)$ is an eigenfunction of $\hat{p}$.

You can create operators with just the `prefix` declaration; the function to perform the operation does not have to be defined.

```
(%i1)   prefix("Aop")$
(%i2)   Aop x^3;
```

$$(\%o1) \quad \text{Aop}\,(x)^3$$

Notice that we get $\text{Aop}\,(x)^3$, and not $\text{Aop}\,(x^3)$. Operators defined with `prefix` are given very high precedence, so they will operate on the x before the exponentiation occurs.

If you'd rather not worry about setting operator precedence for the operators you define, just use parentheses around the operand. For example, you can type `Aop (x^3)` to operate on x^3.

Maxima does provide a mechanism for controlling the precedence of defined operators. A second argument in `prefix` sets the "right binding power" of the operator. By default, the right binding power is set very high (to 180). For comparison, the right binding powers of `^`, `*`, `+`, `−`, and `=` are 139, 120, 100, 134, and 80, respectively. If the right binding power of the operator is less than the right bounding power of an operation, that operation will be performed *before* the operator is applied. For example, using the default right binding power,

```
(%i1)   prefix("Aop")$
(%i2)   Aop A*x^3 + B*sin(x);
```

$$(\%o2) \quad \sin(x)\,B + x^3\,\text{Aop}\,(A)$$

Setting the right binding power below multiplication (120), but above addition (100), gives

```
(%i3)   prefix("Aop",115)$ /* set  */
(%i4)   Aop A*x^3 + B*sin(x);
```

$$(\%o4) \quad \sin(x)\,B + \text{Aop}\,(x^3\,A)$$

Table 12.1 Operator properties that can be declared with `declare("Aop," property)`, where *Aop* is the name of the operator.

Property	Result
`additive`	$\hat{a}(x + y + z)$ simplifies to $\hat{a}(x) + \hat{a}(y) + \hat{a}(z)$
`linear`	Both additive and outative; $\hat{a}(c_1 x + c_2 y + c_3 z)$ simplifies to $c_1\hat{a}(x) + c_2\hat{a}(y) + c_3\hat{a}(z)$ if c_1, c_2, and c_3 are constants.
`multiplicative`	$\hat{a}(xyz)$ simplifies to $\hat{a}(x)\hat{a}(y)\hat{a}(z)$
`outative`	$\hat{a}(cx)$ simplifies to $c\hat{a}(x)$ if c is a constant.

Setting the right binding power below addition (100) gives

```
(%i5)   prefix("Aop",90)$
(%i6)   Aop A*x^3 + B*sin(x);
```

$$(\%o6) \quad \text{Aop}\left(\sin(x)\, B + x^3\, A\right)$$

To undefine a symbol as an operator, use `remove("symbol",operator)` (include the double quotes); to remove all of its properties (including operator status) use `kill("symbol")` as usual.

We can declare properties for operators that determine how expressions containing the operators are simplified (Table 12.1). For example, a linear operator $\hat{a}$ is additive ($\hat{a}(x + y) = \hat{a}(x) + \hat{a}(y)$) and outative ($\hat{a}(cx) = c\hat{a}(x)$, when c is a constant).

```
(%i1)   prefix("Aop")$
(%i2)   Aop (A*x^3 + B*sin(x));
(%i3)   declare([A,B], constant)$
(%i4)   declare("Aop", linear)$
(%i5)   Aop (A*x^3 + B*sin(x));
```

$$(\%o3) \quad \text{Aop}\left(B\sin(x) + A\,x^3\right)$$
$$(\%o6) \quad B\,\text{Aop}(\sin(x)) + A\,\text{Aop}\left(x^3\right)$$

Two operators $\hat{a}$ and $\hat{b}$ are said to *commute* if $\hat{a}(\hat{b}\Psi) = \hat{b}(\hat{a}\Psi)$. The difference between the left- and right-hand sides of the equation is called the *commutator* $\left[\hat{a}, \hat{b}\right]$:

$$[\hat{a}, \hat{b}] = \hat{a}\hat{b} - \hat{b}\hat{a} \qquad \text{The commutator} \qquad (12.5)$$

We can define the commutator in Maxima as a `matchfix` operator, that is, an operator that has left and right delimiters around its operands. For example,

```
(%i1)   matchfix("~ [", "]~ ")$
(%i2)   ~ [a , b]~  :=(a(b(f))-b(a(f)))/f;
```

$$(\%o2) \quad \sim [a,b] \sim := \frac{a\left(b\left(f\right)\right) - b\left(a\left(f\right)\right)}{f}$$

The `matchfix` function defines the delimiters of the operator. We've chosen ~ [and] ~ as delimiters because Maxima already uses square brackets to mark lists and subscripts. The next line defines the operator as a function that operates on a and b. We must use function notation rather than prefix operator notation in the definition.

Let's see if the position operator $\hat{x} = x\times$ commutes with the momentum operator $\hat{p}$. We must tell Maxima that the arbitrary function f depends on x when defining the momentum operator, or df/dx will evaluate to zero.

```
(%i3)   depends(f,x)$
(%i4)   prefix("Xop")$
(%i5)   Xop f := x*f;
(%i6)   prefix("Pop")$
(%i7)   Pop f := (h/(2*%pi*%i))*diff(f,x);
(%i8)   ~ [ "Xop" , "Pop" ]~ ;
(%i9)   % , ratsimp;
```

(%o4) $\text{Xop}\,(f) := x f$

(%o5) $\text{Pop}\,(f) := \dfrac{h}{2\pi i}\,\text{diff}\,(f, x)$

(%o6) $\dfrac{\dfrac{\%i h\left(\left(\frac{d}{dx}f\right)x + f\right)}{2\pi} - \dfrac{\%i\left(\frac{d}{dx}f\right)hx}{2\pi}}{f}$

(%o7) $\dfrac{i h}{2\pi}$

We had to put double quotes around Xop and Pop in the commutator, to keep them from executing without operands before the commutator is computed. The result shows that $\hat{x}$ and $\hat{p}$ don't commute – a fact that is of fundamental importance in quantum mechanics.

 Worksheet 12.0: Operators and operator algebra

In this worksheet we'll define and manipulate several common operators encountered in quantum chemistry.

12.2   Fourier Series

Periodic functions can often be represented by a sum of sine and cosine terms. If the function goes through one complete cycle in $2L$ seconds with a complete cycle defined by $f(x)$ on a finite interval from $-L$ to $+L$, it can be written as a Fourier series expansion

$$f(x) = \frac{a_0}{2} + \sum_{n=1}^{\infty}\left(a_n \cos\frac{n\pi x}{L} + b_n \sin\frac{n\pi x}{L}\right) \qquad \begin{array}{l}\text{Fourier series expansion}\\ \text{of } f(x) \text{ on } (-L, +L)\end{array} \qquad (12.6)$$

where the constant coefficients a_n and b_n are

$$a_n = \frac{1}{L}\int_{-L}^{+L} f(x)\cos\frac{n\pi x}{L}dx, \quad n = 0, 1, 2, \ldots \qquad \begin{array}{l}\text{Fourier coefficients for}\\ \text{cosine terms}\end{array} \qquad (12.7)$$

$$b_n = \frac{1}{L}\int_{-L}^{+L} f(x)\sin\frac{n\pi x}{L}dx, \quad n = 1, 2, \ldots \qquad \begin{array}{l}\text{Fourier coefficients for}\\ \text{sine terms}\end{array} \qquad (12.8)$$

Let's define a function to compute the first N terms of the Fourier series for a function $f(x)$ with a complete cycle on $(-L, +L)$:

```
(%i1)   fourier_expansion(N,f,L)  := block(
            local(n, a, b),
            declare(n,integer),
            assume(n>=0),
            a[n] := (1/L)*integrate(f(x)*cos(n*%pi*x/L), x, -L,
            ➥ L),
            b[n] := (1/L)*integrate(f(x)*sin(n*%pi*x/L), x, -L,
            ➥ L),
            a[0]/2 + sum(a[i]*cos(i*%pi*x/L), i, 1, N) + sum(b[i
            ➥ ]*sin(i*%pi*x/L), i, 1, N)
        )$
```

If the function is odd $(f(-x) = -f(x))$, only the sine terms contribute. For example, with $f(x) = x$,

```
(%i2)   f(x)  := x$
(%i3)   plot2d([f(x), fourier_expansion(1,f,L),
            ➥ fourier_expansion(2,f,L), fourier_expansion(3,f,L)
            ➥ , fourier_expansion(4,f,L)], [x, -L, L], [
            ➥ gnuplot_preamble, "set key bottom right"]), L=%pi$
```

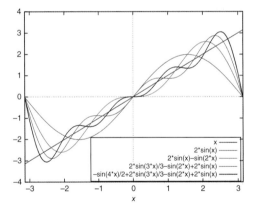

Notice that the Fourier series is a better approximation to the function when more terms are included. The approximation becomes exact with an infinite number of terms.

Notice also that even if $f(x)$ is not periodic, the Fourier series generated is always periodic. In this case, the function $f(x)$ only defines one cycle on $(-\pi, \pi)$; its Fourier series repeats the cycle:

```
(%i4)   plot2d([f(x), fourier_expansion(10,f,%pi)], [x,-10,10],
            ➥ [legend, "f(x) = x", "10-term Fourier series for
            ➥ f(x)"], [gnuplot_preamble, "set key bottom"])$
```

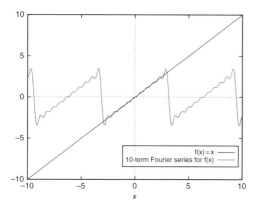

If the function is even (that is, $f(-x) = f(x)$) only the cosine terms contribute. For example, with $f(x) = x^2$,

```
(%i5)   f(x) := x^2;
(%i6)   plot2d([f(x), fourier_expansion(1,f,L),
        ➥ fourier_expansion(2,f,L), fourier_expansion(3,f,L)
        ➥ , fourier_expansion(4,f,L)], [x, -L, L]),L=%pi$
```

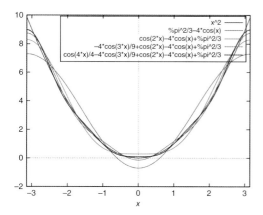

We'll sometimes encounter integrals in Equations (12.7) and (12.8) that `integrate` cannot evaluate symbolically. The `quadpack` package has a function to numerically evaluate this specific integral:

`quad_qawo(f(x), x, a, b, ω, trig)`
Numerically integrates $\int_a^b f(x)\cos(\omega x)\mathrm{d}x$ when `trig` is `'cos`, or $\int_a^b f(x)\sin(\omega x)\mathrm{d}x$ when `trig` is `'sin`. The output is a `quadpack` list (see Section 6.5.4).

This routine is useful for handling discontinuous functions that `integrate` cannot integrate symbolically, for example, a square pulse:

```
(%i1)   fourier_expansion(N,f,L) := block(
        local(n, a, b),
```

```
        declare(n,integer),
        assume(n>=0),
        a[n] := (1/L)*quad_qawo(f(x), x, -L, L, n*%pi/L,
            ➥ 'cos)[1],
        b[n] := (1/L)*quad_qawo(f(x), x, -L, L, n*%pi/L,
            ➥ 'sin)[1],
        a[0]/2 + sum(a[i]*cos(i*%pi*x/L), i, 1, N) + sum(b[i
            ➥ ]*sin(i*%pi*x/L), i, 1, N)
    )$
(%i2)  f(x) := if x<-1 or x>1 then 0 else 1$
(%i3)  plot2d([f(x), fourier_expansion(10,f(x),2)], [x, -6, 6],
            ➥ [legend, "square pulse", "N=10 Fourier series"],
            ➥ [gnuplot_preamble, "set key below"])$
```

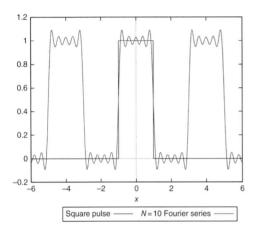

Maxima has its own functions for computing Fourier coefficients and series in the `fourie` package:

`fourier(f(x), x, L)`
Computes Fourier coefficients for the Fourier series expansion of $f(x)$ which defines one cycle on $(-L, L)$. You must `load(fourie)` before using this function.

`fourexpand(coefficients, x, L, nterms)`
Computes *nterms* of the Fourier series expansion with coefficients c previously computed by `fourier` for a function $f(x)$ defining one cycle on $(-L, L)$. You must `load(fourie)` before using this function.

Let's use these functions to write and plot the four-term Fourier expansion of $f(x) = abs(x)$ cycling on $(-\pi, \pi)$:

```
(%i1)  f(x) := abs(x);
(%i2)  load(fourie)$
(%i3)  fourier(f(x), x, %pi);
(%i4)  fourexpand(%, x, %pi, 4);
(%i5)  plot2d([f(x), %], [x, -%pi, %pi])$
```

(%o1) $f(x) := |x|$

(%t3) $a_0 = \dfrac{\pi}{2}$

(%t4) $a_n = \dfrac{2\left(\frac{\pi \sin(\pi n)}{n} + \frac{\cos(\pi n)}{n^2} - \frac{1}{n^2}\right)}{\pi}$

(%t5) $b_n = 0$

(%o5) $[\%t3, \%t4, \%t5]$

(%o6) $-\dfrac{4\cos(3x)}{9\pi} - \dfrac{4\cos(x)}{\pi} + \dfrac{\pi}{2}$

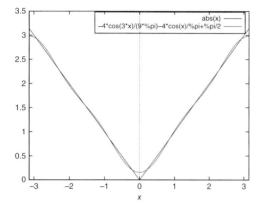

 The `fourie` package uses a slightly different form for the definition of the Fourier series; it gives the value of $a_0/2$ for the value of a_0 as we've written it in Equation (12.6).

The Fourier series can also be written in complex exponential form by applying Euler's formula (Equation (2.4)):

$$
\begin{aligned}
f(x) &= \frac{1}{2} \sum_{n=-\infty}^{\infty} c_n \exp\left(\frac{in\pi}{L}x\right) \\
c_n &= \frac{1}{L} \int_{-L}^{+L} f(x) \exp\left(\frac{-in\pi}{L}x\right) \\
n &= 0, \pm 1, \pm 2, \ldots
\end{aligned}
\qquad
\begin{aligned}
&\text{Fourier series} \\
&\text{in exponential form}
\end{aligned}
\qquad (12.9)
$$

 Worksheet 12.1: Fourier Series

In this worksheet, we'll approximate functions as a summation of harmonic waves using the Fourier series.

12.3   Fourier Transforms

The Fourier series expands a function as a series of harmonic waves, which can be viewed as functions of position or time. Many chemical instruments produce output that is also a function of time $f(t)$. Transforming the output to a series of waves associates frequencies ν with it, which allows the data to be displayed as a spectrum $F(v)$. When the signal and the noise in the signal have different frequencies, the transformation also lets us isolate the signal from the noise.

In Equation (12.9), as L approaches infinity, the coefficients $n\pi/L$ become closer and closer together. The Fourier series becomes an integral called the Fourier transform:

$$F(\omega) = \frac{1}{2\pi} \int_{-\infty}^{+\infty} f(t)\exp(i\omega t)dt \qquad \text{The exponential Fourier transform} \qquad (12.10)$$

where $f(t)$ is the signal in the time domain and $F(\omega)$ is the corresponding signal in the frequency domain. Here, we are using angular frequencies $\omega = 2\pi v$.

The inverse transformation is

$$f(t) = \int_{-\infty}^{+\infty} F(\omega)\exp(-i\omega t)d\omega \qquad \text{The inverse exponential Fourier transform}$$

$$(12.11)$$

Suppose we have an exponentially decaying time signal of the form:

```
(%i1)   f(t)  := cos(5*t)*exp(-abs(t))$
(%i2)   plot2d(f(t),[t,0,10])$
```

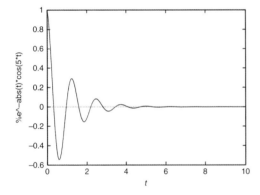

which should have an angular frequency $\omega = 5$. The Fourier transform is

```
(%i3)   load(abs_integrate)$
(%i4)   F(w)  :=  (1/(2*%pi))*(integrate(f(t)*exp(%i*w*t), t, minf
        ➡   , inf) );
(%i5)   F(w), ratsimp;
(%i6)   plot2d(conjugate(F(w))*F(w), [w,0,10], [ylabel, "F(w)^2"
        ➡   ]);
```

$$(\%o4) \quad F(w) := \frac{1}{2\pi} \int_{-\infty}^{\infty} f(t)\exp(iwt)\,dt$$

$$(\%o5) \quad \frac{w^2 + 26}{\pi w^4 - 48\pi w^2 + 676\pi}$$

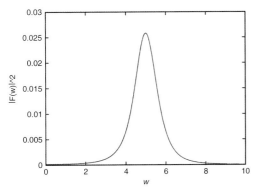

The `abs_integrate` package is needed to extend `integrate` so that it can handle the absolute value function, as we saw in Section 6.5.3. We also plotted $|F(\omega)|^2$ rather than just $F(\omega)$, because in general $F(\omega)$ is a complex function.

We can find the frequency with the maximum value of $|F(\omega)|^2$ with

```
(%i7)    solve(diff(conjugate(F(w))*F(w),w)=0, w), float;
```

(%o7) $[w = -4.99901941743857, w = 4.99901941743857, w = -8.774405685624978\,\%i,$

$w = 8.774405685624978\,\%i, w = -5.099019513592785\,\%i, w = 5.099019513592785\,\%i, w = 0]$

The positive, real, and nonzero solution is $w = 4.99901941743857$, which is close to the expected $w = 5$.

Now let's look at an decaying signal that is composed of two cosine waves, one with $\omega = 5$ and the other with $\omega = 10$. Transformation to the frequency domain should recover the two angular frequencies:

```
(%i1)    f(t) := (cos(5*t)+cos(10*t))*exp(-abs(t));
(%i2)    plot2d(f(t), [t, 0, 6])$
(%i3)    load(abs_integrate)$
(%i4)    F(w) := (1/(2*%pi))*(integrate(f(t)*exp(%i*w*t), t, minf
   ➡  , inf));
(%i5)    F(w), ratsimp;
(%i6)    plot2d(conjugate(%)*%, [w, 0, 15], [ylabel, "F(w)^2"])$
```

(%o1) $f(t) := (\cos(5\,t) + \cos(10\,t))\exp(-|t|)$

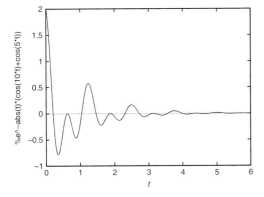

(%o4) $F(w) := \dfrac{1}{2\pi} \displaystyle\int_{-\infty}^{\infty} f(t)\exp(iwt)\,dt$

(%o5) $\dfrac{2w^6 - 119w^4 + 881w^2 + 333502}{\pi w^8 - 246\pi w^6 + 20381\pi w^4 - 623496\pi w^2 + 6895876\pi}$

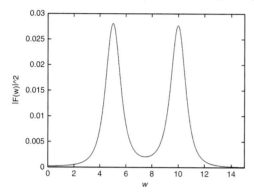

The final plot of the frequency spectrum gives peaks at $\omega = 5$ and $\omega = 10$, as expected:

```
(%i7)   find_root(diff(conjugate(F(w))*F(w), w), w, 4, 6);
(%i8)   find_root(diff(conjugate(F(w))*F(w), w), w, 9,11);
```

(%o7)   5.006151038211477
(%o8)   9.992150776211813

The `fourie` package has a `fourint` function that computes the exponential Fourier transform directly:

fourint(f(t), t)
Computes the exponential Fourier transform of $f(t)$ in terms of two coefficients, a_z and b_z, where the Fourier transform $F(z)$ is given by

$$F(z) = \frac{1}{2}a_z + \frac{i}{2}b_z$$

You must `load(fourie)` before calling this function.

Repeating the transform above, we have

```
(%i1)   load(fourie)$
(%i2)   f(t)  := (cos(5*t)+cos(10*t))*exp(-abs(t));
(%i3)   fourint(f(t),t);
(%i4)   ratsimp(a[z]/2+%i*b[z]/2), %;
```

(%o2) $f(t) := (\cos(5t) + \cos(10t))\exp(-|t|)$

(%t3) $a_z = (2(\dfrac{2z^6}{z^8 - 246z^6 + 20381z^4 - 623496z^2 + 6895876}$

$- \dfrac{119z^4}{z^8 - 246z^6 + 20381z^4 - 623496z^2 + 6895876} +$

$\dfrac{881z^2}{z^8 - 246z^6 + 20381z^4 - 623496z^2 + 6895876} + \dfrac{333502}{z^8 - 246z^6 + 20381z^4 - 623496z^2 + 6895876}))/\pi$

$$(\%t4) \quad b_z = 0$$

$$(\%o4) \quad [\%t3, \%t4]$$

$$(\%o5) \quad \frac{2\,z^6 - 119\,z^4 + 881\,z^2 + 333502}{\pi\,z^8 - 246\,\pi\,z^6 + 20381\,\pi\,z^4 - 623496\,\pi\,z^2 + 6895876\,\pi}$$

The last line is $F(\omega)$, and it is identical to the result we calculated by hand (using z as the symbol for angular frequency rather than ω).

12.3.1 The Fast Fourier Transform

In practice, we have digitized signals rather than continuous functions to transform. Transforming discrete data is computationally intensive and requires a large number of samples, but the Cooley–Tukey algorithm (also called the *fast Fourier transform*) allows the transform to be applied efficiently [82].

Let's apply the fast Fourier transform to the decaying cosines signal we used in the previous section:

```
(%i1)   f(t) := (cos(5*t)+cos(10*t))*exp(-abs(t))$
```

We must select a number of points N that is a power of 2. We also need a sampling interval `tinc` and a maximum time `tmax`.

```
(%i2)   N : 2^6$
(%i3)   tinc: 0.1$
(%i4)   tmax: tinc*(N-1)$
```

Now sample the signal, and plot the sampled points:

```
(%i5)   time : makelist(tinc*t, t, 0, N-1)$
(%i6)   digitized_f: f(time)$
(%i7)   plot2d([f(t), [discrete, time, digitized_f]], [t, 0,
        ➥ tmax], [style, lines, points], [title, "Sampled
        ➥ signal curve"], [legend, "true signal", "samples"
        ➥ ])$
```

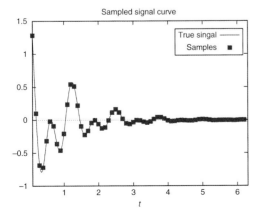

Transform the sampled points from the time domain to the frequency domain using the fast Fourier transform routine `fft`:

```
fft(data)
```
Computes the fast Fourier transform on a list of *data*. The number of elements must be a power of 2. You must `load(fft)` before using this function.

```
(%i8)   load(fft)$
(%i9)   spectrum: fft(digitized_f)$
(%i10)  frequency: makelist(2*%pi*i/tmax, i, 0, N-1)$
(%i11)  plot2d([discrete,frequency, cabs(spectrum)^2], [x, 0,
        ➥ 20], [title, "Frequency spectrum computed by FFT"
        ➥ ],[xlabel, "angular frequency"], [ylabel, "
        ➥ intensity"]);
```

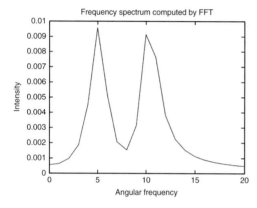

Compare this spectrum with the one we obtained from doing a Fourier transform on $f(t)$. The peaks are around $\omega = 5$ and $\omega = 10$, as they should be.

To check the transform, let's do an inverse Fourier transform with `inverse_fft` and see if we get back our original sampled points:

```
(%i12)  inverse_fft(spectrum)$
(%i13)  plot2d([[discrete, time, realpart(%)], f(t)], [t, tinc,
        ➥ tmax], [style, points, lines], [legend, "inverse
        ➥ FFT points", "true signal"])$
```

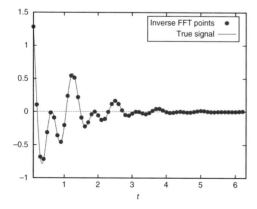

The original points are recovered by the inverse transform, with a small amount of roundoff error.

How much of the time signal do we need to sample? A form of the uncertainty principle applies to time signals and their frequency transforms. If Δt is the duration of the signal $f(t)$ and Δv is the range of frequencies in $F(v)$, then [83]

$$\Delta t \Delta v \geq \frac{1}{4\pi} \qquad \text{The bandwidth theorem} \tag{12.12}$$

In other words, the shorter the duration of the signal we sample, the less certain we can be about the frequencies we can extract from it.

The fast Fourier transform has many other applications in chemistry. In data aquisition, the transform and its inverse is often used to filter noise from a signal. In Worksheet 12.2, we'll add noise to the time domain signal, digitize it again, and transform it to the frequency domain. After filtering out the frequencies that correspond to noise, we'll transform the data back to the time domain to obtain a "clean" signal. The fast Fourier transform can also be used to compute continuous probability density functions from discrete data using a technique called *kernel density estimation* [84] (KDE), which was explored in Worksheet 4.3.

 Worksheet 12.2: Fourier Transforms

In this worksheet, we'll transform both continuous and discrete signals from the time domain to the frequency domain, and vice versa. We'll also use the Fourier transform to separate a signal from random noise.

12.4 The Laplace Transform

The Laplace transform maps an expression in t onto a rational expression in a new variable s. It is mainly used in chemistry to solve PDEs [81] and to convert ODEs that model chemical dynamics into equivalent algebraic equations that are easily solved [85, 86]. In statistical mechanics, Laplace transforms can be used to determine the number and density of states [87].

Applying Laplace transforms in the solution of ODEs has the advantage of building initial values into the equation before the equation is solved [88]. This is the technique used by the des-olve linear differential equation solver we used in Section 11.4, which is available as Equations ⟩ ⟩ Solve ODE with Laplace... from the wxMaxima menu.

For a function $F(t)$ with $t > 0$, the Laplace transform is defined as

$$\mathcal{L}\left[F(t)\right] = f(s) = \int_0^\infty \exp(-st)F(t)\mathrm{d}t \tag{12.13}$$

For example, here is the Laplace transform of $\sin(at)$:

```
(%i1)   assume(s>0, s>a)$
(%i2)   L(F) := integrate(exp(-s*t)*F, t, 0, inf );
(%i3)   L(sin(a*t));
```

$$(\%o2) \quad \mathrm{L}(F) := \int_0^\infty \exp\left((-s)\,t\right)\,F\mathrm{d}t$$

$$(\%o3) \quad \frac{a}{s^2 + a^2}$$

Notice that the Laplace transform uses a real exponential as a kernel, while the Fourier transform uses complex exponentials. If complex variables are used in the Laplace transform, it becomes a version of the Fourier transform, with similar properties [89].

Maxima provides built-in functions `laplace` and `ilt`[3] for the Laplace transform and inverse Laplace transform:

 `laplace(F, t, s)`
Compute the Laplace transform of $F(t)$ in terms of the transform parameter s.

 `ilt(f, s, t)`
Compute the inverse Laplace transform of $f(t)$ in terms of the transform parameter s.

For example, the Laplace transform of $\sin(at)$ is

(%i1)   **laplace(sin(a*t),t,s);**

(%o1) $\dfrac{a}{s^2 + a^2}$

The inverse Laplace transform restores the original function:

(%i2)   **ilt(a/(s^2+a^2), s, t);**

Is a zero or nonzero? nonzero;
  (%o2) $\sin(at)$

The mathematical theory behind Laplace transforms is beyond the scope of this book. For an introduction to the theory, see Ref. [90].

---

3 You can select ⎡Calculus⟩ Laplace transform...⎤ or ⎡Calculus⟩ Inverse Laplace transform...⎤ to insert the `laplace` or `ilt` functions with a dialog.

Glossary

absolute calibration A direct fit of a measurement or instrument response y to some independent variable x, such as the concentration or amount of substance. The fit can then be used to predict x from y.

accuracy Correctness.

antisymmetric matrix A square matrix with elements $M_{ij} = -M_{ji}$.

argument A variable or expression that is passed to a function. For example, x is the argument of $f(x)$.

augmented Lagrangian method A method for finding the minimum of a function subject to equality constraints on the function's variables.

autonomous Describes a differential equation that does not explicitly depend on the independent variable; the independent variable appears only in the derivative.

basis function One of a set of simple functions that can be used in linear combinations to represent more complicated functions.

bias The difference between an estimate or average measurement and the true value.

big float A decimal number with arbitrarily high precision; also called a `bfloat`.

Boolean A variable or expression that has a value of `true` or `false`.

bootstrapping Drawing new samples from the original data and then computing statistics across the samples.

boundary condition A condition that must be satisfied at the boundary of an interval or region where solutions of a differential equation are sought.

calibration Correcting biased measurements.

calibration curve A fit of observed y values to precisely known x values for a series of standards

Cauchy ratio test A series $\sum_i^{\infty} u_i$ is convergent if $\lim_{n \to \infty} |u_{n+1}/u_n| < 1$ and divergent if $\lim_{n \to \infty} |u_{n+1}/u_n| > 1$.

chain rule for derivatives If z is a function of y, and y is a function of x, z is also a function of x and its derivative can be written as

$$\frac{dz}{dx} = \frac{dz}{dy}\frac{dy}{dx}.$$

characteristic polynomial The polynomial $|\mathbf{A} - \lambda\mathbf{I}| = 0$, where $\mathbf{A}$ is a square matrix. The roots of the polynomial in λ are the eigenvalues of $\mathbf{A}$.

chi-squared The figure of merit for the weighted least-squares method, defined by

$$\chi^2 \equiv \sum_{i=1}^{N} \left(\frac{y_i - f(x_i; \alpha)}{\sigma_i} \right)^2$$

closed form Describes an expression that can be written in terms of a finite number of arithmetic, trigonometric, exponential, or logarithmic functions.

column vector A matrix with only one column.

comma-separated value file A spreadsheet file with column values separated by commas or tabs, and rows separated by line breaks.

confidence interval A range of values that contains a specific value of a population parameter with a specific probability. For example, a confidence interval around a sample mean might have a 95% chance of containing the population mean.

confidence level The probability that a measurement lies within a given interval.

context In Maxima, a list of facts established by the `assume` function.

contour plot A plot of a function of more than one variable, drawn by tracing curves with fixed function values.

correlated Describes a pair of experimental variables that are not independent of each other. The covariance of the two variables is nonzero.

correlation matrix A matrix r with elements r_{ij} that are the normalized covariances $C_{ij}/\sqrt{C_{ii}C_{jj}}$ between the ith and jth elements of a vector of random variables.

covariance A measure of how strongly one random variable x changes another (y). The covariance between x and y is the expected (mean) value of $(x - \bar{x})(y - \bar{y})$.

covariance matrix A matrix C with elements C_{ij} that are the covariances of between the ith and jth elements of a vector of random variables.

critical point A point where the first derivative of a function changes sign.

cross product The cross product of two vectors is a vector perpendicular to the two vectors with length equal to the area of a parallelogram with the two vectors as sides. Also called a vector product.

CSV comma-separated value file

cubic spline Smoothly varying piecewise cubic polynomials often used to interpolate, integrate, or differentiate a discrete dataset.

cumulative distribution A function that gives the probability that a random variable X is less than or equal to some particular value x. It is equal to the integral of the probability distribution function from $-\infty$ to x.

curl The cross product of the gradient operator with a vector.

cylindrical coordinates A three-dimensional coordinate system that adds altitude (z) to the two-dimensional polar coordinates r and ϕ.

data reduction Transformation of data to a simpler and more meaningful form.

degrees of freedom The number of values that can be varied freely and independently in a statistical calculation.

dependent variable A variable that changes in response to changes in independent variables.

design matrix A matrix used in statistical analysis that lists independent variables that explain the variation in a dependent variable. In least-squares fitting, each column of the design matrix lists the values of a particular variable, and each row represents a data point in the fit.

determinant A quantity associated with a square matrix that determines its behavior when it is used as a coefficient or transformation matrix. The determinant of a 2×2 matrix $\begin{pmatrix} a & b \\ c & d \end{pmatrix}$ is $ad - bc$; for a 3×3 matrix $\begin{pmatrix} a & b & c \\ d & e & f \\ g & h & i \end{pmatrix}$ the determinant is $aei + bfg + cdh - ceg - bdi - afh$.

determination coefficient The ratio of explained variation to total variation in a dependent variable y around its mean.

diagonal matrix A matrix with nonzero elements only along its major diagonal.

differential equation An equation that relates an independent variable x with derivatives of a dependent variable y.

direction field A geometric representation of a differential equation as a vector field. Each vector in the field is a tangent to a solution of the differential equation.

divergence (1) A power series that does not converge to a single value. (2) The dot product of the gradient operator with a vector.

dot product The sum of the products of corresponding elements in two lists or arrays. Also called the scalar product or the inner product.

dummy argument An unassigned variable used to define the arguments as a function.

e-notation A method for coding numbers in scientific notation; for example, 6.02×10^{23} is 6.02e23 in e-notation.

eigenfunction A function that when operated on by an operator gives a number (called an eigenvalue) times itself.

eigenvalue (1) A number λ that makes $|\mathbf{A} - \lambda\mathbf{I}| = 0$ for a square matrix $\mathbf{A}$. (2) A value of a parameter in a differential equation that yields a nonzero solution.

eigenvector A vector that when multiplied by a matrix or operated on by an operator gives a number (called an eigenvalue) times itself.

equilibrium point A point where $f(x, y) = 0$ for all x for a differential equation $\frac{dy}{dx} = f(x, y)$.

equivalent Two expressions a and b are considered equivalent if rational simplification of a $-$ b is zero.

error The difference between an observed value and the "true" or correct value.

error analysis Estimation of the relative sizes of experimental errors and their effect on a final result. Error analysis guides interpretation of results and suggests ways to improve the design of an experiment.

error function The integral $\mathrm{erf}(x) = \frac{2}{\sqrt{\pi}} \int_0^x e^{-t^2}\,dt$, which often occurs in measurement theory and in the solution of partial differential equations.

evaluation environment In Maxima, an expression that is evaluated in an environment specified by the comma-separated functions, equations, assignments, or switches that follow it. For example, y=m*x+b, b=0, m=a*b, y=A, x=c; yields A = a*b*c.

extrapolation Estimating a variable outside the range of its previously measured values.

feature In Maxima, a mathematical property of a variable, function, or expression.

field A function of position.

figure of merit A measure of the performance or quality of a method or fit, used to compare different methods or fits.

floating point number A number with a decimal point. In Maxima, floating point numbers (or "floats") have about 16 significant figures.

Fourier series An expansion of a periodic function as an infinite series of orthogonal sine and cosine functions.

Fourier transform An integral transform that is used to represent functions of time as functions of frequency.

general solution A solution of a differential equation that doesn't take initial conditions or boundary conditions into account.

gnuplot A powerful open-source graphing engine built into Maxima, and also available as a stand-alone program.

goodness-of-fit A statistical measure that can be used to compute the probability that the data matches the fit purely by chance.

gradient The derivative of a scalar field with respect to a position vector.

hat matrix A matrix **H** that maps a vector of observed values **y** onto a vector of fitted values $\hat{\mathbf{y}}$. Its diagonal elements describe the leverages of the observed values.

Hermite polynomial A particular polynomial solution $y = H_n(x)$ of the equation $\frac{d^2y}{dx^2} - 2x\frac{dy}{dx} + 2ny = 0$ which obeys the recursion relation $-2\,H_n(x)\,x + H_{n+1}(x) + 2\,n\,H_{n-1}(x) = 0$.

Hessian matrix A matrix $\mathcal{H}$ whose elements are the second derivatives of a function $f(x_1, x_2, \ldots)$ with respect to its variables; $\mathcal{H}_{ij}$ is $\partial^2 f / \partial x_i \partial x_j$.

histogram A bar graph with bar widths indicating a range of values for the x variable, and bar heights indicating the number of times the variable was observed in that range.

identity matrix A square matrix **I** that has ones along the diagonal and zeros elsewhere.

ill-conditioned Describes a calculation that rapidly accumulates round-off error or other small errors in its inputs.

implicit equation An equation in several variables that hasn't been explicitly solved in terms of any of its variables. For example, $x^2 + y^2 = 1$ is an implicit equation for a unit circle.

imprecision contour Lines or bands around a fit line that estimate the fit's imprecision.

independent variable A variable that can be controlled in an experiment. Changing one independent variable has no effect on other independent variables.

inflection point A point where the second derivative changes sign.

initial condition A fixed value for a function or its derivatives at a particular point that is required for an acceptable solution of a differential equation.

input cell In Maxima, a cell for editing and executing commands.

integral transform An operator that maps one function $f(x)$ onto another $F(y)$ via an integral of the form $F(y) = \int K(x, y)f(x)dx$. The function $K(x, y)$ is called the *kernel* of the transform.

integrand The expression between $\int$ and dx in an integral, where x is the integration variable.

integration constant An additive constant c that appears in the general solution of an indefinite integral $\int f(x)dx = F(x) + c$.

interpolation Estimating a variable within the range of its previously measured values.

iteration Repetition of a calculation until some set criterion is met.

jackknifing A procedure for estimating the uncertainties in fit parameters. The dataset is "jackknifed" by dropping one data point from the set of N points and refitting. Repeating the procedure for each data point generates a list of jackknifed fit parameters. The standard error of a fit parameter is then $(N-1)/\sqrt{N}$ times the standard deviation of the jackknifed fit parameters.

Jacobian matrix A matrix of all first-order partial derivatives of a vector function **f** with respect to a vector **x**. The elements of the matrix are $\partial f_i / \partial x_j$, where f_i is the ith component of **f** and x_j is the jth component of **x**.

Lagrange interpolation Interpolation using a polynomial of degree n that passes through all $n + 1$ data points.

Laplace transform An integral transform that can reduce a linear differential equation to a more easily solved algebraic equation.

Laplacian The operator ∇^2 which operates on a scalar field $f(x, y, z)$ as

$$\nabla^2 f = \nabla \cdot \nabla f = \frac{\partial^2 f}{\partial x^2} + \frac{\partial^2 f}{\partial y^2} + \frac{\partial^2 f}{\partial z^2}$$

leverage A measure of how far a data point is from the mean x value.

limiting distribution A hypothetical distribution that sample distributions converge towards as sample size becomes larger. The normal distribution and the Poisson distribution are limiting distributions.

limiting error The measurement error that makes the largest contribution to the error in a calculated result.

linear differential equation A type of differential equation that can have solutions that are linear combinations of its other solutions. Linear equations don't contain any products or powers of the dependent variable and its derivatives.

linear equation Linear equations don't contain any products or powers of their variables.

list In Maxima, a list is a series of comma-separated items enclosed in square brackets.

logical expression An expression that evaluates to `true` or `false`.

loop Code for iterative repetition of a calculation.

mapping Applying a function to each element of a list, vector, or part of an expression. In Maxima this can be done using the `map` function.

matrix An array of items arranged in rows and columns.

matrix product The matrix product **A.B** forms a matrix **C** with elements C_{ij} that are the dot product of row i of **A** with column j of **B**.

model equation An equation that predicts a dependent variable from one or more independent variables. It often contains a number of adjustable parameters; for example, a linear model contains a slope and a y-intercept.

multiple linear regression A method for fitting a least-squares line to data when there is more than one independent variable.

nested list A list of lists.

nonlinear differential equation A type of differential equation that includes derivative coefficients that include the dependent variable, or powers of the dependent variable or its derivatives.

normal distribution A theoretical population distribution that is defined in terms of its mean and its standard deviation. When the distribution is graphed, it has a characteristic bell or "Gaussian" shape.

normal equations A system of equations obtained by setting the first derivatives of a figure of merit with respect to its parameters equal to zero. Solving these equations simultaneously gives the values of the parameters that minimize the figure of merit.

normalization (1) Multiplying a function by a constant so that the function's integral over the range of its independent variable is 1. (2) Dividing a vector by its length to obtain a vector of length 1.

normally distributed Describes a random variable that has a bell-shaped "Gaussian" distribution.

noun A symbol that is not evaluated.

null hypothesis A hypothesis used in statistical testing that proposes that there is no significant difference between two quantities; any difference that is observed comes from random error in the data.

null space The set of vectors **x** that satisfy **Cx** = **0** for a matrix **C**.

ODE ordinary differential equation

OLS ordinary least-squares method

operator A mapping of one vector field or function onto another.

ordinary differential equation An equation that contains ordinary derivatives (derivatives of a function of one variable).

ordinary least-squares A method for fitting a function to data with equal error variances; it minimizes the sum of the squared residuals.

orthogonal A pair of vectors or functions are orthogonal if their inner product is zero.

outlier A point that has an unusually large error.

overfitting Refers to a fit that artificially cancels some of the experimental error in the data. Models that overfit the data pass very close to all of the data points, but they describe the space between points poorly.

p-**value** In hypothesis testing, a *p*-value is the probability of obtaining a test statistic at least as large as the one that was observed, if the null hypothesis is true.

package A collection of Maxima function and variable definitions available as an external file. Generally useful packages are loaded automatically when Maxima starts; some special-purpose packages may have to be loaded with the `load` command.

Padé approximation A ratio of polynomials that efficiently approximates a Taylor series.

parametric curve The coordinates of points along a parametric curve can be expressed as functions of a variable, called a parameter.

partial derivative The derivative of a function of more than one variable.

partial differential equation An equation that contains partial derivatives.

PDE partial differential equation

Pearson's *r* A measure of the linear dependence between two variables, obtained by dividing the covariance of the variables by the product of their standard deviations.

Poisson distribution A discrete population distribution that can be used to compute the probability that a random event occurs *n* times, when the mean number of times the event occurs is known.

polar coordinate A system for locating a point in a plane by specifying its distance from the origin (r) and the angle a line from the point to the origin makes with the *x* axis.

population The set of all possible values of a variable of interest.

population distribution A function that gives the exact distribution or scatter of all possible values of a variable around its mean value. Also called the parent distribution.

population mean The average value of a variable taken over all of its possible values.

power series An infinite series of the form $\sum_{i=0}^{\infty} a_i x^i$.

precision Reproducibility.

predicate A function that returns `true` or `false` based on the values of its arguments.

probability density A function that gives the probability that a variable will have a particular value.

random error An error without an identifiable cause that affects the precision of results.

rational expression An expression that can be expressed exactly as a ratio of polynomials.

rational number A number that can be expressed exactly as a ratio of integers.

recursive function A function that calls itself.

relational operator An operator that compares two operands.

remainder term The error in approximating a function with a truncated Taylor series.

residual The difference between an observed and predicted result.

root Values of *x* that make $f(x) = 0$.

round-off error The difference between a number's exact mathematical value and its computed approximation.

row vector A matrix with just one row.

saddle point A point on a surface that is a maximum in one direction and a minimum in another.

scalar A quantity that has only magnitude, without direction.

scalar field A scalar function that depends on x, y, and z.

secular equation An equation in the system $\mathbf{A} - \lambda\mathbf{I} = 0$.

sensitivity coefficient A partial derivative $\partial f / \partial x$ in an error propagation calculation, which shows how sensitive f is to small changes in x.

separation of variables A technique for solving partial differential equations by first rewriting them as a system of one-variable ordinary differential equations.

series A sum of terms that have a fixed form.

Shapiro–Wilk test A statistical test with the null hypothesis that a sample comes from a normally distributed population. If the p-value given by the test is less than some chosen α level (e.g. 0.05 for 95% confidence) then the null hypothesis can be rejected.

singularity A point at which a function takes on an infinite value.

spherical coordinates A coordinate system that specifies the location of points in three-dimensional space using the distance from the origin r, the polar angle θ, and the azimuthal angle ϕ.

square matrix A matrix with the same number of rows and columns.

standard A substance or solution with a precisely known property. For example, a standard solution has a precisely known concentration.

standard deviation A measure of how much data scatters about its mean. When the sample is the entire population, the standard deviation is $\sigma = \sqrt{\dfrac{\sum_{i=1}^{N}(x_i - \mu)^2}{N}}$; for a finite sample from a population, the best estimate of the standard deviation is $s = \sqrt{\dfrac{\sum_{i=1}^{N}(x_i - \bar{x})^2}{N-1}}$.

standard error The theoretical standard deviation of a sample parameter when it is computed over all possible samples taken from a population.

standard error of the regression An estimate of the square root of the residual variance in a regression analysis that assumes all measurements have equal variance. Also called the standard error of the estimate, or the standard error of the fit.

Student's t A limiting distribution for the distribution of sample means for small samples.

studentized Refers to a residual divided by its estimated standard deviation. Studentizing a residual expresses it in units of standard deviation and (usually) makes it easier to interpret.

sum of squared errors The sum of squared differences between data and a true, expected, or modeled value.

symmetric matrix A square matrix with elements $M_{ij} = M_{ji}$.

system variable A variable that controls how Maxima displays results.

systematic error An error with an identifiable cause that affects the accuracy of results.

Taylor series A power series expansion of a function $f(x)$ around a point $x = c$ that has the form:

$$f(x) = \sum_{i=0}^{\infty} \frac{(x-c)^i}{i!}\left(\frac{\mathrm{d}^i f(x)}{\mathrm{d}x^i}\right)_{x=c}$$

TLS total least-squares method

total least-squares A method for fitting a function to data when the variances in both the dependent and independent variables are known and not constant.

transpose The transpose of a matrix $\mathbf{A}$ takes the matrix's rows as columns, and columns as rows. If $\mathbf{A}$ is an $m \times n$ matrix, the transpose $\mathbf{A}^{\mathrm{T}}$ is an $n \times m$ matrix.

type I error Incorrectly rejecting a true null hypothesis.

type II error Incorrectly rejecting a true alternative hypothesis, or incorrectly accepting a false null hypothesis.

underfitting Refers to a fit that doesn't capture all of the variation in the data. The residual variance will be greater than the error variance when we are underfitting the data.

unit round-off error The maximum error that can occur when rounding off to one.

variance The square of the standard deviation.

vector A quantity that has both magnitude and direction.

vector field A function that associates a vector with every point in space.

verb A symbol that is evaluated.

weighted least-squares A method for fitting a function to data that minimizes χ^2.

WLS weighted least-squares method

References

1 Duffy, D.J. and Kienitz, J. (2011). *Monte Carlo Frameworks*, 1e, vol. 1, 601. Wiley.

2 Maxima (2014). Maxima, a computer algebra system. version 5.34.1. http://maxima.sourceforge.net/ (accessed 24 April 2018).

3 Foundation, F.S. (2014). GNU general public license. http://www.gnu.org/copyleft/gpl.html (accessed 24 April 2018).

4 Vodopivec, A. (2014). wxMaxima, an interface for the computer algebra system Maxima. Version 13.04.2. http://wxmaxima.sourceforge.net/ (accessed 24 April 2018).

5 Maxima.sourceforge.net (2009). Xmaxima, an interface for maxima. http://maxima.sourceforge.net/docs/xmaxima/xmaxima.html (accessed 24 April 2018).

6 Grothmann, R. (2014). The Euler Math Toolbox. http://euler.rene-grothmann.de/index.html (accessed 24 April 2018).

7 Stein, W. and The Sage Developers (2014). Sage Mathematics Software. http://www.sagemath.org (accessed 24 April 2018).

8 van der Hoeven, J. (1998). GNU TeXmacs. http://www.texmacs.org (accessed 24 April 2018).

9 Ivashov, A. (2014). SMath Studio Desktop. http://smath.info (accessed 24 April 2018).

10 Rieder, A. (2014). *Cantor*. https://edu.kde.org/cantor/ (accessed 24 April 2018).

11 Maxima.sourceforge.net (2014). Web interfaces using Maxima. http://maxima.sourceforge.net/relatedprojects.html (accessed 24 April 2018).

12 Seibel, P. (2005). *Practical Common Lisp*, 1e. Apress.

13 Office, U.S.G.A. (1992). Patriot missle defense: software problem lead to system failure at Dhahran, Saudi Arabia, GAO Congressional Reports. http://www.gao.gov/assets/220/215614.pdf (accessed 5 April 2014).

14 Fox, R. and Hill, T. (2008). An exact value for Avogadro's number. *ResearchGate* 104–107. doi:10.1511/2007.64.368.

15 Steiner, E. (2008). *The Chemistry Maths Book*, 2e. Oxford University Press.

16 Brinton, W.C. (1939). *Graphic Presentation*. Brinton Associates.

17 Atkins, P., Paula, J.D., and Friedman, R. (2013). *Physical Chemistry: Quanta, Matter, and Change*, 9e, 106–107. Oxford University Press.

18 Huber, P.J. (1987). Experiences with three-dimensional scatterplots. *Journal of the American Statistical Association* 82 (398): 448–453.

19 Khare, P. and Swarup, A. (2009). *Engineering Physics: Fundamentals and Modern Applications*, 319–320. Jones & Bartlett Learning.

20 Jónasson, K. (2012). *Applied Parallel and Scientific Computing, 10th International Conference, PARA 2010*, vol. 1, p. 251. Reykjavík: Springer Science and Business Media.

21 Peet, M. (2008). *Tamar: A Novel*, 1e, 110. Candlewick.

Symbolic Mathematics for Chemists: A Guide for Maxima Users, First Edition. Fred Senese.
© 2019 John Wiley & Sons Ltd. Published 2019 by John Wiley & Sons Ltd.
Companion website: http://booksupport.wiley.com

22 Johnstone, H.F. and Leppla, P.W. (1934). The solubility of sulfur dioxide at low partial pressures. The ionization constant and heat of ionization of sulfurous acid. *Journal of American Chemical Society*, 56 (11): 2233–2238.

23 Wittgenstein, L. (1953). *Philosophical Investigations*, 1e, vol. 1, xlv. Wiley.

24 Banerjee, A., Adams, N., Simons, J., and Shepard, R. (1985). Search for stationary points on surfaces. *The Journal of Physical Chemistry* 89 (1): 52–57.

25 Liu, D. and Nocedal, J. (1989). On the limited memory BFGS method for large scale optimization. *Mathematical Programming B* 45: 503–528.

26 Nocedal, J. and Wright, S. (2006). *Numerical Optimization*. 514–528. Springer.

27 Krommer, A.R. and Ueberhuber, C.W. (1998). *Computational Integration*. SIAM.

28 Piessens, R., de Doncker-Kapenger, E., Ueberhuber, C., and Kahaner, D. (1983). *QUADPACK, A Subroutine Package for Automatic Integration*. Springer-Verlag.

29 Atkins, P., Paula, J.D., and Friedman, R. (2013). *Physical Chemistry: Quanta, Matter, and Change*, 9e, 605–607. Oxford University Press.

30 Sevastyanov, R.M. and Chemyavskaya, R.A. (1987). Virial coefficients of neon, argon, and krypton at temperatures up to 3000 k. *Journal of Engineering Physics* 52 (6): 703–705.

31 Weast, R.C. (ed.) (1977). *CRC Handbook of Chemistry and Physics*, 58e, D–152. Boca Raton, FL. CRC Press.

32 Atkins, P. and Paula, J.D. (2010). *Physical Chemistry*, 9e, 627. W. H. Freeman.

33 Sylvester, J.J. (1851). On the relation between the minor determinants of linearly equivalent quadratic functions. *Philosophical Magazine*, 1 (4): 295–305.

34 Dirac, P.A.M. (1926). On the theory of quantum mechanics. *Proceedings of the Royal Society* A (112): 661–677.

35 Thorne, L.R. (2009). An innovative approach to balancing chemical-reaction equations: a simplified matrix-inversion technique for determining the matrix null space. *Chemical Educator* 15: 304–308.

36 Ford, B. and Hall, G. (1974). The generalized eigenvalue problem in quantum chemistry. *Computer Physics Communications* 8 (5): 337–348.

37 Pinkerton, R.C. (1952). A jacobian method for the rapid evaluation of thermodynamic derivatives, without the use of tables. *The Journal of Physical Chemistry* 56 (6): 799–800.

38 Atkins, P. and Paula, J.D. (2010). *Physical Chemistry*, 9e, 772. W. H. Freeman.

39 Chen-To, T. (1994). A survey of the improper use of ∇ in vector analysis, University of Michigan, College of Engineering Technical Report RL 909. http://deepblue.lib.umich.edu/bitstream/handle/2027.42/7869/bad1475.0001.001.pdf (accessed 6 June 2014).

40 Russell, B. (1931). *The Scientific Outlook. Routledge Classics*, 1e, vol. 1, 42. Routledge.

41 Reich, E.S. (2012). Timing glitches dog neutrino claim. *Nature* 483 (7387): 17.

42 Barrett, H.H. and Myers, K.J. (2013). *Foundations of Image Science*. Wiley.

43 Rutherford, E., Chadwick, J., and Ellis, C.D. (2010). *Radiations from Radioactive Substances*. Cambridge University Press.

44 Adler, F.R. (2012). *Modeling the Dynamics of Life: Calculus and Probability for Life Scientists*. Cengage Learning.

45 Moroi, Y. (1992). *Micelles: Theoretical and Applied Aspects*. Springer.

46 Altman, D.G. (2005). Standard deviations and standard errors. *British Medical Journal*, 331 (7521): 903.

47 Curran-Everett, D. (2008). Explorations in statistics: standard deviations and standard errors. *Advances in Physiology Education*, 32 (3): 203–208.

48 Biau, D.J. (2011). In brief: standard deviation and standard error. *Clinical Orthopaedics and Related Research* 469 (9): 2661–2664.

49 Gosset, W.S. (1908). The probable error of a mean. *Biometrika* 6 (1): 1–25.

50 Johnson, N., Kotz, S., and Balakrishnan, N. (1995). *Continuous Univariate Distributions*, pp. 322–361, no. v. 2 in Wiley series in probability and mathematical statistics: Applied probability and statistics. Wiley. https://books.google.com/books?id=0QzvAAAAMAAJ (accessed 24 April 2018).

51 Box, G.E.P. (1953). Non-normality and tests on variances. *Biometrika* 40 (3/4): 318–335.

52 Bevington, P.R. and Robinson, D.K. (2002). *Data Reduction and Error Analysis for the Physical Sciences*, 3e. McGraw-Hill.

53 Vigen, T. (2014). Spurious Correlations. http://www.tylervigen.com/ (accessed 15 May 2014).

54 Akaike, H. (1976). An information criterion (AIC). *Mathematical Sciences* 14 (153): 5–9.

55 Schwarz, G.E. (1978). Estimating the dimension of a model. *Annals of Statistics* 6 (2): 461–464.

56 Press, W.H., Teukolsky, S.A., Vetterling, W.T., and Flannery, B.P. (1992). *Numerical Recipes in FORTRAN: The Art of Scientific Computing*, 2e. Cambridge University Press.

57 Anscombe, F.J. (1973). Graphs in statistical analysis. *The American Statistician*. 27 (1): 17–21.

58 Woolson, R.F. and Clarke, W.R. (2011). *Methods for the Analysis of Biomedical Data*. Wiley.

59 White, H. (1980). A heteroskedastic-consistent covariance matrix estimator and a direct test of heteroskedasticity. *Econometrica* 48: 817–838.

60 Kvålseth, T.O. (1985). Cautionary note about R^2. *The American Statistician* 39: 279–285.

61 Atkins, P. and Paula, J.D. (2010). *Physical Chemistry*, 9e, 456–459. W. H. Freeman.

62 Herzberg, G. (1950). *Spectra of Diatomic Molecules*. D. Van Nostrand Co.

63 Belsley, D.A., Kuh, E., and Welsch, R.E. (1980). *Regression Diagnostics*. Wiley.

64 Zeng, Q.C., Zhang, E., Dong, H., and Tellinghuisen, J. (2008). Weighted least squares in calibration: estimating data variance functions in high-performance liquid chromatography. *Journal of Chromatography A* 1206 (2): 147–152.

65 Spilker, M.E. and Vicini, P. (2001). An evaluation of extended vs weighted least squares for parameter estimation in physiological modeling. *Journal of Biomedical Informatics* 34 (5): 348–364.

66 Draper, N.R. and Smith, H. (1981). *Applied Regression Analysis*, 85–96. Wiley.

67 Tellinghuisen, J. (2009). Least squares in calibration: weights, nonlinearity, and other nuisances. *Methods Enzymology* 454: 259–285.

68 National institute of Standards and Technology (2012). NIST/SEMATECH e-Handbook of Statistical Methods. http://www.itl.nist.gov/div898/handbook/pmd/section4/pmd452.htm (accessed 12 June 2013).

69 Willet, J.B. and Singer, J.D. (1988). Another cautionary note about R^2: its use in weighted least-squares regression analysis. *The American Statistician* 42 (3): 236–238.

70 Irvin, J.A. and Quickenden, T.I. (1983). Linear least squares treatement when there are errors in both x and y. *Journal of Chemical Education*, 60 (9): 711–712.

71 Pearson, K. (1901). On lines and planes of closest fit to systems of points in space. *Philosophical Magazine* 2: 559–572.

72 York, D. (1966). Least-square fitting of a straight line. *Canadian Journal of Physics* 44: 1079–1086.

73 Zeng, Q.C., Zhang, E., and Tellinghuisen, J. (2008). Univariate calibration by reversed regression of heteroscedastic data: a case study. *Analyst* 133: 1649–1655.

74 Martin, Y.C. and Hackbarth, J.J. (1977). *Chemometrics: Theory and Application: Examples of the Application of Nonlinear Regression Analysis to Chemical Data*. ACS Symposium Series, 153–164. American Chemical Society.

75 Efron, B. and Gong, G. (1983). A leisurely look at the bootstrap, the jackknife, and cross-validation. *American Statistician* 37 (1): 36–48.

76 Caceci, M.S. (1989). Estimating error limits in parametric curve fitting. *Analytical Chemistry*, 61 (20): 2324–2327.

77 Russell, B. (1948). *Human Knowledge: Its Scope and Limits*, Routledge Classics, 1e, vol. 1, 166. Routledge.

78 Isenberg, C. (1992). *The Science of Soap Films and Soap Bubbles*, 151. Dover.

79 Constanda, C. (2013). *Differential Equations: A Primer for Scientists and Engineers*. Springer.

80 Atkinson, K., Han, W., and Stewart, D.E. (2011). *Numerical Solution of Ordinary Differential Equations*, 9e, 70–89. Wiley.

81 Loney, N.W. (2006). *Applied Mathematical Methods for Chemical Engineers*, 224–231. CRC Press.

82 Brigham, E.O. (1988). *The Fast Fourier Transform and Its Applications*. Prentice Hall.

83 Gatti, P.L. (2014). *Applied Structural and Mechanical Vibrations: Theory and Methods*, 2e, 39–40. CRC Press.

84 Silverman, B.W. (1986). *Density Estimation for Statistics and Data Analysis*, 1e. Chapman and Hall.

85 Barthel, J. (1998). *Physical Chemistry of Electrolyte Solutions: Modern Aspects*, 62–88. Springer Science and Business Media.

86 Rogers, D.W. (2011). *Concise Physical Chemistry*. Wiley chap. 10.

87 Hase, W.L. (1996). *Unimolecular Reaction Dynamics: Theory and Experiments*. Oxford University Press, chap. 4.

88 Jenson, V.G. and Jeffreys, G.V. (1977). *Mathematical Methods in Chemical Engineering*, 15–160. Elsevier.

89 Feshbach, H. and Morse, P.M. (1953). *Methods of Theoretical Physics*, 1e, vol. 1, 467. McGraw-Hill.

90 Spiegel, M.R. (1965). *Schaum's Outlines: Laplace Transforms*, 1e. McGraw-Hill.

Index